MICROBIAL BIOFILMS: FORMATION AND CONTROL

A complete list of titles in the Society for
Applied Bacteriology Technical Series appears at
the end of this volume

THE SOCIETY FOR APPLIED BACTERIOLOGY
TECHNICAL SERIES NO 30

Microbial Biofilms: Formation and Control

Edited by
S.P. DENYER
Department of Pharmacy
University of Brighton

S.P. GORMAN
School of Pharmacy
The Queen's University of Belfast

M. SUSSMAN
Department of Microbiology
Medical School
Newcastle upon Tyne

OXFORD
BLACKWELL SCIENTIFIC PUBLICATIONS
LONDON EDINBURGH BOSTON
MELBOURNE PARIS BERLIN VIENNA

© 1993 by the Society for Applied Bacteriology
and published for them by
Blackwell Scientific Publications
Editorial Offices:
Osney Mead, Oxford OX2 0EL
25 John Street, London WC1N 2BL
23 Ainslie Place, Edinburgh EH3 6AJ
238 Main Street, Cambridge
  Massachusetts 02142, USA
54 University Street, Carlton
  Victoria 3053, Australia

Other Editorial Offices:
Librairie Arnette SA
1, rue de Lille
75007 Paris
France

Blackwell Wissenschafts-Verlag GmbH
Düsseldorfer Str. 38
D-10707 Berlin
Germany

Blackwell MZV
Feldgasse 13
A-1238 Wien
Austria

First published 1993

Set by Setrite Typesetters, Hong Kong
Printed and bound in Great Britain
at the University Press, Cambridge

DISTRIBUTORS

Marston Book Services Ltd
PO Box 87, Oxford OX2 0DT
(*Orders*: Tel: 0865 791155
        Fax: 0865 791927
        Telex: 837515)

USA
Blackwell Scientific Publications, Inc.
238 Main Street, Cambridge, MA 02142
(*Orders*: Tel: 800 759-6102
        617 876−7000)

Canada
Oxford University Press
70 Wynford Drive
Don Mills, Ontario M3C 1J9
(*Orders*: Tel: 416 441-2941)

Australia
Blackwell Scientific Publications Pty Ltd
54 University Street, Carlton, Victoria 3053
(*Orders*: Tel: 03 347-5552)

A catalogue record for this title
is available from the British Library

ISBN 0-632-03753-9

Library of Congress
Cataloging-in-Publication Data

Microbial biofilms: formation and control/
    edited by S.P. Denyer,
    S.P. Gorman, M. Sussman.
            p.   cm. −
    (Society for Applied Bacteriology
    technical series; no. 30)
      Includes bibliographical references
      and index.
      ISBN 0-632-03753-9
      1. Biofilms — Congresses. I. Denyer, S.P.
    II. Gorman, S.P. III. Sussman, Max.
    IV. Series: Technical series
    (Society for Applied Bacteriology); no. 30.
      [DNLM: 1. Bacterial Adhesion —
    congresses.   2. Microbiology — congresses.
    3. Microscopy — methods — congresses.
    W1 SO851F no. 30
    1993/QW 52 M62353 1993]
    QR100.8.B55M53   1993
    576′.119 − dc20
    DNLM/DLC for Library of Congress

Contents

Contributors   ix

Preface   xii

1   **Microbial Biofilm Formation and Characterization**   1
    H.M. LAPPIN-SCOTT, J. JASS AND J.W. COSTERTON
    Formation of Biofilms, 2
    Methods for Study of Biofilm Formation and Control, 6
    Conclusions, 11
    References, 11

2   **Mechanisms of Microbial Adherence**   13
    S.P. DENYER, G.W. HANLON AND M.C. DAVIES
    Strength of Attachment, 13
    Interfacial Forces and Interactions, 15
    Methods for Studying Surface Physicochemistry, 16
    Conclusions, 24
    References, 25

3   **Laboratory Methods for Biofilm Production**   29
    P. GILBERT AND D.G. ALLISON
    Closed Culture Models of Biofilm Development, 31
    Open Continuous Culture of Biofilms, 37
    References, 46

4   **The Physiology and Biochemistry of Biofilm**   51
    J.W.T. WIMPENNY, S.L. KINNIMENT AND M.A. SCOURFIELD
    Biofilm Structure and Function, 51
    Biofilm Physiology, 59
    Constant-Depth Biofilm: Rationalization for the Development of a Model
       Growth System, 63
    Use of the CDFF to Investigate the Physiology of Biofilm, 69
    Future Challenges, 84
    References, 88

5 **Confocal Laser Scanning Microscopy of Adherent Microorganisms, Biofilms and Surfaces** 95
   S.P. GORMAN, W.M. MAWHINNEY AND C.G. ADAIR
   Confocal Laser Scanning Microscopy, 96
   Method, 98
   Discussion, 101
   References, 106

6 **Attachment in Disease** 109
   S. PATRICK AND M.J. LARKIN
   Characterization of Surface Structures by Microscopy, 110
   Immunological Characterization of the Surface Structures, 116
   Preparation of Homogeneous Populations, 117
   Study of Attachment, 121
   Conclusions, 128
   References, 129

7 **Medical Device-Associated Adhesion** 133
   M.H. WILCOX
   Laboratory Diagnosis of Device-Associated Infection, 134
   Slime Production by Coagulase-Negative Staphylococci, 136
   Experimental Variables of *in vitro* Adherence, 140
   Conclusions, 144
   References, 144

8 **Antimicrobial and Other Methods for Controlling Microbial Adhesion in Infection** 147
   S.P. DENYER, G.W. HANLON, M.C. DAVIES AND S.P. GORMAN
   Regulation of Adhesion to Epithelial Cells by Antimicrobial Chemotherapy, 147
   Control of Adhesion to Model Biomaterial Surfaces, 152
   Control of Adhesion to Modified Medical Devices, 155
   Conclusions, 163
   References, 163

9 **Methods for the Study of Dental Plaque Formation and Control** 167
   M. ADDY, M.A. SLAYNE AND W.G. WADE
   Models of Bacterial Attachment and Plaque Accumulation on Teeth, 167
   Laboratory Methods for the Bacteriological Examination of Plaque, 170
   Clinical Methods for the Measurement of Plaque Formation, 171
   Clinical Measurement of Gingivitis, 173
   Methods to Evaluate Plaque Control Agents, 175
   References, 182

10 **Sensitivity of Bacteria in Biofilms to Antibacterial Agents** ... 187
W.W. NICHOLS

Colony on Filter: A Simple Method for Exposure of Bacterial Colonies to
Antibacterial Agents, 187
The Modified Robbins Device, 189
Batch-Culture Studies of Coagulase-Negative Staphylococcal Biofilms
on Biomedical Materials, 190
Chemostat-Grown Biofilms under Low-Iron Conditions, 192
Growth Rate-Controlled Continuously Perfused Biofilms, 193
Glass Slides and Animal Models, 195
Methods Predicted to be of Importance in Future Studies, 197
References, 198

11 *Legionella* **Biofilms and their Control** 201
C.W. KEEVIL, A.B. DOWSETT AND J. ROGERS

Development of the Chemostat Biofilm Model, 202
Uses for the Model, 205
References, 213

12 **Formation and Control of Coliform Biofilms in Drinking
Water Distribution Systems** 217
C.W. MACKERNESS, J.S. COLBOURNE, P.L.J. DENNIS,
T. RACHWAL AND C.W. KEEVIL

Distribution System Biofilm Formation, 219
Continuous-Culture Model, 219
Biofilm Development, 222
Strategies for Biofilm Control, 224
References, 225

13 **The Control of Biofilm in Recreational Waters** 227
T.D. WYATT

Recreational Water Facilities, 227
Features of Recreational Waters, 229
Microorganisms Associated with Recreational Water, 231
Physiological Aspects of Biofilm-Associated Organisms, 233
Biocides Used in Recreational Waters, 234
Methods for Investigating Water Quality, 236
Methods for Investigating the Effects of Biocides on Organisms, 240
Methods of Overcoming the Problems of Biofilms, 242
Conclusions, 243
References, 244

14 **Adhesion of Biofilms in Flowing Systems** 247
M.E. CALLOW, R. SANTOS AND T.R. BOTT

Flow Apparatus, 248
Radial Flow Chamber, 252

Flow Cell, 254
Conclusions, 257
References, 257

15 **Plasmid Exchange between Soil Bacteria in Continuous-
    Flow Laboratory Microcosms** 259
    L. SUN, M.J. BAZIN AND J.M. LYNCH
    Column Reactor Design and Operation, 260
    Plasmid Exchange in the Column Reactors, 262
    Conclusions, 263
    References, 265

16 **Microbial Films in the Light Engineering Industry** 267
    P.E. COOK AND C.C. GAYLARDE
    Formation and Examination of Biofilms in Engineering Fluids, 268
    Control of Biofilms in the Light Engineering Industry, 280
    References, 281

17 **Adherent Growth of *Staphylococcus aureus* from Poultry
    Processing Plants** 285
    C.E.R. DODD, B.J. CHAFFEY, K. DAEMS AND W.M. WAITES
    Inoculum Preparation, 286
    Impedance, 286
    Radiolabel Incorporation into DNA, 287
    Conclusions, 290
    References, 291

18 **Microbial Adherence to Food Contact Surfaces** 293
    A. GILMOUR, A.B. WILSON AND T.W. FRASER
    Removal by Rinsing, 293
    In situ Epifluorescence Microscopy, 297
    In situ Bioluminescence Measurements, 300
    Scanning Electron Microscopy, 302
    References, 312

19 **The Statistical Evaluation of Adherence Assays** 315
    A.D. WOOLFSON
    Parametric and Non-Parametric Hypothesis Testing, 316
    Adherence Assays: Comparison of Results in Two Independent Groups by
        Parametric and Non-Parametric Tests, 318
    Statistical Analysis of Multiple Comparisons, 321
    Selection and Computation Procedures, 324
    References, 324

    Index 327

Contributors

C.G. ADAIR, *School of Pharmacy, Medical Biology Centre, The Queen's University of Belfast, Belfast BT9 7BL, UK*

M. ADDY, *Department of Periodontology, Dental School, University of Wales College of Medicine, Heath Park, Cardiff CF4 4XN, UK*

D.G. ALLISON, *Department of Pharmacy, University of Manchester, Oxford Road, Manchester M13 9PL, UK*

M.J. BAZIN, *Life Sciences Division, King's College London, Campden Hill Road, London W8 7AH, UK*

T.R. BOTT, *School of Chemical Engineering, University of Birmingham, Edgbaston, Birmingham B15 2TT, UK*

M.E. CALLOW, *School of Biological Sciences, University of Birmingham, Edgbaston, Birmingham B15 2TT, UK*

B.J. CHAFFEY, *Department of Applied Biochemistry and Food Science, University of Nottingham, Sutton Bonington Campus, Loughborough LE12 5RD, UK*

J.S. COLBOURNE, *Thames Water Utilities PLC, Water and Environmental Sciences, Newgent House, Vastern Road, Reading RG1 2GB, UK*

P.E. COOK, *Department of Food Science and Technology, University of Reading, Reading RG6 2AP, UK*

J.W. COSTERTON, *Department of Biological Sciences, University of Calgary, Calgary, Alberta, Canada T2N 1N4*

K. DAEMS, *Department of Applied Biochemistry and Food Science, University of Nottingham, Sutton Bonington Campus, Loughborough LE12 5RD, UK*

M.C. DAVIES, *Department of Pharmaceutical Sciences, University of Nottingham, Nottingham NG7 2RD, UK*

P.L.J. DENNIS, *Thames Water Utilities PLC, Water and Environmental Sciences, Newgent House, Vastern Road, Reading RG1 2GB, UK*

S.P. DENYER, *Department of Pharmacy, University of Brighton, Moulsecoomb, Brighton BN2 4GJ, UK*

C.E.R. DODD, *Department of Applied Biochemistry and Food Science, University of Nottingham, Sutton Bonington Campus, Loughborough LE12 5RD, UK*

A.B. DOWSETT, *Pathology Division, PHLS Centre for Applied Microbiology and Research, Porton Down, Salisbury SP4 0JG, UK*

T.W. FRASER, *Department of Agriculture for Northern Ireland, Agriculture and Food Science Centre, Newforge Lane, Belfast BT9 5PX, UK*

C.C. GAYLARDE, *Department of Soils, University of Rio Grande do Sul, Porto Alegre, RS, Brazil*

ix

P. GILBERT, *Department of Pharmacy, University of Manchester, Oxford Road, Manchester M13 9PL, UK*

A. GILMOUR, *The Queen's University of Belfast and Department of Agriculture for Northern Ireland, Agriculture and Food Science Centre, Newforge Lane, Belfast BT9 5PX, UK*

S.P. GORMAN, *School of Pharmacy, Medical Biology Centre, The Queen's University of Belfast, Belfast BT9 7BL, UK*

G.W. HANLON, *Department of Pharmacy, University of Brighton, Moulsecoomb, Brighton BN2 4GJ, UK*

J. JASS, *Department of Biological Sciences, Hatherly Laboratories, University of Exeter, Exeter EX4 4PS, UK*

C.W. KEEVIL, *Pathology Division, PHLS Centre for Applied Microbiology and Research, Porton Down, Salisbury SP4 0JG, UK*

S.L. KINNIMENT, *School of Pure and Applied Biology, University of Wales College of Cardiff, Cardiff CF1 3TL, UK*

H.M. LAPPIN-SCOTT, *Department of Biological Sciences, Hatherly Laboratories, University of Exeter, Exeter EX4 4PS, UK*

M.J. LARKIN, *School of Biology and Biochemistry, The Queen's University of Belfast, Belfast BT9 5AG, UK*

J.M. LYNCH, *Horticulture Research International, Worthing Road, Littlehampton BN17 6LP, UK*

C.W. MACKERNESS, *Pathology Division, PHLS Centre for Applied Microbiology and Research, Porton Down, Salisbury SP4 0JG, UK*

W.M. MAWHINNEY, *School of Pharmacy, Medical Biology Centre, The Queen's University of Belfast, Belfast BT9 7BL, UK*

W.W. NICHOLS, *Department of Infection Research, ZENECA Pharmaceuticals, Mereside, Alderley Park, Macclesfield SK10 4TG, UK*

S. PATRICK, *Department of Microbiology and Immunobiology, The Queen's University of Belfast, Grosvenor Road, Belfast BT12 6BN, UK*

T. RACHWAL, *Thames Water Utilities PLC, Water and Environmental Sciences, Newgent House, Vastern Road, Reading RG1 2GB, UK*

J. ROGERS, *Pathology Division, PHLS Centre for Applied Microbiology and Research, Porton Down, Salisbury SP4 0JG, UK*

R. SANTOS, *School of Chemical Engineering, University of Birmingham, Edgbaston, Birmingham B15 2TT, UK*

M.A. SCOURFIELD, *Department of Dermatology, University of Wales College of Medicine, Heath Park, Cardiff CF4 4XN, UK*

M.A. SLAYNE, *Department of Periodontology, Dental School, University of Wales College of Medicine, Heath Park, Cardiff CF4 4XN, UK*

L. SUN, *Life Sciences Division, King's College London, Campden Hill Road, London W8 7AH, UK*

W.G. WADE, *Department of Periodontology, Dental School, University of Wales College of Medicine, Heath Park, Cardiff CF4 4XN, UK*

W.M. WAITES, *Department of Applied Biochemistry and Food Science, University of Nottingham, Sutton Bonington Campus, Loughborough LE12 5RD, UK*

M.H. WILCOX, *Department of Experimental and Clinical Microbiology, University of Sheffield Medical School, Beech Hill Road, Sheffield S10 2RX, UK*

A.B. WILSON, *The Queen's University of Belfast, Agriculture and Food Science Centre, Newforge Lane, Belfast BT9 5PX, UK*

J.W.T. WIMPENNY, *School of Pure and Applied Biology, University of Wales College of Cardiff, Cardiff CF1 3TL, UK*

A.D. WOOLFSON, *School of Pharmacy, Medical Biology Centre, The Queen's University of Belfast, Belfast BT9 7BL, UK*

T.D. WYATT, *Microbiology Department, Mater Infirmorum Hospital, Crumlin Road, Belfast BT14 6AB, UK*

Preface

The consequences, beneficial and deleterious, of the association between microbes and surfaces have long been recognized. As far back as the 14th century, Guy de Chauliac, a French surgeon, recorded the relationship between the presence of foreign bodies and delayed wound healing (Voorhees, 1985, *Archives of Surgery*, **120**, 289−295), while just over a century ago the symbiosis between *Rhizobium* and the roots of leguminous plants was first recorded (Beijerinck, 1888, *Botanische Zeitung*, **46**−**50**, 725−804). Although the first detailed description of microbial attachment to surfaces appeared exactly 50 years ago (Zobell, 1943, *Journal of Bacteriology*, **46**, 39−56), it was not until the late 1970s that the word *biofilm* made its appearance in the scientific literature.

The almost universal association between microorganisms and surfaces is now widely accepted. Mature biofilms may contain as many as 10^{16} cells/m^3, which is equivalent to 10^{10} cells/ml, and considerably more than usually arise in suspension. Indeed, many regard surfaces as the preferred site for microbial growth. It is hardly surprising, therefore, that the presence of a biofilm on a surface can have profound effects, including those on health, water-dependent industrial processes, energy efficiency and water quality. It has been estimated that, in Britain alone, the cost of biofouling represents some 0.5% of the Gross National Product (Pritchard, 1981, in *Fouling of Heat Transfer Equipment*, edited by Somerscales & Knudsen, pp. 513−523. Hemisphere, Washington). Moreover, considerable morbidity and mortality results from the refractory nature of the microbial biofilms that arise on implanted medical devices during infections.

The study of microbial attachment and biofilm formation and growth draws on many scientific disciplines, including microbiology, biochemistry, chemistry, medicine, dentistry, water technology, engineering and food science. We were fortunate to recruit colleagues from all these disciplines to contribute to a Demonstration Meeting of the Society for Applied Bacteriology, held at the Department of Food Science and Technology, University of Reading, on 25 September 1991. This volume, the 30th in the Technical Series of the Society, brings together as a permanent record some new techniques as well as some older established methods that were presented on that occasion. We hope that they will prove useful, not only for the exploration and characterization of biofilms, but also in the search for means to control their adverse activities. As in other branches of scientific effort, it is the careful application of reproducible methods that provides the best hope of a fuller understanding.

We should like to thank all those who presented the original demonstrations in Reading and also turned them into the chapters that make up this book. The smooth running of the meeting owed much to Mr A.J. Reynolds, the local organizer, and Dr Susan Passmore, the Society's Meetings Secretary.

S.P. Denyer
S.P. Gorman
M. Sussman

1: Microbial Biofilm Formation and Characterization

H.M. LAPPIN-SCOTT[1], J. JASS[1] AND J.W. COSTERTON[2]

[1]*Department of Biological Sciences, Hatherly Laboratories, University of Exeter, Exeter EX4 4PS, UK; and* [2]*Department of Biological Sciences, University of Calgary, Calgary, Alberta, Canada T2N 1N4*

Biofilms form on available surfaces in virtually all aquatic ecosystems that can support microbial growth. The growth of bacteria on surfaces is now an interdisciplinary research topic of interest to microbiologists, engineers, ecologists and chemists. In order to increase our understanding of the formation and control of biofilms, techniques to study these processes have been developed. Some of these methods involve disrupting the biofilm from the surface before growth parameters are measured. Examples of such parameters include the heterotrophic potential, viable cell counts and total counts obtained with acridine orange staining and epifluorescence microscopy. There are clearly limitations to such approaches, because biofilms disrupted from surfaces do not have the same properties as those still intact on a surface (Murray *et al.*, 1987). However, many of the more recent developments in this field have included direct, non-disruptive methods that allow observations of biofilms without removing them from the surface to which they are attached. Such techniques include Fourier transform infrared (FT-IR) spectroscopy (Nichols *et al.*, 1985; Nivens *et al.*, 1986; Jolley *et al.*, 1989; Geesey & Bremer, 1990), confocal laser microscopy (Caldwell & Lawrence, 1989a) and nuclear magnetic resonance (NMR) (Hoyle *et al.*, 1990). This chapter includes some of these techniques to monitor the formation and control of biofilms. The reader is directed to reviews on the formation and control of biofilms by Costerton *et al.* (1987) and Lappin-Scott & Costerton (1989). Throughout this chapter, the attached microorganisms are termed the sessile population and those living in the aqueous phase are termed the planktonic population.

Formation of Biofilms

The Robbins device

In order to study biofilms it was first necessary to develop a suitable apparatus to control the flow of microorganisms over a surface and provide samples of the biofilms at various stages of development. The first such apparatus — the Robbins device — was constructed in our laboratory at the University of Calgary by Jim Robbins more than 10 years ago. It was composed of a 92.5 cm long hollow steel pipe with 12 clusters of four sample ports orientated at 90° to each other around the pipe, providing a total of 48 sampling ports (McCoy *et al.*, 1981; Ruseska *et al.*, 1982). Each sample stud was made from a metal screw to which a surface could be attached. The device was used to study both changes in biofilm formation with time, and methods of controlling biofilm growth with antimicrobial agents. Since this prototype was developed as an *in situ* sampling device for monitoring biofilms in industrial pipelines it was rather cumbersome, and not directly applicable for use in many medical systems. Some modifications were later made to develop the modified Robbins device (MRD). The MRD has usefully been applied in the study of biofilms in areas as diverse as medical, environmental and industrial systems.

The MRD is constructed from a 41.5 cm long perspex block, 20 mm high and 26 mm across with a rectangular lumen approximately 2 mm deep and 10 mm wide, containing 25 evenly spaced sample ports each 11 mm in diameter, with polypropylene connectors at either end for attaching the tubing from the culture vessel (Fig. 1.1, A). The sample studs, also made of perspex, are 26 mm in length and are inserted into the sample ports. The stud design includes a bevelled upper section (14 mm in depth) for easy removal from the MRD. The lower section of the stud (12 mm in depth) is designed to fit tightly into

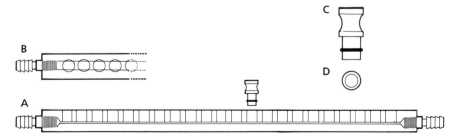

FIG. 1.1. The modified Robbins device showing the sample ports and the stud design. A, Cross-section showing the lumen within the perspex block, the sample ports and a stud. B, View from above. C, Cross-section of a stud. D, View of the 1 cm rim of the stud within which the discs are inserted.

the MRD and this is assisted by a rubber O-ring which effectively seals the port to prevent leakage. The O-ring is positioned in a narrow groove 21 mm from the top of the stud. The bottom surface of the sample stud has a 1 mm rim and the surface to be colonized is cut into a disc (termed the surface disc) and placed inside this rim, together with a backing disc.

The procedure for attaching the surface disc is as follows: black rubber backing discs are cut from 1.6 mm thick silicone rubber sheets (Esco Rubber, Bibby Sterilin Ltd, Stone, Staffordshire, UK). The backing is cut from the sheet with a hammer and an 8.5 mm punch which cuts surface and backing discs to fit exactly onto the stud. The backing disc is glued on to the surface disc with cyanoacrylate glue (Loctite Super Glue 3) and then both are pushed firmly into the stud, with the surface disc uppermost. The surface of the disc is then gently cleaned with alcohol to remove any contaminants before the stud is inserted into the MRD. When all of the studs are in place the discs are adjusted until they lie flush with the lumen to minimize current eddies around the surface. Before use the MRD is sterilized with ethylene oxide gas.

The Robbins device and batch culture growth

After the MRD has been prepared and sterilized it can be attached to either a batch culture or a chemostat culture of microorganisms. The batch apparatus consists of a growth medium reservoir connected to a peristaltic pump and the MRD via silicone rubber tubing. After the culture has been pumped through the MRD it is either recirculated back to the reservoir or pumped directly into an effluent container. For studies of the formation of biofilms the growth medium is inoculated with an exponentially growing culture, maintained at the required temperature by a water bath or in a constant temperature room. The culture is continuously pumped through the MRD at 50–60 ml/h. The duration of the run is determined by the operator, but is usually of the order of 48 h. Studs are removed aseptically at various time intervals and examined by a combination of techniques (Fig. 1.2), including viable cell counts, epifluorescence microscopy, scanning electron microscopy, the heterotrophic potential, total carbohydrate and total protein.

For studies of the control of biofilms, the above method for the formation of the biofilm is undertaken and then the growth reservoir is aseptically removed and replaced by a vessel containing one or more antimicrobial agents. For example, Nickel *et al.* (1985) and Jass (1990) pumped tobramycin and β-lactams respectively through the MRD to determine the concentration required to control biofilm growth. At various time intervals the studs are removed and sampled, as described in Fig. 1.2. Changes in the viable cell counts indicate the effectiveness of the antimicrobial agent on sessile populations.

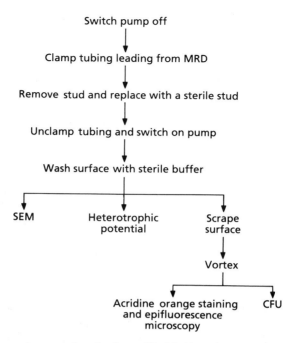

Fɪɢ. 1.2. The sampling procedure for the modified Robbins device. SEM, Scanning electron microscopy.

The Robbins device and chemostat growth

The initial systems introduced for biofilm studies used microorganisms growing in batch culture. However, an improved understanding of the importance of the growth rate on biofilm formation (Keevil *et al.*, 1987; Brown *et al.*, 1988) led to the development of chemostat models to study sessile growth. Several methods have been reported to generate biofilms at growth rates below μ_{max} (Keevil *et al.*, 1987; Gilbert *et al.*, 1989). Work in our laboratories has focused on methods to attach an MRD to a chemostat. The benefits of such an approach are that it offers a method for studying biofilm formation at different growth rates; that it is an easily sampled, reproducible system, and that the 25 sample ports allow biofilm formation and control to be accurately monitored over time. The different types of continuous culture systems, including chemostats, have been reviewed by Gottschal (1990).

There are two methods of attaching an MRD to a chemostat. First, it can be attached closely to the chemostat on the effluent line and the planktonic population pumped out at a flow rate controlled by the dilution rate, through the MRD and over the surface discs. This is termed the 'flow-through' system and is used to study biofilms where intermittent flow conditions prevail — that is, where the surfaces are not continuously submerged in planktonic microorganisms. With this system the flow rate through the MRD cannot be controlled separately from the growth rate, and a mixture of air and culture is dispensed from the chemostat over the surfaces (Jass *et al.*, 1992). With this approach in our laboratories, *Pseudomonas fluorescens* has been shown to form a dense biofilm (Fig. 1.3; Jass *et al.*, 1992). This system has the major advantages that contamination of the planktonic population via the MRD is minimized, and that the MRD (and biofilm) is separated from the planktonic population. This also has the advantage that, after the biofilm has formed, the MRD can be disconnected from the chemostat and used for further studies, or it can be attached to a second chemostat operating under a different set of growth conditions.

Secondly, the MRD can be attached to a loop of silicone tubing that is connected from the chemostat, through a peristaltic pump, to the MRD and

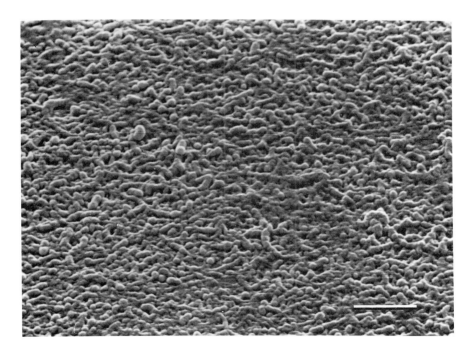

FIG. 1.3. Scanning electron micrograph showing a biofilm of *Pseudomonas fluorescens* growing on silastic rubber. The modified Robbins device was attached to a chemostat with the flow-through system. Bar represents 5 μm.

then back into the chemostat. The culture is constantly pumped through the MRD and the studs are continuously immersed in culture. This is termed the *recirculating system* and is used to study biofilms where the planktonic cells are growing at a specific growth rate and the flow rate over the MRD can be separately controlled.

The MRD is a very versatile apparatus for studying both the formation and control of biofilms. By changing the surface disc it is possible to investigate the colonization of a variety of materials, and to investigate the best methods of removing the biofilm by exploring different combinations and concentrations of antimicrobial agents. The choice of establishing biofilms on the MRD from batch or chemostat cultures extends these facilities to situations where it is relevant to work below μ_{max}. The MRD is particularly useful for studying biofilm formation in tubular devices such as catheters or industrial pipelines.

Methods for Study of Biofilm Formation and Control

Fourier transform infrared spectroscopy

Fourier transform infrared spectroscopy (FT-IR) is a powerful, non-destructive technique which can continuously analyse the chemical nature of bacterial attachment and biofilm formation (Nichols *et al.*, 1985) by applying attenuated total reflectance (ATR) optics. In ATR spectroscopy, infrared radiation is directed through an internal reflectance element (IRE), such as germanium or zinc selenide crystals. Bacteria are attached to an IRE by pumping microorganisms grown either in a batch culture (Jass, 1990) or a chemostat (Bremer & Geesey, 1991) over the surface (Fig. 1.4(a)). Since the IRE is of a higher refractive index than the biofilm, the infrared radiation, which is directed into the IRE at an angle of incidence greater than the critical angle, will be totally internally reflected along the length of the IRE (Fig. 1.4(b)). Under these conditions some radiation leaves the IRE at the crystal−liquid interface, and is termed the *evanescent wave*. The distance the evanescent wave travels into the film depends on the refactive index of the IRE, the wavelength and the angle of incidence. For example, the depth of the evanescent wave for germanium is $0.3-0.7\,\mu m$, and for zinc selenide is $1-2\,\mu m$ in the region of $2000-1000/cm$ (Fig. 1.4(c)). The resultant spectra are then due to the attenuation of the absorbed radiation by the molecules present within the range of the evanescent wave, at the immediate surface of the IRE. Thus, when the IRE is colonized by microorganisms the infrared radiation will be reduced by the absorbance of molecules from the biofilm. One type of IRE is a cylindrical internal reflectance element (Spectra Tech, Stanford, USA) in a macrocircle cell assembly (Fig. 1.4(a)).

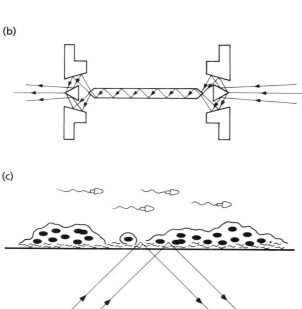

FIG. 1.4. (a) The attenuated total reflectance circle cell apparatus, including the circle cell, the cylindrical internal reflectance element and the mirrors. See text for details. (b) A schematic diagram of the optics of the cylindrical IRE. (c) A schematic diagram of the beam penetrating into the biofilm on the surface of the IRE.

The application of interferometric instrumentation and spectral enhancement techniques to infrared spectroscopy allows the use of aqueous solvents in biological systems. Many molecules within biological systems have absorption bands in the same region (around 1600/cm) as the broad water band in aqueous systems. By use of the Fourier transform instrumentation the water band can effectively be subtracted from the spectrum. Therefore, a Fourier transform infrared spectrometer employing a liquid nitrogen-cooled mercury—cadmium—teluride detector in the mid-range (4000—400/cm) must be used in conjunction with a computer dedicated to the task.

The use of single ATR FT-IR to monitor events of real-time analysis of a biological film formed in seawater was developed by Nichols *et al.* (1985) at Florida State University. They used a germanium IRE and collected 100 interferograms at a resolution of 8/cm, and the resulting spectra were calculated with a separate dry IRE cell as a standard. The absorbance contribution from the water was then subtracted and the resultant spectra were characteristic of the biological film that accumulated on the surface of the IRE. A band of increasing intensity over time at 1100/cm was observed, which was predominantly the C—O stretch of the carbohydrates. This technique was further adapted by Geesey *et al.* (1986) to monitor the real-time events of the deterioration of copper by exopolymers and sessile populations. By coating a germanium circular IRE with a fine film of copper, its deterioration due to bacterial polysaccharides and biofilms was observed by its appearance and changes to the water absorption band at 1640/cm (Geesey *et al.*, 1987; Jolley *et al.*, 1989). Geesey and his colleagues also used zinc selenide IRE to sample further into the aqueous phase (Geesey *et al.*, 1986, 1987).

Bremer & Geesey (1991) developed a double ATR FT-IR system that collected concurrent spectra for the reference and the sample. The modelling of a double-beam spectrometer eliminates any outside changes in the instrumentation during the experiment, thus producing spectra with potentially no artefacts. Spectra were collected in the range of 2000—1000/cm, with a resolution of 4/cm on a Perkin Elmer 1800 Fourier transform infrared spectrometer. The double-sided interferograms were apodized with a weak Beer—Norton function prior to the Fourier transformation, and the single-beam spectra were calculated with either a control spectrum obtained prior to inoculation with bacteria, or the spectrum from the reference cell. The amide I (1638/cm), II (1578 and 1536/cm) and III (1275/cm) absorbance bands, attributed to the protein components of the biofilm, were monitored over 188 h. Although Bremer & Geesey (1991) appreciated that the double-beam system could eliminate artefacts and so improve the interpretation of the spectra, a significant difference between the two systems was not apparent in this experiment.

ATR FT-IR was also developed to monitor the penetration of β-lactams through biofilms to the crystal interface (Jass, 1990). A biofilm of *Ps. aeruginosa* was formed on a cylindrical germanium IRE and then challenged by β-lactam antibiotics. A Digilab FTS-60 Fourier transform infrared spectrometer with a mid-range detector was used to collect 400 interferograms at a 4/cm resolution. The interferograms were double-sided and apodized with a triangular function prior to Fourier transformation. The spectra were then calculated with a background spectrum obtained at zero time. The increase in the band at 1768/cm demonstrated the presence of the antibiotic β-lactam carbonyl bond within 0.6 μm of the surface of the germanium IRE.

FT-IR spectroscopy is ideal for observing real-time events of a chemical nature in biofilms on a surface of a germanium or zinc selenide IRE. ATR FT-IR requires the use of very sensitive computer-controlled equipment by a skilled operator. Data interpretation can be laborious and time-consuming, especially when large numbers of spectra are obtained. Care must be taken to ensure that artefacts in the spectra are not introduced, because these can be mistaken for chemical changes. The greatest difficulty with this technique is the interpretation of the spectra, since aqueous systems are so chemically complex.

^{13}C Nuclear magnetic resonance

This method has been used to monitor the uptake and dissimilation of radiolabelled glucose by biofilm inhabitants (Hoyle *et al.*, 1990). The biofilm was generated in a novel manner by incorporating sterile glass wool into growth medium. For example, Hoyle *et al.* (1990) grew *Ps. aeruginosa* 579 or *Escherichia coli* MER 600 in brain−heart infusion broth containing glass wool and incubated at 36°C for 24 h. The glass wool was recovered from the broth and the planktonic bacteria were removed by washing the wool five times in 20 ml of 10 mmol/l hydroxyethylpiperazine ethanesulphonic acid (HEPES), pH 7.0. It was reported that the washed glass wool contained up to 10^{10} cfu/g, and this dense biofilm was observed by scanning electron microscopy. After the biofilm was established within the glass wool, the wool was placed in an NMR widebore tube (1 cm diameter) containing 2−5 ml HEPES buffer and 50 mM of ^{13}C glucose labelled at the C-1, C-2 or C-6 position. The biofilm was aerated by the introduction of sterile air through a Pasteur pipette inserted into the widebore tube, and the temperature was maintained at 25°C. While the sessile microorganisms utilized the ^{13}C glucose, the NMR spectra were collected on a 50 MHz Fourier transform JEOL FX-270 NMR spectrometer. Each spectrum was composed of 700-1000 scans at 1 s intervals. The chemical shifts were assigned relative to an external standard of trimethylsilane. Hoyle *et al.* (1990) used this technique to monitor the conversion of ^{13}C glucose to ethanol and acetate by *E. coli*.

This technique can be used to monitor *in situ* physiological events that occur within biofilms. To date such physiological studies have focused on the planktonic population, so the techniques described in this chapter extend these facilities to sessile populations. The radiolabelled carbon can be incorporated into a range of different carbon sources to follow the utilization of these by biofilm inhabitants. This technique requires a complex, well developed, mature biofilm. High concentrations of sugars — in the milligram range — are required, and this does not always reflect conditions *in situ*.

Scanning confocal laser microscopy and computer-enhanced microscopy

Scanning confocal laser microscopy (SCLM) and computer-enhanced light microscopy were developed by Caldwell and his colleagues at the University of Saskatchewan to study attachment rates, detachment rates, exopolymer production, cell metabolism and viability and responses to antimicrobial agents within biofilms (Caldwell, 1985; Caldwell & Germida, 1985; Lawrence *et al*., 1987; Caldwell & Lawrence, 1989b; Delaquis *et al*., 1989; Lawrence *et al*., 1989). The computer enhancement of microscopy images increases the depth of field of a normal light microscope. For example, Caldwell & Lawrence (1989a) developed a continuous-flow slide culture apparatus to observe microbial attachment and biofilm formation on the surface of the flow cell wall, and used computer-enhanced microscopy and image processing to study the growth kinetics of an attached population of *Ps. fluorescens*. By digitizing and storing a microscopic image, the dynamic nature of bacterial adhesion and biofilm formation can be observed by direct comparison and image subtraction. This technique is limited by the fact that the complete cell or microcolony is not all in the same focal plane, hence only a two-dimensional image can be obtained. This limitation was overcome by the development of SCLM for bacterial adhesion and biofilm studies (Caldwell *et al*., 1992). SCLM is used to produce thin horizontal or vertical sections of biofilms (0.2 μm thick) and eliminates out-of-focus haze within biofilms. The haze is removed because the SCLM uses pinhole apertures at the laser and the detector, which effectively remove any stray light. Since only the light in the plane of focus is directed to the detector and all other light is filtered out, the in-focus image will consist of a very thin section of the specimen. Thus, as the beam scans the specimen, it will obtain a two-dimensional image of this very thin section. To obtain a three-dimensional image the beam has to be focused at a different depth of the specimen, and a three-dimensional composition may be constructed. By scanning the image at time intervals a dynamic impression of the events can be monitored. This technique may be used in conjunction with various stains, such as a negative fluorescent stain (resazurin or fluorescein) used to study microcolony formation (see Chapter 5).

Conclusions

There have been many exciting developments in the use of techniques to study events occurring within biofilms, and this chapter has dealt with only a few. The combinations of biofilm sampling devices, such as the MRD and the non-destructive techniques described here, should improve our understanding of the formation of biofilms and, in turn, improve the control of biofilm formation.

References

BREMER, P.J. & GEESEY, G.G. (1991) An evaluation of biofilm development utilizing attenuated total reflectance Fourier transform infrared spectroscopy. *Biofouling*, 3, 89−100.

BROWN, M.R.W., ALLISON, D.G. & GILBERT, P. (1988) Resistance of bacterial biofilms to antibiotics: a growth rate related effect? *Journal of Antimicrobial Chemotherapy*, 22, 777−780.

CALDWELL, D.E. (1985) New developments in computer-enhanced microscopy (CEM). *Journal of Microbiological Methods*, 4, 117−125.

CALDWELL, D.E. & GERMIDA, J.J. (1985) Evaluation of difference imagery for visualizing and quantitating microbial growth. *Canadian Journal of Microbiology*, 31, 35−44.

CALDWELL, D.E. & LAWRENCE, J.R. (1989a) Microbial growth and behaviour within surface microenvironments. In Hattori, T., Ishida, Y., Maruyama, Y., Morita, R.Y. & Uchida, A. (eds.) *Recent Advances in Microbial Ecology*, pp. 140−145. Japan Scientific Societies Press, Tokyo.

CALDWELL, D.E. & LAWRENCE, J.R. (1989b) Image analysis and computer modelling of microbial growth on surfaces. *Binary*, 1, 147−150.

CALDWELL, D.E., CORBER, D.R. & LAWRENCE, J.R. (1992) Confocal laser microscopy and computer image analysis in microbial ecology. *Advances in Microbial Ecology*, 12, 1−67.

COSTERTON, J.W., CHENG, K.-J., GEESEY, G.G., LADD, T.I., NICKEL, J.C., DASGUPTA, M. & MARRIE, T.J. (1987) Bacterial biofilms in nature and disease. *Annual Review of Microbiology*, 41, 435−464.

DELAQUIS, P.J., CALDWELL, D.E., LAWRENCE, J.R. & McCURDY, A.R. (1989) Detachment of *Pseudomonas fluorescens* from biofilm on glass surfaces in response to nutrient stress. *Microbial Ecology*, 18, 199−210.

GEESEY, G.G. & BREMER, P.J. (1990) Applications of Fourier transform infrared spectroscopy to studies of copper corrosion under bacterial biofilms. *Marine Technology Society Journal*, 24, 36−43.

GEESEY, G.G., MITTELMAN, M.W., IWAOKA, T. & GRIFFITHS, P.R. (1986) Role of bacterial exopolymers in the deterioration of metallic copper surfaces. *Materials Performance*, 25, 37−40.

GEESEY, G.G., IWAOKA, T. & GRIFFITHS, P.R. (1987) Characterization of interfacial phenomena occurring during exposure of a thin copper film to an aqueous suspension of an acidic polysaccharide. *Journal of Colloid and Interface Science*, 120, 370−376.

GILBERT, P., ALLISON, D.G., EVANS, D.J., HANDLEY, P.S. & BROWN, M.R.W. (1989) Growth rate control of adherent bacterial populations. *Applied and Environmental Microbiology*, 55, 1308−1311.

GOTTSCHAL, J.C. (1990) Different types of continuous culture in ecological studies. In Grigorova, R. & Norris, J.R. (eds.) *Methods in Microbiology*, Vol. 22, pp. 87−125. Academic Press, London.

HOYLE, D.B., EZRA, F.S. & RUSSELL, A.F. (1990) ^{13}C *Nuclear Magnetic Resonance of Bacterial*

Biofilms. American Society for Microbiology Abstract Programme, Abstract I-61, 90th Annual Meeting of the American Society for Microbiology, May 1990. American Society for Microbiology, Washington.

JASS, J. (1990) β-*Lactam Penetration through Pseudomonas Aeruginosa Biofilms Monitored by ATR-FTIR Spectroscopy.* MSc Thesis, University of Calgary, Canada.

JASS, J., SHARP, E.V., COSTERTON, J.W. & LAPPIN-SCOTT, H.M. (1992) *Colonization of Silastic Rubber by Pseudomonas fluorescens and Pseudomonas putida Using a Chemostat and a Modified Robbins Device.* American Society for Microbiology Programme, Abstract D266. 92nd Annual Meeting of the American Society for Microbiology, May 1992. American Society for Microbiology, Washington.

JOLLEY, J.G., GEESEY, G.G., HANKINS, M.R., WRIGHT, R.B. & WICHLACZ, P.L. (1989) *In situ*, real-time FT-IR/CIR/ATR study of the biocorrosion of copper by gum arabic, alginic acid, bacterial culture supernatant and *Pseudomonas atlantica* exopolymer. *Applied Spectroscopy*, **43**, 1062–1067.

KEEVIL, C.W., BRADSHAW, D.J., DOWSETT, A.B. & FEARY, T.W. (1987) Microbial film formation: dental plaque deposition on acrylic tiles using continuous culture techniques. *Journal of Applied Bacteriology*, **62**, 129–138.

LAPPIN-SCOTT, H.M. & COSTERTON, J.W. (1989) Bacterial biofilms and surface fouling. *Biofouling*, **1**, 323–342.

LAWRENCE, J.R., DELAQUIS, P.J., KORBER, D.R. & CALDWELL, D.E. (1987) Behavior of *Pseudomonas fluorescens* within the hydrodynamic boundary layers of surface microenvironments. *Microbial Ecology*, **14**, 1–14.

LAWRENCE, J.R., MALONE, J.A., KORBER, D.R. & CALDWELL, D.E. (1989) Computer image enhancement to increase depth of field in phase contrast microscopy. *Binary*, **1**, 181–185.

McCOY, W.F., BRYERS, J.D., ROBBINS, J. & COSTERTON, J.W. (1981) Observations of fouling biofilm formation. *Canadian Journal of Microbiology*, **27**, 910–917.

MURRAY, R.E., COOKSEY, K.E. & PRISCU, J.C. (1987) Influence of physical disruption on growth of attached bacteria. *Applied and Environmental Microbiology*, **53**, 2997–2999.

NICHOLS, P.D., HENSON, J.M., GUCKERT, J.B., NIVENS, D.E. & WHITE, D.C. (1985) Fourier transform–infrared spectroscopic methods for microbial ecology: analysis of bacteria, bacteria–polymer mixtures and biofilms. *Journal of Microbiological Methods*, **4**, 79–94.

NICKEL, J.C., RUSESKA, I., WRIGHT, J.B. & COSTERTON, J.W. (1985) Tobramycin resistance of *Pseudomonas aeruginosa* cells growing as a biofilm on urinary catheter material. *Antimicrobial Agents and Chemotherapy*, **27**, 619–624.

NIVENS, D.E., NICHOLS, P.D., HENSON, J.M. GEESEY, G.G. & WHITE, D.C. (1986) Reversible acceleration of the corrosion of AISI 304 stainless steel exposed to seawater induced by growth and secretions of the marine bacterium *Vibrio natriegens*. *Corrosion*, **42**, 204–210.

RUSESKA, I., ROBBINS, J., COSTERTON, J.W. & LASHEN, E.S. (1982) Biocide testing against corrosion-causing oil-field bacteria helps control plugging. *Oil & Gas Journal*, 8 March, 153–164.

2: Mechanisms of Microbial Adherence

S.P. DENYER[1], G.W. HANLON[1] AND M.C. DAVIES[2]

[1]*Department of Pharmacy, University of Brighton, Moulsecoomb, Brighton BN2 4GJ; and* [2]*Department of Pharmaceutical Sciences, University of Nottingham, Nottingham NG7 2RD, UK*

Most microorganisms can attach to surfaces; the rate of attachment and the avidity of the interaction reflect an interplay between microbial and substratum surface characteristics, previous conditioning of the substratum and fluid shear stress. This attachment is the first step in the colonization of surfaces and precedes the process of consolidation, during which the initially weak adhesive forces are strengthened by exopolymer formation and, finally, by growth to form an established biofilm.

The pattern and extent of attachment can be described by the adsorption isotherm (Fig. 2.1), which may be of the type I (Langmuir) or type II (sigmoid) form. Type I uptake shows a rapid rise in the number of attached cells, with increasing concentration up to a limiting value; such isotherms indicate that the adsorption is restricted to a monolayer. The sigmoid type II isotherm represents multilayer adsorption to non-porous surfaces, and the inflection point represents the formation of a complete monolayer. Instances of apparently constant uptake, where surface saturation kinetics do not apply, have been reported, but it is likely that these reflect a type II isotherm with poorly defined inflection.

Strength of Attachment

Methods for assessing the strength of bacterial attachment may be based on either *cell distraction* or *dynamic adhesion* (Fowler & McKay, 1980), both of which are amenable to mathematical interpretation (Escher & Characklis, 1990).

Cell distraction methods have been reviewed by Fowler & McKay (1980), who described *adhesion number* or *critical force* tests. The former distinguish

Microbial Biofilms:
Formation and Control

13

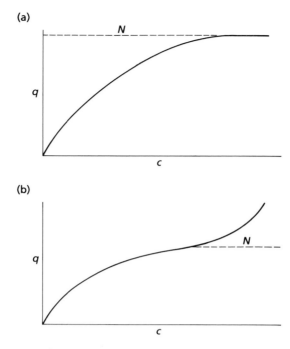

FIG. 2.1. Adsorption isotherms. (a) Type I (Langmuir) isotherm. (b) Type II (multilayer) isotherm. q, Number of adherent bacteria; c, bacterial concentration; N, number of binding sites per unit area.

between adherent and non-adherent cells by the application of a fixed distraction force to a population of cells which have interacted with a surface. Various modifications of this type of test have been described in which accurately definable distractive forces are achieved, or where the shear rate is quantifiable. The limitation of the adhesion number test, however, is that it is a study at one particular shear force.

The critical force test examines a population of adherent cells under a steadily increasing fluid shear stress, until detachment occurs. Mohandas *et al.* (1974) used both a parallel-plate flow system and a rotating-disc system to find a minimum critical shear stress above which cells would eventually detach, but which must be reached for distraction to occur. This critical shear value was related to the surface characteristics of the substratum.

Dynamic adhesion methods take into account the initial physicochemical adsorption stage of adhesion, as opposed to cell distraction methods, which tend to concentrate on the secondary stabilization stage. Fowler & McKay (1980) described a radial flow chamber which consisted of two parallel plates separated by a narrow spacing. Culture fluid containing bacteria was pumped at a constant volumetric flow rate through the centre of one disc and then through the gap between the discs in a radial fashion. This resulted in a radial shear stress gradient across the space between the discs. Under these conditions, at a critical radius the fluid shear is too high to permit bacterial attachment; outside this radius the bacteria can adhere and form a biofilm. The critical radius can be used to calculate the critical shear stress, which is a measure of the initial adhesive forces involved in attachment.

Interfacial Forces and Interactions

A study of the attachment of bacteria to surfaces requires an understanding of the physicochemical characteristics of the two surfaces and the interactions between them (Oliveira, 1992). In general, both surfaces possess an overall negative charge and, for attachment to take place, the resulting electrostatic repulsion barrier must be overcome by attractive forces (Table 2.1). The eventual outcome of this interplay between competing forces is governed by thermodynamic principles (Absolom *et al.*, 1983) and is described by the DLVO theory, which derives its name from the authors who proposed it (Derjaguin & Landau, 1941; Verwey & Overbeek, 1948). The shorter the distance between substratum and bacterial cell, adhesive forces begin to predominate, and this is favoured by the presence of cellular appendages and extracellular polymers (Oliveira, 1992).

Characklis (1990) differentiated between *reversible adsorption*, arising primarily from long-range interaction forces, and *irreversible adhesion*, generally considered to result from a more definitive interaction, such as that mediated

TABLE 2.1. *Forces involved in microbial adhesion to surfaces*

Type of interaction	Interaction forces	Approximate interaction energy (kJ/mol)
Reversible Long range, weak, low specificity	van der Waals Electrostatic	20–50
Irreversible Short range, generally high specificity	Dipole–dipole Dipole-induced dipole Ion–dipole Ionic Hydrogen bonds Hydrophobic	40–400

by a specific adhesin. The latter interactions rely largely on the short-range physical forces of attraction summarized in Table 2.1, and are optimized by the close spatial arrangement of the interacting component groups in an adhesin−receptor link (Hjerten & Wadstrom, 1990).

Bacterial adhesins are generally protein in nature, whereas host cell receptors are typically specific carbohydrate residues on the eukaryotic cell surface (Beachey *et al.*, 1988; Findlay & Falkow, 1989). Adhesion may be mediated by fimbrial appendages, such as the mannose-binding fimbriae of the Entero-bacteriaceae, or by binding proteins anchored directly to the microbial surface. For example, *Staphylococcus aureus* has fibronectin-binding proteins that attach to a glycoprotein (fibronectin) found in large amounts on epithelial cell surfaces (Lindberg *et al.*, 1990). Bacteria also react specifically with other connective-tissue and serum proteins, such as fibrinogen (Lindberg *et al.*, 1990), collagen, laminin and vitronectin (Wadstrom *et al.*, 1990). These proteins may, if they are involved in the conditioning of implanted medical devices, serve to mediate bacterial attachment to their surfaces (Christensen *et al.*, 1989).

Methods for Studying Surface Physicochemistry

The principal physicochemical forces that mediate attraction, in particular to inert non-biological surfaces, are those associated with surface charge and hydrophobicity, which are themselves a function of the structure and chemical composition of the surface.

The following section describes some of the methods used to study these physicochemical parameters. There are many modifications to these methods and this diversity, with the inherent plasticity of the microbial cell surface, has led to considerable differences in experimental findings. It is important to remember that each technique has intrinsic advantages and disadvantages which should be taken into account in the interpretation of the results. In general, the techniques described have found the widest application in com-paring different strains of related organisms, in order to predict adherence. Even in this limited application, it is important to stress that tight control is necessary during preparation of the inoculum, including the selection of growth medium, culture age and growth phase, metabolic status, cell pre-paration and suspending environment (Rozgonyi *et al.*, 1990), in order to minimize the influence of phenotypic variation on surface character.

Surface hydrophobicity

Hydrophobic interactions are believed to play a major part in the adherence of commensal and pathogenic organisms to their target tissues (Rozgonyi *et al.*, 1990) and similar mechanisms are likely to be responsible for adhesion to inanimate substrata. Hydrophobic interactions are brought about because water molecules situated around a non-polar solute are more highly ordered than the bulk water, and hence the establishment of a hydrophobic interaction displaces the water and thus increases the disorder of the system; in other words, there is an increase in entropy.

A variety of methods has been devised to probe the relative hydrophobicity of microbial and, in particular, bacterial cells. These methods may give differing results, because they are based on different principles and therefore measure hydrophobicity at different depths of the bacterial surface.

Partition between aqueous and hydrocarbon phases

Bacterial adherence to hydrocarbons (BATH) is a simple quantitative test for microbial hydrophobicity, first described by Rosenberg *et al.* (1980) and further developed in later studies (Rosenberg, 1984; Sharon *et al.*, 1986). Bacteria are harvested from growth medium and washed in phosphate−urea−magnesium (PUM) buffer (pH 7.1) comprising (g/l): K_2HPO_4, 16.87; KH_2PO_4, 7.26; $MgSO_4 . 7H_2O$, 0.2; urea, 1.8. The cells are resuspended in PUM buffer to an initial absorbance of approximately 1.0, measured at a suitable wavelength in the range 400−550 nm, and 1.2 ml is dispensed into each of a series of test tubes. Volumes of hydrocarbons of 0.2, 0.15, 0.1 and 0.05 ml are added to the cell suspensions, which are then held at room temperature for 10 min, vortex-mixed for 2 min and allowed to stand at room temperature to facilitate hydrocarbon separation; a short period of slow centrifugation can sometimes be of assistance. Liquid hydrocarbons that may be used include, *n*-hexane, *p*-xylene, *n*-octane and *n*-hexadecane. The absorbance of the aqueous phase after separation is measured and the results are recorded as the percentage absorbance of the aqueous phase after treatment, compared with the absorbance of the initial bacterial suspension.

The more hydrophilic bacteria will remain in the aqueous phase, whereas hydrophobic cells will migrate to the water−hydrocarbon interface, leading to significantly lowered absorbance readings. An example for *Bacteroides fragilis* can be found in Chapter 6.

The BATH technique is rapid and the results are easy to interpret; key factors that affect reproducibility and performance are discussed in Rosenberg (1984). Hexadecane is the favoured hydrocarbon for this assay, but it is important to be aware that components in the bacterial surface may be altered when in contact with any of the recommended hydrocarbons.

Hydrophobic interaction assay with octyl or phenyl sepharose gel

This procedure was developed from the technique of hydrophobic interaction chromatography (HIC) first used by Smyth *et al.* (1978). HIC measures microbial adsorption to octyl or phenyl sepharose beads packed into small

Pasteur pipette columns. A known number or concentration of bacterial cells is passed through the column and the number of cells retained after several washes is a measure of the hydrophobicity of the cell surface. Studies using this technique are described in Chapter 8. Various problems have been associated with this method, including mechanical trapping of cells within the gel (Hazen, 1990), which is a particular problem with clumped cells. To overcome this, an alternative method has been developed which involves the simple mixing of cells with the gel, a method known as hydrophobic interaction assay (HIA).

In HIA, phenyl sepharose gel is washed four times in triple-distilled water to remove any preservative and to ensure full hydration to its swollen form. It is then washed once in 0.02 mol/l phosphate buffer with 2 mol/l sodium chloride (pH 7.2; PBSS). The swollen gel (30 g) is then resuspended to a final volume of 100 ml in PBSS and maintained at 20°C with continuous stirring. Volumes of 1 ml of this gel are then placed in chromic acid-cleaned glass test tubes. To these are added $0.05-0.08$ ml of a bacterial suspension $(5 \times 10^9$ cells/ml in PBSS) and the volume is made up to 2 ml with PBSS. A similar set of control mixtures is prepared by substituting buffer for gel. All the tubes are vortex-mixed for 60 s, followed by centrifugation at $150 g$ for 2 min to sediment the gel. The absorbance of the supernatant at 420 nm is determined in each tube. The proportion of cells adsorbed to the gel can be expressed as the percentage difference in optical density between the test and control. The greater the retention on the gel the more hydrophobic the bacterial surface. Clearly, in this technique only the outer surface of the bacterium interacts with the gel, particularly when the gel ligand molecules are short (e.g. octyl sepharose).

Contact angle measurements on bacterial lawns

The hydrophobic characteristics of bacterial surfaces can be measured by contact-angle determinations on cell layers (van Oss, 1978). Current developments involve deposition of a bacterial lawn on cellulose nitrate or polycarbonate membrane filters, followed by contact-angle measurements with the sessile drop technique (Neumann & Good, 1979). Such measurements can be used directly as a measure of surface hydrophobicity: the greater the contact angle with water the more hydrophobic the surface; or, by using liquids with different surface tensions, such as water and diiodomethane, surface free energy calculations can be made with the geometric mean equation (Busscher *et al.*, 1984). A typical experimental methodology is summarized below.

Prewashed 25 mm diameter cellulose nitrate membrane filters (Millipore), with a pore size of 0.45 μm, are fixed in a filtration unit (Sartorius, UK) and further washed by passing 500 ml of triple-distilled water under vacuum. The bacterial lawns for measuring contact angles are prepared by filtering a washed suspension of cells in triple-distilled water to a final surface density of 10^8 cells/mm^2. The filters are then carefully removed from the filtration unit and placed on glass microscope slides lying on moistened filter paper in a closed Petri dish. The filters must be maintained under these constant moisture conditions until required, when they are divided into two halves, mounted and fixed on to aluminium sample discs with double-sided adhesive tape.

Aqueous contact angles are determined at 20°C with triple-distilled water of surface tension 72.6 mN/m. A small drop of *ca.* 5 μl is placed on the bacterial lawn and the contact angle is determined with a telegoniometer, or by projecting a magnified image on to a screen at a distance of 1.5 m with a modified projector. Measurements are made with an accuracy of ±2° on both halves of the filter, with at least six drops per lawn. Readings are averaged from four sample surfaces. During this process some dehydration of the filters through water evaporation may be expected, and contact angles are therefore either determined after a fixed drying period or as a function of drying time.

Recently, axisymmetric drop shape analysis has been described, which is believed to overcome some of the problems inherent in contact-angle measurements on biological substrates (Duncan-Hewitt *et al.*, 1989).

Salt aggregation tests (SAT)

First described by Lindahl *et al.* (1981), this test is based on the precipitation of cells from suspension ('salting out'), as a measure of the relative hydrophobicity of microorganisms. The procedure employs serial dilutions of ammonium sulphate (0.2–4 mol/l, in 0.2 mol/l increments) in sodium phosphate buffer (0.002 mol/l, pH 6.8), which are dispensed as 25 μl volumes into 24-well tissue culture trays or glass depression slides. To each well is added an equal volume of the bacterial suspension (5×10^9 cells/ml in sodium phosphate buffer, pH 6.8) which is mixed with gentle agitation at 20°C for 2 min. The SAT value is taken as the lowest concentration of ammonium sulphate to cause bacterial aggregation. Aggregation is indicated when the bacteria coalesce to give a clear solution, when viewed against a dark background. Hydrophobic bacteria tend to aggregate at lower concentrations of ammonium sulphate than do hydrophilic bacteria.

A modification to this test employs ammonium sulphate solution stained with methylene blue (Rozgonyi *et al.*, 1990). The test is performed on white hydrophobic paper card and the aggregation read immediately after mixing, while the card is gently rocked. The reaction is easily read with the naked eye, and the paper cards can be dried and retained for record purposes.

In the SAT, aggregation is believed to arise through reduction of repulsive charges and the subsequent dominance of hydrophobic interactions at the outer surface layer of the bacteria.

Attachment to hydrophobic polystyrene

Several authors have used polystyrene to follow the tendency of bacteria to adhere to hydrophobic surfaces (e.g. Harber *et al.*, 1985). Interactions are assumed to be mediated by hydrophobic–hydrophobic attraction, which affords a potential technique to compare the hydrophobicity of the outermost layers of microbial surfaces.

Polystyrene, either spin- or dip-cast as a thin film on to glass slides (Khan *et al.*, 1988), as polystyrene microspheres (Hazen & Hazen, 1987), or microtitre trays (Denyer *et al.*, 1990a) can be used as substrata for attachment. Experimental conditions vary, but generally either phosphate-buffered saline (PBS) or PUM suspending media and microbial suspensions of 1×10^6 cells/ml (yeasts) or 5×10^8 cells/ml (bacteria) are used. Contact times are from 1 to 4 h, often with gentle shaking, after which unattached bacteria can be determined by viable counting. Alternatively, the adherent population can be estimated by ATP extraction and bioluminescence determination (Harber *et al.*, 1985) or by staining and colorimetric or image analysis (Khan *et al.*, 1988; Denyer *et al.*, 1990a).

Other methods

Other techniques, which have found less widespread application, include partition and binding of fatty acids to cells from aqueous buffers (Kjelleberg *et al.*, 1980; Malmqvist, 1983) and cell partitioning between aqueous/polymer and polymer/polymer two-phase systems (Magnusson *et al.*, 1977). In the former technique, the low molecular weight probes are likely to penetrate the surface layer of bacteria more deeply than the macromolecules, e.g. polyethylene glycol, used in the two-phase systems. It is also possible that the partitioning materials themselves may affect the nature of the microbial surface.

Surface charge

Both bacteria and substrata acquire a surface charge — usually negative — as the result of the adsorption of ions or the ionization of existing surface groups. These charged surfaces will then attract counter-ions from the surrounding aqueous phase to give rise to a measurable electrical double layer of counter-ions associated with each surface. Therefore, as the bacterium approaches the surface of the substrate a repulsive interaction comes into play, resulting from the overlaping ionic atmospheres around the two surfaces. The magnitude of this repulsion will depend on the potentials of the two surfaces, the ionic strength and dielectric constant of the surrounding medium and the distance between the bacterium and substrate.

A variety of methods has been devised to determine microbial surface charge, including colloid titration (Watanabe & Takesue, 1976), attachment to charge-modified polystyrene (Shea & Williamson, 1990), fluorescent probes, ion exchange resins and electrophoretic mobility. The latter two methods have found the widest application. Particularly valuable is electrophoretic mobility, which allows an estimation of the zeta potential of bacterial cells. In addition, isoelectric focusing has been used to differentiate between species and within populations on the basis of charge. Electrophoretic mobility as a function of pH has aided in the identification of specific charged surface constituents.

Zeta potentials

The measurement of bacterial cell electrophoretic mobility and zeta potential can be made by the technique of quasielastic laser light scattering. This involves the application of a known electric field across a suspension of the microbial sample contained in a cylindrical sample cell. The sample is illuminated with a laser light source (15 mW helium–neon laser) which is split into two beams of equal intensity. The split beams are made to cross at a stationary level (zero electroosmotic flow) in the sample cell, which forms an ellipsoid measuring volume. In this region a pattern of interference fringes is formed as a result of the coherence of the two laser beams. The spacing of these light and dark bands is an exact function of the beam-crossing angle and laser frequency. Particles moving through the fringe system will scatter light with a frequency different from that of the incident beam, as a result of the Doppler effect.

To determine the charge and velocity of the particles under investigation, one of the laser beams is modulated by a frequency of 250 Hz. This causes the interference fringes to drift in a direction parallel to the particle motion. Charged particles will move either with the fringe system or against it, depending on the polarity of the applied electric field. Synchronization of the modulating frequency and the field polarity with the particle movement then provides information about the sign and magnitude of the particle surface charge. The Malvern Zetasizer II (Malvern Instruments, UK) is a system which determines the Doppler shift frequency by generating a correlation function of the scattered light intensity and converting this to a frequency spectrum via a Fourier transform.

For electrophoretic measurements, washed bacterial cells are suspended in phosphate buffer (50 mmol/l, pH 7.2) to a cell density of 10^8 cells/ml by vortexing for 120 s. A total of 15 measurements is performed on each sample at 25°C, and the electrophoretic mobility and zeta potential are calculated by reference to the Doppler shift frequency and the particle velocity.

Retention on anionic ion exchange resin

Harkes *et al.* (1992) described a technique by means of which the relative surface charge of bacteria can be determined by measuring their retention on an anionic ion exchange resin, such as Dowex 1×8 mesh size 100/200 (80–150 µm). Pasteur pipettes are filled with 0.5 g of resin and rinsed with 3 ml of PBS, pH 7. Bacterial suspensions are adjusted to an OD_{540} of 1, and 1 ml of suspension is added to the column. The column is eluted with PBS and the absorbance measured. The relative surface charge of the bacteria is expressed as the percentage of bacteria bound to the resin. This technique can be modified in a manner similar to that of the HIA method described above.

Surface chemical analysis

The physicochemical interaction forces associated with microbial adhesion are determined by the surface chemistry of the organism. X-ray photoelectron spectroscopy (XPS), also known as ESCA (electron spectroscopy for chemical analysis) makes possible a chemical analysis of the composition of microbial cell surfaces, generally at depths of between 2 and 5 nm (Mozes *et al.*, 1991).

Bacterial cells for analysis are washed three times in distilled water and resuspended to 20 ml at a cell density of 1×10^{10} cells/ml before overnight freeze-drying. Freeze-dried cells are then carefully mounted with double-sided adhesive tape on to metal stubs and placed into an ultra-high vacuum environment in an electron spectrometer (Escalab Mk II, VG Scientific, UK). The sample is then exposed to a low-energy monochromatic X-ray source, which causes emission of photoelectrons from the atomic shells of the elements present in the surface (Paynter, 1988). These electrons possess an energy characteristic of the element and molecular orbital from which they are emitted. The electrons are then detected and counted according to this energy. By counting the number of electrons at each energy value, a spectrum of peaks corresponding to the elements of the surface is generated. The area under each peak is a measure of the relative amount of each element, while the shape and position of the peaks reflect the chemical state of each element (Ratner, 1983; Sherwood, 1985).

Typical analytical conditions include (i) an excitation source of Mg-K α X-rays; (ii) fixed analysis transmission mode at a pass energy of 50 eV for the survey scan and, at 20 eV, for the high-resolution scans of the C1s, N1s,

O1s and P spectra; (iii) an electron take-off angle of 25° at a vacuum of 10^{-8} mmHg.

Correlation between methods

This section describes the surface characterization of a range of clinical isolates of *Staphylococcus epidermidis* by HIA, contact-angle measurement, zeta potential determination, XPS and adhesion to polystyrene.

HIA, contact-angle determinations and adhesion to polystyrene differentiated between seven staphylococcal strains examined, and were in agreement with respect to the hydrophobicity rank order. A similar trend was not revealed with zeta potentials, and only strain 905 was distinguishable from the rest (Table 2.2; see also Rosenberg & Doyle, 1990). Surface elemental analysis was performed on three representative strains of staphylococci, chosen on the basis of their relative hydrophobicity. The results suggested the presence of surface exposed protein, carbohydrate and, possibly, lipoteichoic acid (van der Mei et al., 1988) (Table 2.3). Although all these components might be expected to contribute to the overall surface characteristics of the cells, there was no clear correlation with either hydrophobicity or surface charge. Nevertheless, XPS has been used to describe a number of determinant factors relevant to physicochemical behaviour (Mozes et al., 1989).

Several methods used to measure hydrophobicity show reasonable correlation, although some discrepancies may occur in bacterial strains that exhibit less pronounced cell-surface hydrophobicity (van Loosdrecht et al., 1987; Wadstrom et al., 1987; Rosenberg & Doyle, 1990; Rozgonyi et al., 1990). It is not altogether surprising that some variations arise, because the

TABLE 2.2. *Surface characteristics of clinical isolates of Staphylococcus epidermidis*

Strain	Water contact angle	Zeta potential (mV)	HIA percentage of cells adsorbed	Adhesion to polystyrene (relative adherence)
900	15 ± 2*	-41.5 ± 0.2	36.0	4.6
901	38 ± 2	-40.0 ± 0.7	70.0	26.9
902	25 ± 2	-43.2 ± 0.4	51.0	12.0
903	25 ± 2	-43.6 ± 0.2	46.0	6.5
904	26 ± 2	-44.6 ± 1.0	54.0	9.1
905	23 ± 2	-48.7 ± 2.0	41.0	8.1
906	26 ± 2	-44.2 ± 0.3	48.0	8.1

* All figures ±SD.

TABLE 2.3. *XPS surface elemental analysis of
staphylococcal cells grown in nutrient broth*

	Elemental ratio		
Strain	N/C	O/C	P/C
900	0.180	0.360	0.039
901	0.176	0.335	0.033
904	0.167	0.345	0.021

methodologies used probe different components of the cell surface. The small molecular weight hydrocarbons used in the BATH test would be expected to penetrate the surface of the bacterial cell wall to a greater extent than high molecular weight polymers. In addition, artefacts may be introduced with the BATH test if the hydrocarbon damages the cell wall, because the test may then be measuring the hydrophobicity of a denatured surface. HIA and the SAT method measure hydrophobicity at the outer surface of the cell envelope, and so tend to give comparable hydrophobicity values. The reader is referred to Rozgonyi *et al.* (1990) for a list of surface components of bacterial cells which are thought to predominate in each of the test methods.

Discrepancies can also arise because the tests themselves are subject to variations in methodology which will influence the final results. It has been shown that bacterial cell density can affect the results obtained for the BATH and SAT tests, and it is also possible to overload the gel in the HIA test. Therefore, the effect of cell loading must be evaluated at the outset for each of the tests. The ionic strength of the suspending medium will have a dramatic effect on the apparent hydrophobicity for each of the tests described, and effects such as pH and the presence of surfactants will also affect the results obtained.

Conclusions

Any experimental approach to the evaluation of the interactive processes between bacterium and surface must pay particular attention to the effect of the environment on both surfaces. In particular, for such studies to be relevant to the *in vivo* situation, the organisms must be clearly representative of those found in the natural environment. Few studies have appreciated the importance of this, and many have approached the investigation of microbial adherence by using laboratory-grown cultures. However, recent studies have demonstrated the plasticity of the microbial cell surface and the intimate relationship between the phenotypic expression of surface characteristics and the environment. Normal methods of laboratory culture, with artificial culture media, are alien to the situation in nature, where nutrient limitation, reduced growth rate and environmental challenge serve continuously to affect the growth characteristics of the microorganism. Artificial laboratory conditions of culture lead to the development of organisms with few of the characteristics seen *in vivo* (Denyer *et al.*, 1990a,b).

References

ABSOLOM, D.R., LAMBERTI, F.V., POLICOVA, Z., ZINGG, W., VAN OSS, C.J. & NEUMANN, A.W. (1983) Surface thermodynamics of bacterial adhesion. *Applied and Environmental Microbiology*, **46**, 90−97.

BEACHEY, E.H., GIAMPAPA, C.S. & ABRAHAM, S.N. (1988) Bacterial adherence. Adhesion receptor-mediated attachment of pathogenic bacteria to mucosal surfaces. *American Review of Respiratory Diseases*, **138**, S45−S48.

BUSSCHER, H.J., WEERKAMP, A.H., VAN DER MEI, H.C., VAN PELT, A.W.J., DE JONG, H.P. & ARENDS, J. (1984) Measurement of the surface free energy of bacterial cell surfaces and its relevance for adhesion. *Applied and Environmental Microbiology*, **48**, 980−983.

CHARACKLIS, W.G. (1990) Biofilm processes. In Characklis, W.G. & Marshall, K.C. (eds.) *Biofilms*, p. 195−231. John Wiley, New York.

CHRISTENSEN, G.D., BADDOUR, L.M., HASTY, D.L., LAWRENCE, J.H. & SIMPSON, W.A. (1989) Microbial and foreign body factors in the pathogenesis of medical device infections. In Bisno, A.L. & Waldvogel, F.A. (eds.) *Infections Associated with Indwelling Medical Devices*, pp. 27−59. American Society for Microbiology, Washington.

DENYER, S.P., DAVIES, M.C., EVANS, J.A., FINCH, R.G., SMITH, D.G.E., WILCOX, M.H. & WILLIAMS, P. (1990a) Influence of carbon dioxide on the surface characteristics and adherence potential of coagulase-negative staphylococci. *Journal of Clinical Microbiology*, **28**, 1813−1817.

DENYER, S.P., DAVIES, M.C., EVANS, J.A., FINCH, R.G., SMITH, G.D.E., WILCOX, M.H. & WILLIAMS, P. (1990b) Phenotypic changes in staphylococcal cell surface characteristics associated with growth in human peritoneal dialysis fluid. In Wadstrom, T., Eliasson, I., Holder, I. & Ljungh, A. (eds.) *Pathogenesis of Wound and Biomaterial-Associated Infections*, pp. 233−244. Springer-Verlag, London.

DERJAGUIN, B.V. & LANDAU, V. (1941) Theory of the stability of strongly charged lyophobic sols and the adhesion of strongly charged particles in solutions of electrolytes. *Acta Physicochemica USSR*, **14**, 633−662.

DUNCAN-HEWITT, W.C., POLICOVA, Z., CHENG, P., VARGHA-BUTLER, E.I. & NEUMANN, A.W. (1989) Semi-automatic measurement of contact angles on cell layers by modified axisymmetric drop shape analysis. *Colloids and Surfaces*, **42**, 391−403.

ESCHER, A. & CHARACKLIS, W.G. (1990) Modeling the initial events in biofilm accumulation. In Characklis, W.G. & Marshall, K.C. (eds.) *Biofilms*, pp. 445−486. John Wiley & Sons, New York.

FINDLAY, B.B. & FALKOW, S. (1989) Common themes in microbial pathogenicity. *Microbiological Reviews*, **53**, 210−230.

FOWLER, H.W. & McKAY, A.J. (1980) The measurement of microbial adhesion. In Berkeley, R.C.W., Lynch, J.M., Melling, J., Rutter, P.R. & Vincent, B. (eds.) *Microbial Adhesion to Surfaces*, pp. 143−162. Ellis Horwood, Chichester.

HARBER, M.J., MACKENZIE, R. & ASSCHER, A.W. (1985) Polystyrene tube bioluminescence assay for bacterial adherence. In Sussman, M. (ed.) *The Virulence of Escherichia coli*, pp. 389–393. Academic Press, London.

HARKES, G., DANKERT, J. & FEIJEN, J. (1992) Growth of uropathogenic *Escherichia coli* strains at solid surfaces. *Journal of Biomaterial Science, Polymer Edition*, **3**, 403–418.

HAZEN, K.C. (1990) Cell surface hydrophobicity of medically important fungi, especially *Candida* species. In Doyle, R.J. & Rosenberg, M. (eds.) *Microbial Cell Surface Hydrophobicity*, pp. 249–295. American Society for Microbiology, Washington DC.

HAZEN, K.C. & HAZEN, B.W. (1987) A polystyrene microsphere assay for detecting surface hydrophobicity variations within *Candida albicans* populations. *Journal of Microbiological Methods*, **6**, 289–299.

HJERTEN, S. & WADSTROM, T. (1990) What types of bonds are responsible for the adhesion of bacteria and viruses to native and artificial surfaces? In Wadstrom, T., Eliasson, I., Holder, I. & Ljungh, A. (eds.) *Pathogenesis of Wound and Biomaterial-Associated Infections*, pp. 55–63. Springer-Verlag, London.

KHAN, M.A., DENYER, S.P., DAVIES, M.C. & DAVIS, S.S. (1988) Adhesion of *Staphylococcus epidermidis* to modified polystyrene surfaces. *Journal of Pharmacy and Pharmacology*, **40**, 17P.

KJELLEBERG, S., LAGERCRANTZ, C. & LARSSON, TH. (1980) Quantitative analysis of bacterial hydrophobicity by the binding of dodecanoic acid. *Acta Pathologica Microbiologica Scandinavica, Section B*, **91**, 69–73.

LINDAHL, M., FARIS, A., WADSTROM, T. & HJERTEN, S. (1981) A new test based on 'salting out' to measure relative surface hydrophobicity of bacterial cells. *Biochimica Biophysica Acta*, **667**, 471–476.

LINDBERG, M., JONSSON, K., MULLER, H.-P., SIGNAS, C. & HOOK, M. (1990) Fibronectin binding proteins from *Staphylococcus aureus*. In Wadstrom, T., Eliasson, I., Holder, I. & Ljungh, A. (eds.) *Pathogenesis of Wound and Biomaterial-Associated Infections*, pp. 55–63. Springer-Verlag, London.

MAGNUSSON, K.-E., STENDAHL, O., TAGESSON, C., EDEBO, L. & JOHANSSON, G. (1977) The tendency of smooth and rough *Salmonella typhimurium* bacteria and lipopolysaccharide to hydrophobic and ionic interaction as studied in aqueous polymer two-phase systems. *Acta Pathologica Microbiologica Scandinavica, Section B*, **85**, 212–218.

MALMQVIST, T. (1983) Bacterial hydrophobicity measured as partition of palmitic acid between the two immiscible phases of cell surface and buffer. *Acta Pathologica Microbiologica Scandinavica, Section B*, **91**, 69–73.

MOHANDAS, N., HOCHMUTH, R.M. & SPAETH, E.E. (1974) Adhesion of red cells to foreign surfaces in the presence of flow. *Journal of Biomedical Material Research*, **8**, 119–136.

MOZES, N., AMORY, D.E., LEONARD, A.J. & ROUXHET, P.G. (1989) Surface properties of microbial cells and their role in adhesion and flocculation. *Colloids and Surfaces*, **42**, 313–329.

MOZES, N., HANDLEY, P.S., BUSSCHER, H.J. & ROUXHET, P.G. (1991) *Microbial Cell Surface Analysis: Structural and Physicochemical Methods*. VHC, New York.

NEUMANN, A.W. & GOOD, R.J. (1979) Techniques of measuring contact angles. In Good, R.J. & Stromberg, R.R. (eds.) *Surface and Colloid Science*, Vol II, pp. 31–91. Plenum Press, New York.

OLIVEIRA, D.R. (1992) Physico-chemical aspects of adhesion. In Melo, L.F., Bott, T.R., Fletcher, M. & Capdeville, B. (eds.) *Biofilms: Science and Technology*, pp. 45–58. Kluwer Academic Publishers, Dordrecht.

PAYNTER, R.W. (1988) Introduction to X-ray photoelectron spectroscopy. In Ratner, B.D. (ed.) *Surface Characterisation of Biomaterials*, pp. 40–57. Elsevier Science Publishers, Amsterdam.

RATNER, B.D. (1983) Surface characterisation of biomaterials by electron spectroscopy for chemical analysis. *Annals of Biomedical Engineering*, **11**, 313–316.

ROSENBERG, M. (1984) Bacterial adherence to hydrocarbons: a useful technique for studying cell surface hydrophobicity. *FEMS Microbiology Letters*, 22, 289–295.

ROSENBERG, M. & DOYLE, R.J. (1990) Microbial cell surface hydrophobicity: history, measurement and significance. In Doyle, R.J. & Rosenberg, M. (eds.) *Microbial Cell Surface Hydrophobicity*, pp. 1–37. American Society for Microbiology, Washington.

ROSENBERG, M., GUTNICK, D. & ROSENBERG, E. (1980) Adherence of bacteria to hydrocarbons: a simple method for measuring cell surface hydrophobicity. *FEMS Microbiology Letters*, 9, 29–33.

ROZGONYI, F., LJUNGH, A., MAMO, W., HJERTEN, S. & WADSTROM, T. (1990) Bacterial cell-surface hydrophobicity. In Wadstrom, T., Eliasson, I., Holder, I. & Ljungh, A. (eds.) *Pathogenesis of Wound and Biomaterial-Associated Infections*, pp. 233–244. Springer-Verlag, London.

SHARON, D., BARR-NESS, R. & ROSENBERG, M. (1986) Measurement of the kinetics of bacterial adherence to hexadecane in polystyrene cuvettes. *FEMS Microbiology Letters*, 36, 115–118.

SHEA, C. & WILLIAMSON, J.C. (1990) Rapid analysis of bacterial adhesion in a microplate assay. *Biotechniques*, 8, 610–611.

SHERWOOD, P.M.A. (1985) Data analysis in photoelectron spectroscopy. In Briggs, D. & Seah, M.P. (eds.) *Practical Surface Analysis by Auger Photoelectron Spectroscopy*, pp. 445–463. John Wiley, New York.

SMYTH, C.J., JONSSON, P., OLSSON, E., SODERLIND, O., ROSENGREN, S., HJERTEN, S. & WADSTROM, T. (1978) Differences in hydrophobic surface characteristics of porcine entero-pathogenic *Escherichia coli* with or without K88 antigen as revealed by hydrophobic interaction chromatography. *Infection and Immunity*, 22, 462–472.

VAN DER MEI, H.C., LEONARD, A.J., WEERKAMPA, H., ROUXHET, P.G. & BUSSCHER, H.J. (1988) Properties of oral streptococci relevant for adherence: zeta potential, surface free energy and elemental composition. *Colloids and Surface*, 32, 297–305.

VAN LOOSDRECHT, M.C.M., LYKTEN, J., NORDE, W., SCHRAA, G. & ZEHNDER, A.J.B. (1987) The role of bacterial cell wall hydrophobicity in adhesion. *Applied and Environmental Microbiology*, 53, 1893–1897.

VAN, OSS, C.J. (1978) Phagocytosis is a surface phenomenon. *Annual Review of Microbiology*, 32, 19–39.

VERWEY, E.J.W. & OVERBEEK, J.T.G. (1948) *Theory of Stability of Lyophobic Colloids*. Elsevier, Amsterdam.

WADSTROM, T., ANDERSON, K., SYDOW, M., AXELSSON, L., LINDGREN, S. & GULLMAR, B. (1987) Surface properties of lactobacilli isolated from the small intestine of pigs. *Journal of Applied Bacteriology*, 62, 513–520.

WADSTROM, T., ERDEI, J., PAULSSON, M. & LJUNGH, A. (1990) Fibronectin, collagen and vitronectin binding of coagulase-negative staphylococci. In Wadstrom, T., Eliasson, I., Holder, I. & Ljungh, A. (eds.) *Pathogenesis of Wound and Biomaterial-Associated Infections*, pp. 339–349. Springer-Verlag, London.

WATANABE, K. & TAKESUE, S. (1976) Colloid titration for determining the surface charge of bacterial spores. *Journal of General Microbiology*, 96, 221–223.

3: Laboratory Methods for Biofilm Production

P. GILBERT AND D.G. ALLISON

Department of Pharmacy, University of Manchester, Oxford Road, Manchester M13 9PL, UK

In nature and in disease biofilms are the most frequently encountered physiological form adopted by microorganisms (Costerton *et al.*, 1987). Biofilms are functional consortia of microbial cells organized within extracellular polymer matrices and associated with surfaces. Their physiology, metabolism and organization are greatly dependent on the nature of these surfaces, and also the prevailing physicochemical environment. Thus, biofilms associated with the inside of pipework (Costerton & Lashen, 1984; Shaw *et al.*, 1985), cooling towers (Lee & West, 1991), infected indwelling medical devices (Gristina *et al.*, 1985; Jacques *et al.*, 1986; Costerton *et al.*, 1987), soft animal and plant tissues (Lippincott & Lippincott, 1969), bone joints (Marrie & Costerton, 1985), soil aggregates and river and sea beds (Paerl, 1975) differ significantly with respect to their physiological properties, organization and the metabolism of their component cells. Accordingly, numerous laboratory methods have evolved in attempts to model such systems. These models vary from simple systems based on solid media to complex biofermenters. Individual models reproduce particular physiological constraints upon biofilm growth and development, which are pertinent to different ecosystems. Choice of an appropriate model system must therefore be tempered not only by technical constraints but also by an understanding of the *in vivo/in situ* organization of the biofilm.

Nutrient availability within natural ecosystems, such as biofilms, governs not only the growth rate of the individual cells — itself a modulator of cell physiology (Brown *et al.*, 1990; Gilbert *et al.*, 1990) — but also the physiological status of the population with respect to metabolism of the rate-limiting nutrient, cell-envelope composition and expression of extracellular and surface-exposed antigens (Brown & Williams, 1985). Each of these factors will affect the nature of the developed biofilms, and also their susceptibility to treatment with antibiotics of biocides (Brown *et al.*, 1988). Since an important consequence

Microbial Biofilms:
Formation and Control

of biofilm growth is the establishment of nutrient, physicochemical and gaseous gradients within the cell population, factors such as growth rate, nutrient availability and biocide penetration will differ spatially within the biofilm and produce mixed populations with heterogeneous properties. Facultative anaerobes such as *Escherichia coli* may establish gradients of pH and secondary metabolites within the biofilm, where low oxygen tension in the depths of the biofilm may induce fermentative growth and acid production, whereas at the interface between the biofilm and the liquid phase, aerobic growth may dominate. If biofilms are disaggregated before study, these different physiologies and the consequences of consortial organization will be lost. The properties of an intact biofilm may, therefore, differ significantly from those of dispersed cells derived from them (Allison *et al.*, 1990a, b; Evans *et al.*, 1991a). The formation and stability of biofilms will be markedly affected by the flow of liquids around or through them. Thus, biofilms associated with flowing systems such as pipes, intravascular catheters, etc. will be subject to shearing forces associated with the movement of liquids, whereas the dispersive effects on wet or moist surfaces will be minimal. Where growth rate-limiting nutrients are sequestered by organisms by extracellular mechanisms, such as siderophores, growth at water/solid interfaces is likely to disperse and reduce the efficiency of nutrient acquisition, and also to perturb cell−cell interactions. For any given microbial species, therefore, the location and site of biofilm development will affect its expressed properties. Generally biofilms will occur at one of the following.

1   *Solid/air interfaces* (e.g. lung infections, wet surfaces). Nutrients and moisture are derived from the solid surface and oxygen is derived from the air. Opposing nutrient and oxygen gradients will therefore be established in the biofilm. The properties of the biofilm will be affected by leaching rates for nutrient from the substratum, and also the physicochemical properties of the substratum *per se*.

2   *Inert solid/liquid interfaces* (e.g. pipework, indwelling medical devices, etc.). Oxygen and available nutrients are derived from the liquid environment. Gradients decrease with increasing depth of biofilm. The properties of the biofilm will, in this instance, be affected not only by the physical properties of the inert supports but also by the shearing effects of liquid movement over them.

3   *Solid nutrient/liquid interfaces* (e.g. soft-tissue infection). Nutrients and oxygen may be derived from either the solid or the liquid medium. Accordingly, many different physicochemical gradients may be established within the population. Once again, the nature of the substratum and the shear forces imposed by liquid movement will influence biofilm establishment, development, dispersal and physiology.

Growth in nature is often at rates much lower than those typically obtained in laboratory systems. In the latter, media are devised that optimize growth of the test organism, but in the real world deficiencies in particular nutrients impose growth-rate regulation (Brown, 1977). Such nutrient deficiencies and growth rates must be replicated in any laboratory model of biofilm if it is to be appropriate. Although it is difficult to predict the nutrient-limiting factors and growth rates associated with each and every environment, some generalizations can be made. Animal infections are generally iron-limited (Brown et al., 1984; Shand et al., 1985), whereas freshwater and marine environments are likely to be carbon- or nitrogen-limited (Brock, 1971). The choice of appropriate defined media for use in laboratory studies can therefore increase the relevance of the test inocula generated (Gilbert et al., 1987).

The physicochemical environment and hence the growth rate, constantly change in closed growth systems (batch culture). Such changes reflect consumption of nutrients, production of waste materials and increasing oxygen demand during growth of the population. Thus, whereas batch culture offers simplicity and convenience and is readily adapted to different test organisms, it often lacks relevance to the real world it attempts to model, since such factors are difficult to control. Open growth systems (continuous culture), on the other hand, by providing a constant supply of fresh nutrient and removal of the byproducts of growth, enables not only the imposition of appropriate nutrient deficiencies on the culture but also the possibility, as in a chemostat, of growth-rate control. Experiments that combine planktonic growth in a chemostat with growth-rate control of appropriate model biofilm are essential if the effects of adherence and growth rate are to be differentiated from one another.

This review will consider the various approaches to the laboratory culture of bacterial biofilms, and will characterize these according to the constraints and considerations detailed above. Where possible, some indication will be given as to the appropriateness of each model to particular environments.

Closed Culture Models of Biofilm Development

By far the most convenient culture methods for bacterial cells are closed growth systems represented by shake flasks and solid media. Although it is possible, to some extent, to control the nature of the nutrient which ultimately causes such cultures to enter their stationary phases of growth, growth rates vary significantly throughout, according to the age of the cultures, the build-up of toxic metabolites and the consumption of nutrients. Provided that consideration is given to medium composition, closed culture presents a simple way in which biofilms may be modelled in the laboratory. In these, either the medium directly presents a surface for colonization (agar surface) or the substratum is placed into direct contact with a liquid nutrient.

Solid/air interfaces (agar surfaces)

The simplest *in vitro* method of producing a biofilm population is to inoculate the surface of an agar plate to produce a confluent lawn culture. Volumes (1 ml) of cell suspensions (*ca.* 10^6 cells/ml) are spread evenly over the surfaces of suitable chemically defined media solidified with bacteriological agar No. 1 (Oxoid) and placed in 30 cm diameter Petri dishes. The defined medium should, if possible, reproduce the nutrient environment and pH of the ecosystem to be modelled. Thus, if soft-tissue infections are to be modelled, a suitable iron-deficient medium should be used (Brown *et al.*, 1984). Typically, a single lawn culture of Gram-negative species such as *E. coli* or *Pseudomonas aeruginosa* generated in this way and incubated overnight at $30-35°C$ will yield sufficient cells (10^{14}) for metabolic and physiological analysis. Often the size of bacterial colonies produced on nutrient agar surfaces depends on the overall colony density on the plate. Thus, single isolated colonies are often large, whereas closely packed ones are small. Since nutrient and oxygen gradients will be established within each individual colony and the magnitude of such gradients will depend on colony size, care should be taken that a confluent lawn is generated without the presence of discrete colonies. This will ensure homogeneous cell suspensions and reproducible results (Al-Hiti & Gilbert, 1983). Indeed, in an elegant series of scanning electron micrographs, Shapiro (1987) demonstrated the high degree of morphological and biochemical organization within a single agar-grown colony of *E. coli*. As the colonies aged, morphologically distinguishable zones involving cells of different shapes, sizes and orientation were identified. Similarly Al-Hiti & Gilbert (1983) demonstrated that, for agar-grown cultures, not only the reproducibility of the responses of inocula to antimicrobials, but also the susceptibility to a number of antimicrobial substances, depended on colony density. With these provisos, semisolid supports may mimic bacteria isolated from soft-tissue infections and, although not fully duplicating conditions of growth *in vivo*, they adequately model the close proximity of individual cells to one another and to a nutrient gradient generated from the substratum. In this respect, colonies grown on agar may be representative of biofilms at solid/air interfaces, and provide samples for biochemical analysis. Such effects on cell properties, related to the physical state of the medium, have been reviewed by Lorian (1989). When grown in liquid media, many *E. coli* cells were transformed to a rounded shape as the growth temperature decreased, whereas few rounded forms were observed on solid medium (Iwaya *et al.*, 1978). Significant variations in the ability of *Ps. aeruginosa* strains, isolated from cystic fibrosis patients, to produce highly mucoid colonies on agar but relatively non-mucoid suspensions in liquid culture, have been reported (Chan *et al.*, 1984). Furthermore, variation in the expression of cell-wall antigens during growth in liquid media or on agar has been demonstrated in a number of organisms, including *E. coli*, *Ps. aeruginosa* (Critchley & Basker, 1988), *Staphylococcus aureus* (Cheung & Fischetti, 1988) and *Vibrio parahaemolyticus* (Silverman *et al.*, 1984).

Agar is polyanionic in nature and will interact with and reduce the bioavailability of many cationic agents. In order to assess antimicrobial susceptibility, therefore, it is often necessary to harvest the cells from the agar surface. Although this will clearly negate any protective effects gained by the cell from organization within a glycocalyx, significant variations in minimum inhibitory concentrations (MIC) for *Haemophilus influenza* (Bergeron *et al.*, 1987), *Ps. aeruginosa* (DeMatteo *et al.*, 1981), *E. coli* (Hohl & Felber, 1988) and *Staphylococcus epidermidis* (Kurian & Lorian, 1980) have been reported when cultures harvested from solid media are compared with those grown in the corresponding liquid media. Ideally, cells should be removed from the agar surface before sensitivity testing, in order to eliminate the effect of the agar polysaccharide on antimicrobial potency while an intact biofilm is maintained. This is often achieved by inoculating the cultures on to bacteria-proof cellulose acetate or nitrocellulose membranes (0.45 μm porosity) placed directly over the agar (Millward & Wilson, 1989; Nichols *et al.*, 1989). Exposure to the antimicrobial agent is subsequently achieved by carefully removing the filter from the agar and transferring it to a sterile receiver containing the antibacterial agent. After a further period of incubation, cells are dispersed from the filter by vortexing, serial dilution in physiological saline containing an appropriate neutralizer,and plating out in triplicate on predried agar plates. Residues from the agar may be removed by floating the membrane on several washing solutions before susceptibility testing. Comparisons can thus be made between intact biofilms and those dispersed before antibiotic treatment.

Factors that may affect the properties of agar-grown microcolonies produced in the above manner include the accumulation of metabolites released during growth in the immediate vicinity of the cells; nutrient accessibility only from the underside of the colony, to create differential growth rates throughout a colony; rate of gas diffusion; and a general lack of dispersive shearing forces which may occur at solid/liquid interfaces.

Solid/liquid interfaces (submerged surfaces)

Solid/liquid interfaces probably represent the most frequently encountered sites of biofilm formation. At such interfaces the solid surface may be nutritionally inert, as at the wall of a pipe or the surface of an implanted medical device, or it may provide nutrients/cations for microbial utilization. The liquid phase may be static and allow only diffusive exchange of materials with the developing biofilm, or it may be dynamic. In the latter case, waste products will constantly be diluted away from the biofilm, and localized depletion of nutrients in the liquid phase will be minimized. A dynamic liquid phase will also impose shear stresses on the developing biofilm, and cause cells to be shed and possibly exert a direct influence on exopolymer deposition. Many of these attributes of the biofilm may be modelled in closed growth systems.

Wall growth

The walls of glass vessels containing broth cultures of bacteria will eventually develop adherent biofilms. The extent of wall growth will reflect the degree of agitation of the flask, which may be static or vigorously shaken, the adhesive properties of the microorganism and the nutrient properties of the growth medium. Strongly aerobic organisms will additionally form pellicles of biofilm at the liquid/air interface when the culture vessels are not agitated. Pellicular growth may be harvested directly on to glass slides pushed under the liquid surface. Biofilms produced on the walls of the vessels may be sampled by gently decanting the culture fluid and either rinsing the flasks with sterile buffers and scraping the sides of the flask, or by gentle sonication of the vessel and resuspending fluid. Alternatively, after decanting the culture, biofilm growth may be resumed by placing fresh medium in the flask and continuing the incubation. Repeated replacement of the growth medium in this manner eventually allows substantial numbers of attached cells to be recovered. The physiological properties of the cells obtained in this manner are, however, likely to alter with each successive passage in fresh medium.

Biofilms derived as wall and pellicular growth are poorly controlled with respect to growth rate and nutrient availability. Growth rates in the biofilm are likely to be substantially different from those in the bulk of the culture. In addition, the biofilm population will be in equilibrium with the planktonic dispersed cells. Thus, cells are constantly exchanged between the two phases, and the shear forces of liquid movement may cause segments of the adherent biofilm to slough from the vessel walls and disperse within the bulk phase. Direct comparisons between the planktonic and adherent populations in such studies are, therefore likely to reveal not only differences attributable to adhesion and biofilm *per se*, but also growth-rate differences. Unlike agar-grown colonies, however, these biofilms have not been artificially established on the surface, and as such are representative of biofilms found on the walls of closed vessels, such as water cooling systems and storage tanks.

Submerged sample discs

Bateria have been shown to adhere to many different surfaces, including activated charcoal (LeChevallier *et al.*, 1988), wood (Cundell & Mitchell, 1977), stainless steel (Wilkinson & Hamer, 1974), polystyrene (Fletcher, 1986) and glass (Dempsey 1981; Fletcher, 1986; Allison & Sutherland, 1987).

A range of methods for cultivating biofilms at solid/liquid interfaces allow growth at such surfaces to be modelled (Prosser *et al.*, 1987; Gristina *et al.*, 1987). In the technique of Prosser *et al.* (1987), washed cell pellets are resuspended in buffer to a fixed concentration (*ca.* E_{540} 0.8) and are then dispensed as 80 μl volumes on to presterilized discs (7 mm diameter) composed of various test materials (e.g. silicone). Incubation for 1 h allows organisms to adhere to the test surfaces. Discs are then immersed in liquid nutrient medium in a Petri dish and incubated for varying times in order to allow the biofilms to develop. In the method described by Gristina *et al.* (1987), the initial incubation step is omitted and sample discs are placed directly into inoculated media in the wells of tissue culture plates. After development of the biofilms, biofilm from sample discs can be harvested, either by gentle sonication or by scraping into sterile diluent. Alternatively, if susceptibility tests are warranted, the discs may be placed directly in solutions of antimicrobial agent.

An alternative to these approaches is to suspend the test pieces on cotton or silk threads into culture flasks containing inoculated media. After various periods of incubation, the test pieces can be removed, rinsed and transferred to fresh sterile media. Continual passage in this manner allows substantial biofilms to be developed on the test surfaces. The latter approach represents an improvement over the static incubation of discs, in that gentle shaking will remove loosely attached cells and actively promote the selection of a truly adherent population. It should be noted, however, that organisms will also adhere to the threads, which should therefore be removed before harvesting.

Wetted air/solid interfaces (rotating disc)

Biofilms may often develop on moist, wet surfaces. Such environments may be modelled by the rotating disc method. This consists of a rotating disc of sample material, partially immersed in inoculated growth medium (Fig. 3.1). The surface of the rotating disc is wetted and supplied with nutrients as it passes through the growth medium. Gaseous exchange is likely to occur out of the liquid through a thin liquid film. Such systems have been used to study the effects of fluid shear stress on biofilm development (Abbott *et al.*, 1983), because the shear stress increases with distance from the centre of the disc. Such systems can be operated as closed or open growth systems, according to whether or not the medium reservoir is exchanged with fresh medium.

Summary

The methods described above provide rapid, simple, reproducible methods for establishing thick biofilms on various test materials in closed culture. When glass or other transparent test surfaces are used, the methods allow

(a)

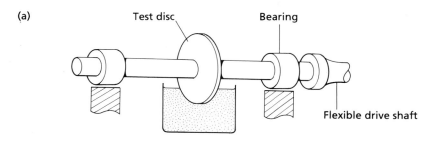

(b)

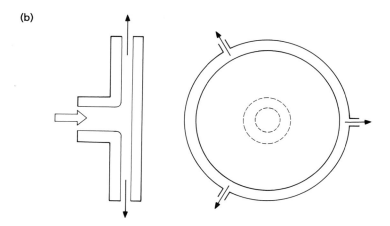

FIG. 3.1. (a) Rotating disc biofilm reactor (after Loeb *et al.*, 1984). (b) The radial flow biofilm reactor consists of two parallel discs with a small distance (*ca.* 500 μm) of separation (after Fowler & McKay, 1980).

staining and light microscopy to be carried out on the biofilm (Zobell & Allen, 1935; Allison & Sutherland, 1987). In these respects, many laboratories have used such approaches to evaluate extracellular polymer production (Allison & Sutherland, 1984), adhesion *per se* (Fletcher, 1986) and antimicrobial susceptibility (Prosser *et al.*, 1987; Dix *et al.*, 1988; Sticker *et al.*, 1989). In the latter studies the amidinocillin, cefamandole and chlorhexidine susceptibility of *E. coli* is substantially reduced for biofilms, compared with equivalent planktonic populations. However, as noted by Brown *et al.*, (1988), these experimental systems are uncontrolled with respect to growth rate and often make comparisons between slow-growing biofilm cells and fast, exponentially growing planktonic cells. As such, the effects of growth rate cannot be distingushed from those associated with adhesion. Nevertheless, biofilms constructed in this manner can be used as a primary screen when there are many antibiotics to be tested (Anwar *et al.*, 1989a, b). With the use of appropriate controls, such techniques can provide valuable information on the properties of sessile bacterial populations embedded in the biofilm without the problems associated with inactivation of antimicrobial agents by heavy concentrations of planktonic bacteria.

Open Continuous Culture of Biofilms

The majority of natural habitats are continuously perfused with fresh supplies of nutrients. Thus, unlike closed (batch) cultures, steady states are often achieved and maintained for long periods of time. Many laboratory models of biofilm development are also continuously perfused in this manner and provide useful, if complex, approaches to the study of biofilms. Evaluation of the mass balance of nutrients and waste products in such continuously perfused systems allows the study of the metabolic activity of the contained biofilms. The inclusion of antimicrobial agents in the medium permits the susceptibility of the developed biofilms to treatment to be evaluated. For antibiotics, pharmacokinetic drug concentration profiles may be modelled by careful manipulation of two or more pumps supplying dosed and undosed media to the reaction vessel (Gilbert, 1985).

Solid/liquid interfaces (submerged open systems)

As for the closed growth systems considered, biofilms may be formed on the surfaces of submerged test pieces suspended in appropriate cultures, or they may be formed on the walls of the culture vessels themselves.

Submerged test pieces

The simplest approach, adopted and described by Keevil *et al.* (1987), and modified by Anwar *et al.* (1989a, b), is to suspend small flat tiles (*ca.* 1 cm^2 × 0.2 mm) of the test material in continuous culture fermenters. This can be facilitated by fastening the tiles with silk or cotton thread (Keevil *et al.*, 1987) and attaching the thread to an entrance port of the fermenter. Suitable fermenters are those described by Gilbert & Stuart (1977) and incorporate wide-necked (B27) ground-glass stoppers into the head of the fermenter. These ports are sufficiently large to allow the sample tiles to be lowered into the fermenter and trap the thread conveniently between the stopper and the vessel wall. Tiles are added to the fermenters when they are at their steady state (usually after 4−5 complete volume changes of medium) and hang freely in the culture. Media should be chosen to establish growth-rate limitation through the depletion of appropriate critical nutrients (i.e. iron depletion for modelling of infections; Brown, 1977). The media should support the growth of the test organism(s) to a stationary-phase cell density of not less than 10^9 cells/ml. Mixing of the medium to achieve aeration and nutrient distribution, contributes to the shearing stresses imposed on the developing biofilm. Initially, cells from the bulk planktonic phase attach to the test surface and initiate the formation of a biofilm. Thereafter an equilibrium is established, with growth of the adherent population, sequestration of planktonic cells to the biofilm and dispersal of biofilm cells from the test piece to the liquid phase. Over a period of several days, depending on the growth rate of the fermenter and the biomass supported by the medium, there is a net development of a substantial biofilm on the immersed tile. For steady-state planktonic populations in the chemostat of $ca.$ 5×10^9/ml, cell densities in the biofilm of $ca.$ 2×10^9/cm^2 may be achieved after 2−3 days. When the development of such populations is studied by periodic removal of the tiles, resuspension of the biofilm and viable counting (Fig. 3.2(a)), it can be seen that in the early stages the net rate of accumulation of cells to the biofilm exceeds the specific growth rate of the planktonic population. After 2−3 days' incubation the net growth of the biofilm is much reduced, and probably equates to a semistationary phase culture. The properties of the biofilm thus formed are, therefore, likely to change substantially with incubation time in the chemostat. Indeed, Anwar et al. (1990) reported substantial differences between antibiotic susceptibility of 'old' and 'young' biofilms generated in this manner. Although chemostats can maintain and control growth rates, a prerequisite is that cultures are homogeneous and well mixed. This ensures that fresh medium is equally distributed to all cells in the population, and also that each individual cell has an equal chance of being washed away (Herbert et al., 1956). This is not the case when adherent populations are included in a fermenter. First, the adherent cells are less likely to be washed out so creating a pressure in favour of adherence which, on prolonged incubation may select for populations of cells with substantially different phenotypes to the original inoculum. Second, fresh nutrient will be distributed, through mixing, only among the homogeneous planktonic population. Nutrient will be available to the attached biofilm only by diffusion. Thus, although cells at the interface between biofilm and bulk liquid phase may be capable of growing at rates equal to the planktonic cells, those within the biofilm must grow more slowly, if at all. Growth rates are also likely to be reduced with increasing depth of the biofilm. It is therefore tempting to suppose that the use of a chemostat gives a growth-rate control throughout the culture, but this is not the case. Moreover, differences in the physiology of adherent and suspended cells are likely to increase significantly with time.

This approach has been used with success to model the formation of

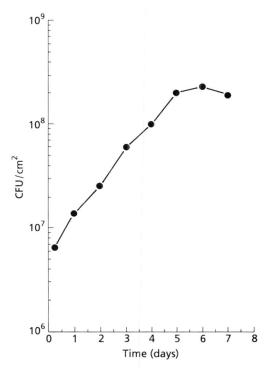

FIG. 3.2. Kinetics of development of *Pseudomonas aeruginosa* biofilms cultured under iron-limited conditions on methacrylate tiles immersed in chemostats at dilution rates of 0.05/h (from Anwar *et al.*, 1989a).

dental plaque on methacrylate dental reins with mixed inocula that represent organisms of the oral cavity (Keevil *et al.*, 1987), the growth of *Legionella* biofilms in cooling systems, where the medium deployed was treated sterilized tapwater (West *et al.*, 1989), and also for a variety of investigations of antibiotic susceptibility (Anwar *et al.*, 1989a, b).

Fixed test surfaces

In the submerged systems described above, sample tiles are suspended freely in the culture medium and it is difficult to control and duplicate fluid shear forces imposed on the developing biofilm. Such forces can be better regulated either by making the sample form part of, or attaching it to, the wall of the fermentation vessel and agitating or flowing the culture medium over it in a controlled, non-turbulent manner. The Roto Torque system (Bakke *et al.*,

1984) epitomizes such an approach. In this device the sample tiles are attached to walls of a continuous fermenter, which is stirred and continuously mixed by means of a rotating inner cylinder. Fluid frictional resistance can then be calculated from the rotational speed of the mixing cylinder and torque measurements. The system permits studies of the engineering effects of biofouling but is too complex for routine microbiological use. An alternative approach is the radial flow reactor (Fig. 3.1(b); Fowler & McKay, 1980). In this system medium is flowed through the centre of a stationary disc against which is placed a second disc composed of sample material. The distance between the two discs is *ca*. 500 μm. Once the sample disc is inoculated, a biofilm develops which is subject to fluid shear forces which decrease with distance from the centre as the cross-sectional area for flow increases. This geometry is particularly useful for studying the initial events of biofilm accumulation.

A simpler approach is to flow sterile medium at various controlled rates through a pipe composed of the test material, or containing samples of test material embedded within it. One such approach, the Robbins device (Fig. 3.3(a) McCoy *et al.*, 1981; see also Chapter 1) is now regarded as classic. In this system a chemostat feeds spent medium and culture directly to a second reactor, designed to simulate an industrial biofouling situation. The second reactor, constructed from 1.2 cm inside-diameter PVC, contains flanges into which $0.5\,cm^2$ circular test pieces can be inserted and locked into place. Once located, the test piece forms part of the inner wall of the reactor, and is available for bacterial colonization. Test pieces may be removed periodically and the biofilms on them subjected to various analyses. Measurements of changes in the resistance to fluid flow through the reactor, due to biofouling, can easily be assessed by measurement of pressure drops across it. The apparatus provides a most effective model of biofouling in tubular flow systems, such as heat exchangers, wastewater pipes etc., and makes possible evaluation of the kinetics of biofilm development and the efficacy of antifouling strategies in such systems.

Many experimental systems utilize the basic principles of the Robbins device. Of particular note are systems that incorporate glass viewing panels (Banks & Bryers, 1991), which allow image analysis of adhesion events and biofilm development within the reactor.

A modified Robbins device (see Chapter 1) has been used effectively to model infections of soft tissues and indwelling medical devices (Nickel *et al.*, 1985; Evans & Holmes, 1987). In this device (Fig. 3.3(b)) retractable pistons form the inside of a rectangular-section pipe constructed in perspex. Plugs of sample material are glued to the ends of the pistons and are inserted into the device, which is sterilized, either by passing hypochlorite solution through it or by ethylene oxide gas. During use, sterile medium is passed from a reservoir through the device at a speed of 1 ml/min. Inoculation is achieved by initially

(a)

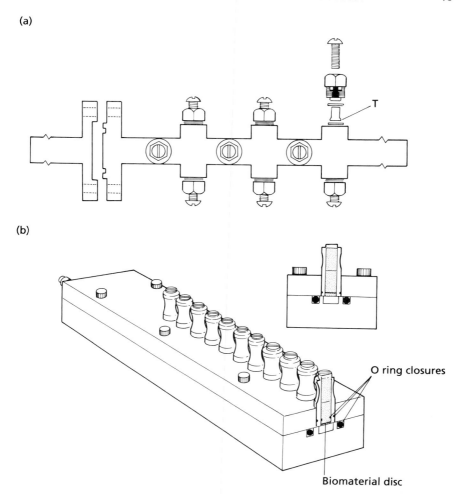

(b)

O ring closures

Biomaterial disc

FIG. 3.3. (a) Robbins device biofilm sampler. Removable test surfaces (T) have 0.5 cm^2 areas exposed to circulating fluid (from McCoy *et al.*, 1981). (b) Cross-sectional view of a modified Robbins device (after Evans & Holmes, 1987).

passing a midexponential phase culture of test organisms through the reactor for several hours. Biofilms develop on the test pieces, which can be sampled by removal of the pistons. Alternatively, if antifouling treatments are to be assessed, agents may be included within the medium.

Since organisms will be shed to the liquid phase it is not possible to predict or control the growth rates which, as for submerged surfaces in chemostat culture (above), will alter with time as the thickness of the biofilm increases and nutrient gradients are established (Fig. 3.4). Typical studies (Nickel *et al.*, 1985; Costerton *et al.*, 1987; Evans & Holmes, 1987; Gristina *et al.*, 1987) with such tubular models compare the properties of adherent cells with those of the equivalent planktonic population of cells passing through the device. Although this technique has provided much valuable information about bacterial biofilms, the interpretation of such data has recently been brought into question (Brown *et al.*, 1988). Although the adherent cells in such models are growing extremely slowly, they are comparable to planktonic controls in a state of rapid growth. Since systems such as the Robbins device used in this way lack effective growth-rate control, they do not differentiate between properties attributable to growth rate and those associated with adherence. There is, however, strong evidence to suggest that many of the properties currently attributed to the growth of cells as biofilms may actually be related to slow growth (Brown *et al.*, 1988, 1991; Gilbert *et al.*, 1990). Care must therefore be taken in the choice of appropriate planktonic controls, if conclusions are to

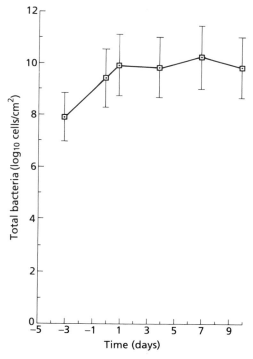

FIG. 3.4. Kinetics of development of a staphylococcal biofilm upon silicone elastomer discs obtained with a modified Robbins device perfused with synthetic dialysis effluent (from Evans & Holmes, 1987).

be drawn about properties of adherent cells *per se*. Many of these potential problems are overcome in the approach adopted by Gilbert *et al.* (1989) (see below).

Growth rate-controlled biofilm fermenter

A method for the selection of synchronous populations of bacteria has been described by Helmstetter & Cummings (1963), in which cells attached to a filter membrane are perfused with fresh medium. In such systems the majority of the cells remain attached to the filter matrix, whereas their progeny are lost to the eluate. Although the method is widely used for selective division synchronization, it has until recently been overlooked that the parent population represents a biofilm, and that the continuous flow of fresh medium offers the potential, as in a chemostat, to control growth rate. By suitable adaptation, the potential of the Helmstetter & Cummings technique to control growth rate in a biofilm has been addressed (Gilbert *et al.*, 1989). The technique provides for a growth rate-controlled biofilm formed on a wet membrane support exposed to the atmosphere. As such, it provides an appropriate model of lung infection and burns.

Midexponential phase cultures (60 ml, *ca.* 10^8 cfu/ml) are pressure-filtered (35 kN m^{-2}) through a prewashed 47 mm diameter cellulose acetate membrane filter (0.22 µm, Oxoid) with a Millipore pressure-filtration unit. The cell-impregnated filter membrane is removed and carefully inverted into the base of a modified, jacketed continuous fermentation apparatus (Fig. 3.5). When fresh medium is passed into the fermentation chamber via the peristaltic pump (1 ml/min), a hydrostatic head develops above the membrane filter which, at steady state, perfuses the filter at the rate of medium addition to the vessel. Viable counts performed on the eluate passing through the filter (Fig. 3.6(a)) show that the number of eluted cells decreases rapidly and reaches steady state after approximately 100 min. This steady state is maintained for up to 5–14 days, according to organism type and flow rates. During this steady-state period the numbers of cells attached to the filter increases only slightly. As in the Helmstetter & Cummings (1963) technique, the initial decrease, represented by the shaded area in Fig. 3.5(a), corresponds to the removal of loosely attached cells from the filter bed and the cells eluted at steady state correspond to newly formed daughter cells, dividing synchronously when transferred to fresh medium (Gilbert *et al.*, 1989; Allison *et al.*, 1990b).

When rates of flow of fresh medium through the filter bed are altered after achieving the initial steady state, and viable counts in the eluate are estimated 40–48 h later, rates of elution of cells from the filter bed can be seen to be directly related to medium flow rate up to a critical flow rate (Fig. 3.6(b)). Thereafter, the rate of elution of cells from the filter bed

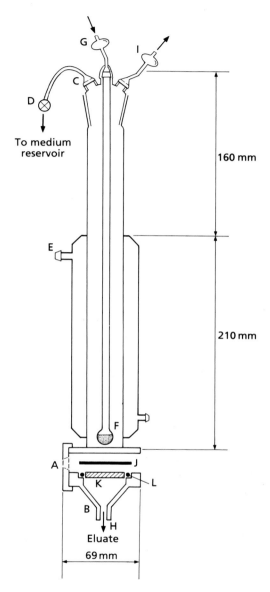

FIG. 3.5. Continuous perfused biofilm fermenter. Biofilm device consists of a glass fermenter (A) and teflon base (B). Fresh medium in (C), via a peristaltic pump (D), maintained at incubation temperatures by water jacket (E). Aeration through glass sinter (F) connected to air filter (G) and pump. Medium outlet (H), air outlet (I). Filter membrane (0.22 μm) (J), stainless steel sintered support (K) and O-ring seal (L) (after Gilbert *et al.*, 1989).

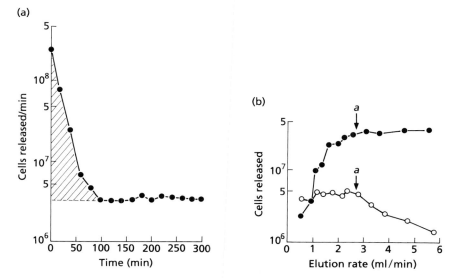

FIG. 3.6. (a) Elution of bacteria from a 0.22 μm filter at a flow rate of 1 ml/min. All loosely bound cells were removed within the first 100 min as represented by the shaded area.
(b) Relationship between the rate of elution of the filter membrane and either (○) viable count/ml of the eluate or (●) rate of organisms released from the membrane/min. From the graph it is possible to calculate the critical medium flow rate a, which at steady state is equivalent to μ_{max}. Growth-rate control is exerted in the period up to a. The rate of cell division in the biofilm at slow elution rates is regulated by the rate of flow of fresh medium (after Gilbert et al., 1989).

remains constant. These results bear a strong similarity to chemostat cultures, where the critical flow rate corresponds to the achievement of μ_{max}, and where the adherent population is under a form of growth-rate control brought about by the rate of perfusion of the filter. Sample biofilms, prepared by low-temperature stage-freeze techniques for scanning electron microscopy, showed dispersed cells attached to the filter matrix yet embedded within an extracellular polymer matrix which strongly resemble samples *in vivo*. Thus, although the initial method of cellular attachment by pressure filtration is artificial, the biofilms so produced at steady state resemble those isolated *in vivo*.

It is important to note that in these systems, the biofilm populations are perfused with nutrients from below. Although this is atypical of biofilms *in situ* on inanimate surfaces, it is representative of bacterial surface infections of soft tissues. The method described offers the ability to control the rate of growth of adherent populations of microorganisms, and has proved a useful model of infection *in vivo*. It is relevant to note, however, that in this growth system the residual adherent population will be enriched in older dividing cells, compared with batch cultures. This may, however, also be true *in vivo*, where the less adherent, newly formed cells are able to relocate and eventually colonize new surfaces (Allison *et al.*, 1990a, b), thereby contributing to the spread of infection. By using control populations grown in a chemostat with media and nutrient deprivations identical to the adherent populations, this method offers the possibility of separating the physiological effects of growth rate and adhesion. The technique has now been sucessfully used for a number of bacterial species, including clinical mucoid isolates of *Ps. aeruginosa* (Evans *et al.*, 1991a) and Gram-positive organisms such as *S. epidermidis* (Evans *et al.*, 1991b; Duguid *et al.*, 1991). The effects of growth rate on cell-surface properties such as hydrophobicity and charge (Allison *et al.*, 1990a, b), biocide (Evans *et al.*, 1990a) and antibiotic (Evans *et al.*, 1990b; Duguid *et al.*, 1991; Evans *et al.*, 1991a) susceptibility of bacterial biofilms, have been investigated with such models. Perfusion of the intact biofilms *in situ* with antibiotics in such instances has allowed evaluation of the effectiveness of chemotherapeutic regimens. Although techniques such as the Robbins device more closely model natural biofilms on impervious surfaces, they do not offer such possibilities for accurate modelling *in vivo*.

References

ABBOT, A., RUTTER, P.R. & BERKELEY, R.C.W. (1983) The influence of ionic strength, pH and protein layer on the interaction between *Streptococcus mutans* and glass surfaces. *Journal of General Microbiology*, **129**, 439–445.

AL-HITI, M.M.A. & GILBERT, P. (1983) A note on inoculum reproducibility: solid versus liquid culture. *Journal of Applied Bacteriology*, **55**, 173–176.

ALLISON, D.G., EVANS, D.J., BROWN, M.R.W. & GILBERT, P. (1990a) Surface hydrophobicity and dispersal of *Pseudomonas aeruginosa* from biofilms. *FEMS Microbiology Letters*, **71**, 101–104.

ALLISON, D.G., EVANS, D.J., BROWN, M.R.W. & GILBERT, P. (1990b) Possible involvement of the division cycle in dispersal of *Escherichia coli* from biofilms. *Journal of Bacteriology*, **172**, 1667–1669.

ALLISON, D.G. & SUTHERLAND, I.W. (1984) A staining technique for attached bacteria and its correlation to extracellular carbohydrate production. *Journal of Microbiological Methods*, **2**, 93–99.

ALLISON, D.G. & SUTHERLAND, I.W. (1987) The role of exopolysaccharides in adhesion of freshwater bacteria. *Journal of General Microbiology*, **133**, 1319–1327.

ANWAR, H., VAN BIESEN, T., DASGUPTA, M.K., LAM, K. & COSTERTON, J.W. (1989a) Interaction of biofilm bacteria with antibiotics in a novel in vitro chemostat system. *Antimicrobial Agents and Chemotherapy*, **33**, 1824–1826.

ANWAR, H., DASGUPTA, M.K., LAM, K. & COSTERTON, J.W. (1989b) Tobramycin resistance of mucoid *Pseudomonas aeruginosa* biofilm grown under iron-limitation. *Journal of Antimicrobial Chemotherapy*, **24**, 647–655.

ANWAR, H., DASGUPTA, M.K. & COSTERTON, J.W. (1990) Testing the susceptibility of bacteria in biofilms to antibacterial agents. *Antimicrobial Agents and Chemotherapy*, **34**, 2043–2046.

BAKKE, R., TRULEAR, M.G., ROBINSON, J.A. & CHARACKLIS, W.G. (1984) Activity of *Pseudomonas aeruginosa* biofilms in steady state. *Biotechnology and Bioengineering*, **26**, 1418–1424.

BANKS, M.K. & BRYERS, J.D. (1991) Bacterial species dominance within a binary culture biofilm. *Applied and Environmental Microbiology*, **57**, 1974–1979.

BERGERON, M.G., SIMARD, P. & PROVENCHER, P. (1987) Influence of growth medium and supplement on growth of *Haemophilus influenzae* and on antibacterial activity of several antibiotics. *Journal of Clinical Microbiology*, **25**, 650–655.

BROCK, T.D. (1971) Microbial growth rates in nature. *Bacteriological Reviews*, **35**, 39–58.

BROWN, M.R.W. (1977) Nutrient depletion and antibiotic susceptibility. *Journal of Antimicrobial Chemotherapy*, **3**, 198–201.

BROWN, M.R.W., ANWAR, H. & LAMBERT, P.A. (1984) Evidence that mucoid *Pseudomonas aeruginosa* in the cystic fibrosis lung grows under iron-restricted conditions. *FEMS Microbiology Letters*, **21**, 113–117.

BROWN, M.R.W., ALLISON, D.G. & GILBERT, P. (1988) Resistance of bacterial biofilms to antibiotics: a growth rate related effect. *Journal of Antimicrobial Chemotherapy*, **22**, 777–789.

BROWN, M.R.W., COLLIER, P.J. & GILBERT, P. (1990) Influence of growth rate on susceptibility to antimicrobial agents: modifications of the cell envelope and batch and continuous culture studies. *Antimicrobial Agents and Chemotherapy*, **34**, 1623–1628.

BROWN, M.R.W., COSTERTON, J.W. & GILBERT, P. (1991) Extrapolation to life outside the test tube. *Journal of Antimicrobial Chemotherapy*, **27**, 565–567.

BROWN, M.R.W. & WILLIAMS, P. (1985) The influence of environment on envelope properties affecting survival of bacteria in infections. *Annual Review of Microbiology*, **39**, 527–556.

CHAN, R.J., LAM, S., LAM, K. & COSTERTON, J.W. (1984) Influence of culture conditions on expression of the mucoid mode of growth of *Pseudomonas aeruginosa*. *Journal of Clinical Microbiology*, **19**, 8–10.

CHEUNG, A.L. & FISCHETTI, V.A. (1988) Variation in the expression of cell wall proteins of *Staphylococcus aureus* grown on solid and liquid media. *Infeciton and Immunity*, **56**, 1061–1065.

COSTERTON, J.W. & LASHEN, E.S. (1984) Influence of biofilm efficacy of biocides on corrosion-causing bacteria. *Materials Performance*, **23**, 34–37.

COSTERTON, J.W., CHENG, K.J., GEESEY, G.G., LADD, T.I., NICKEL, J.C., DASGUPTA, M. & MARRIE, T.J. (1987) Bacterial biofilms in nature and disease. *Annual Review of Microbiology*, **41**, 435–464.

CRITCHLEY, I.A. & BASKER, M.J. (1988) Conventional laboratory agar media provide an iron-limited environment for bacterial growth. *FEMS Microbiology Letters*, **50**, 35–39.

CUNDELL, A.M. & MITCHELL, R. (1977) Microbial succession on a wooden surface exposed to the sea. *International Deterioration Bulletin*, **13**, 67–73.

DeMATTEO, C.S., HAMMER, M.C., BALTCH, A.L., SMITH, R.P., SUTPHEN, N.T. & MICHELSEN, P.B. (1981) Susceptibility of *Pseudomonas aeruginosa* to serium bactericidal activity. *Journal of Laboratory and Clinical Medicine*, **98**, 511–518.

DEMPSEY, M.J. (1981) Colonisation of antifouling paints by marine bacteria. *Botanica Marina*, **24**, 185–191.

DIX, B.A., COHEN, P.S., LAUX, D.C. & CLEELAND, R. (1988) Radiochemical method for evaluating the effects of antibiotics on *Escherichia coli* biofilms. *Antimicrobial Agents and Chemotherapy*, **32**, 770–772.

DUGUID, I.G., EVANS, E., BROWN, M.R.W. & GILBERT, P. (1991) Tobramycin and ciprofloxacin susceptibility of *Staphylococcus epidermidis* in batch and continuous culture. *Abstracts of the 91st Annual Meeting of the American Society for Microbiology*, **A-104**.

EVANS, D.J., ALLISON, D.G., BROWN, M.R.W. & GILBERT P. (1990a) Effect of growth-rate on resistance of Gram-negative biofilms to cetrimide. *Journal of Antimicrobial Chemotherapy*, **26**, 473–478.

EVANS, D.J., BROWN, M.R.W., ALLISON, D.G. & GLIBERT, P. (1990b) Susceptibility of bacterial biofilms to tobramycin: role of specific growth rate and phase in division cycle. *Journal of Antimicrobial Chemotherapy*, 25, 585–591.

EVANS, D.J., ALLISON, D.G., BROWN, M.R.W. & GILBERT, P. (1991b) Susceptibility of *Pseudomonas aeruginosa* and *Escherichia coli* biofilms towards ciprofloxacin: effect of specific growth rate. *Journal of Antimicrobial Chemotherapy*, 27, 177–184.

EVANS, E., DUGUID, I.G., BROWN, M.R.W. & GILBERT, P. (1991a) Surface properties and adhesion of *Staphylococcus epidermidis* in batch and continuous culture. *Abstracts of the 91st Annual Meeting of the American Society for Microbiology*, D-53, Dallas.

EVANS, R.C. & HOLMES, C.J. (1987) Effect of vancomycin hydrochloride on *Staphylococcus epidermidis* biofilm associated with silicone elastomer. *Antimicrobial Agents and Chemotherapy*, 31, 889–894.

FLETCHER, M. (1986) Measurement of glucose utilisation by *Pseudomonas fluorescens* that are free-living and that are attached to surfaces. *Applied and Environmental Microbiology*, 52, 672–676.

FOWLER, H.W. & McKAY, A.J. (1980) The measurement of microbial adhesion. In Berkeley, R.W., Lynch, J.M., Melling, J., Rutter, P.R. & Vincent, B. (eds.) *Microbial Adhesion to Surfaces*, pp. 143–156. Ellis Horwood, Chichester.

GILBERT, P. (1985) Theory and relevance of continuous culture to *in-vitro* models of antibiotic dosing. *Journal of Antimicrobial Chemotherapy*, 15, (Suppl. A), 1–6.

GILBERT, P. & STUART, A. (1977) Small-scale chemostat for the growth of mesophilic and thermophilic microorganisms. *Laboratory Practice*, 26, 627–628.

GILBERT, P., BROWN, M.R.W. & COSTERTON, J.W. (1987) Inocula for antimicrobial sensitivity testing: a critical review. *Journal of Antimicrobial Chemotherapy*, 20, 147–154.

GILBERT, P., COLLIER, P.J. & BROWN, M.R.W. (1990) Influence of growth rate on susceptibility to antimicrobial agents: biofilms, cell cycle, dormancy, and stringent response. *Antimicrobial Agents and Chemotherapy*, 34, 1865–1868.

GILBERT, P., ALLISON, D.G., EVANS, D.J., HANDLEY, P.S. & BROWN, M.R.W. (1989) Growth rate control of adherent bacterial populations. *Applied and Environmental Microbiology*, 55, 1308–1311.

GRISTINA, A.G., OGA, M., WEBB, L.W. & HOBGOOD, C.D. (1985) Adherent bacterial colonisation in the pathogenesis of osteomyelitis. *Science*, 228, 990–993.

GRISTINA, A.G., HOBGOOD, C.D., WEBB, L.X. & MYRVIK, Q.N. (1987) Adhesive colonization of biomaterials and antibiotic resistance. *Biomaterials*, 8, 423–426.

HELMSTETTER, C.E. & CUMMINGS, D.J. (1963) An improved method for the selection of bacterial cells at division. *Biochimica et Biophysica Acta*, 82, 608–610.

HERBERT, D., ELSWORTH, R. & TELLING, R.C. (1956) The continuous culture of bacteria: a theoretical and experimental study. *Journal of General Microbiology*, 14, 601–622.

HOHL, P. & FELBER, A.M. (1988) Effect of method, medium, pH and inoculum on the *in-vitro* antibacterial activities of fleroxacin and norfloxacin. *Journal of Antimicrobial Chemotherapy*, 22 (Suppl. D), 71–80.

IWAYA, M., GOLDMAN, R., TIPPER, D.J., FEINGOLD, B. & STROMINGER, J.L. (1978) Morphology of an *Escherichia coli* mutant with a temperature-dependent round cell shape. *Journal of Bacteriology*, 136, 1143–1158.

JACQUES, M., MARRIE, T.J. & COSTERTON, J.W. (1986) In vitro quantitative adherence of microorganisms to intrauterine contraceptive devices. *Current Microbiology*, 13, 133–137.

KEEVIL, C.W., BRADSHAW, D.J., DOWSETT, A.B. & FEARY, T.W. (1987) Microbial film formation: dental plaque deposition on acrylic tiles using continuous culture techniques. *Journal of Applied Bacteriology*, 62, 129–138.

KURIAN, S. & LORIAN, V. (1980) Discrepancies between results obtained by agar and broth techniques in testing of drug combinations. *Journal of Clinical Microbiology*, 11, 527–529.

LeChavallier, M.W., Cawthon, C.D. & Lee, K.G. (1988) Inactivation of biofilm bacteria. *Applied and Environmental Microbiology*, **54**, 2492–2499.

Lee, J.V. & West, A.A. (1991) Survival and growth of *Legionella* species in the environment. *Journal of Applied Bacteriology*, **70** (Suppl.), 121S–130S.

Lippincott, B.B. & Lippincott, J.A. (1969) Bacterial attachment to a specific wound site as an essential stage in tumour initiation by *Agrobacterium tumefaciens. Journal of Bacteriology*, **97**, 620–628.

Loeb, G.I., Laster, D. & Gracik, T. (1984) The influence of microbial fouling films on hydrodynamic drag of rotating discs. In Costelow, J.D. & Tipper, R.C. (eds.) *Marine Biodeterioration: an Interdisciplinary Study*, pp. 88–100. Naval Institute Press, Annapolis.

Lorian, V. (1989) In vitro simulation of in vivo conditions: physical state of the culture medium. *Journal of Clinical Microbiology*, **27**, 2403–2406.

McCoy, W.F., Bryers, J.D., Robbins, J. & Costerton, J.W. (1981) Observations in fouling biofilm formation. *Canadian Journal of Microbiology*, **27**, 910–917.

Marrie, T.J. & Costerton, J.W. (1985) Mode of growth of bacterial pathogens in chronic polymicrobial human osteomyelitis. *Journal of Clinical Microbiology*, **22**, 924–933.

Millward, T.A. & Wilson, M. (1989) The effect of chlorhexidine on *Streptococcus sangius* biofilms. *Microbios*, **58**, 155–164.

Nichols, W.M., Evans, M.J., Slack, M.P.E. & Walmsley, H.L. (1989) The penetration of antibiotics into aggregates of mucoid and non-mucoid *Pseudomonas aeruginosa. Journal of General Microbiology*, **135**, 1291–1303.

Nickel, J.C., Ruseska, I., Wright, J.B., & Costerton, J.W. (1985) Tobramycin resistance of cells of *Pseudomonas aeruginosa* growing as a biofilm on urinary catheter material. *Antimicrobial Agents and Chemotherapy*, **27**, 619–624.

Paerl, H.W. (1975) Microbial attachment to particles in marine and freshwater ecosystems. *Microbial Ecology*, **2**, 73–83.

Prosser, B.L.T., Taylor, D., Dix, B.A. & Cleeland, R. (1987) Method of evaluating effects of antibiotics on bacterial biofilm. *Antimicrobial Agents and Chemotherapy*, **31**, 1502–1506.

Shand, G.H., Anwar, H., Kadurugamuwa, J., Brown, M.R.W., Silverman, S.H. & Melling, J. (1985) *In vivo* evidence that bacteria in urinary tract infections grow under iron restricted conditions. *Infection and Immunity*, **48**, 35–39.

Shapiro, J.A. (1987) Organisation of developing *Escherichia coli* colonies viewed by scanning electron microscopy. *Journal of Bacteriology*, **169**, 142–156.

Shaw, J.C., Bramhill, B., Wardlaw, N.C. & Costerton, J.W. (1985) Bacterial fouling in a model core system. *Applied and Environmental Microbiology*, **49**, 693–701.

Silverman, M., Belas, R. & Simon, M. (1984) Genetic control of bacterial adhesion. In Marshall, K.C. (ed.) *Microbial Adhesion and Aggregation*, pp. 95–107. Springer-Verlag, New York.

Sticker, D., Dolman, J., Rolfe, S. & Chawla, J. (1989) Activity of antiseptics against *Escherichia coli* growing as biofilms on silicone surfaces. *European Journal of Clinical Microbiology and Infectious Diseases*, **8**, 974–978.

West, A.A., Araujo, R., Dennis, P.J.L., Lee, J.V. & Keevil, C.W. (1989) Chemostat models of *Legionella pneumophila*. In Flannigan B. (ed.) *Airborne Deteriogens and Pathogens*, pp. 107–116. Biodeterioration Society, London.

Wilkinson, T.G. & Hamer, G. (1974) Wall growth in mixed bacterial cultures growing on methane. *Biotechnology and Bioengineering*, **16**, 251–260.

Zobell., C.E. & Allen, E.C. (1935) The significance of marine bacteria on the fouling of submerged surfaces. *Journal of Bacteriology*, **29**, 230–251.

4: The Physiology and Biochemistry of Biofilm

J.W.T. Wimpenny[1], S.L. Kinniment[1] and M.A. Scourfield[2]

[1]*School of Pure and Applied Biology, University of Wales College of Cardiff, Cardiff CF1 3TL; and* [2]*Department of Dermatology, University of Wales College of Medicine, Heath Park, Cardiff CF4 4XN, UK*

In their natural habitats bacteria can encounter changing, competitive and often hostile conditions. In such environments the organism must either adapt its physiology to that appropriate to the prevailing conditions, or die. It is now generally accepted that the majority of bacteria in most ecosystems are attached to surfaces and not suspended in the aqueous phase as are planktonic bacteria. As such they apparently derive benefit from this mode of growth (Lappin-Scott & Costerton, 1989). Biofilm growth is interesting in that, to a certain extent, these sessile organisms create their own environment. Thus, film growth influences the microenvironment, which in turn influences the physiology of its component cells (Characklis *et al.*, 1989). Phenotypic variations in physiology and biochemistry occur in response to growth within a diffusion gradient system with continuously varying concentrations of substrates, products and other solute molecules. Such phenotypic changes may, for example, be due to the deployment of antimicrobials and, as a general rule, the biofilm mode of growth leads to a lowered sensitivity to antimicrobial agents compared with their freely dispersed counterparts (Nichols, 1989). A general discussion of the physiology and biochemistry of the biofilm mode of growth and, where possible, its variation from the planktonic mode of growth, will be presented, followed by an outline of the approach we have taken to investigate the structure and function of biofilm with the constant-depth film fermenter (CDFF).

Biofilm Structure and Function

Elemental composition of biofilms

The elemental composition of various biofilms has been reviewed by Christensen & Characklis (1989), who reported the results of a number of investigations

Microbial Biofilms:
Formation and Control

(Kornegay & Andrews, 1967; Characklis, 1980; Turakhia, 1986). The relative amounts of organic and inorganic biofilm components can be determined by combustion. Volatile and fixed solids generally reflect the organic and inorganic fraction, respectively. The volatile fraction of a suspended microbial population is greater than 90%; for biofilms it is often considerably less, primarily because of the added mass of inorganic constituents adsorbed, entrapped or precipitated within the biofilm matrix. However, in laboratory experiments where biotic components dominate, the volatile fraction of a biofilm may be as high as 80% of the biofilm dry weight (Kornegay & Andrews, 1967). The carbon:nitrogen ratios in some biofilms are considerably higher — approximately five times — than in microbial cells. This probably reflects the large proportion of extracellular polymer substances (EPS) (generally low in nitrogen), if the total carbon rather than organic carbon is measured (Christensen & Characklis, 1989).

The inorganic fraction is especially high in biofilms that accumulate in natural aquatic ecosystems, where silt, clays and sediments are trapped in the matrix, from where they influence the physical properties of the biofilm matrix.

The presence of biofilm can lead to favourable conditions for metal corrosion that would not occur in its absence. These conditions may favour localized corrosion by the formation of differential aeration cells, due to a patchy distribution of the biofilm, as well as by the formation of anaerobic sites at the base of the film due to microbial respiration. This activity in the biofilm often generates conditions favourable for the growth of sulphate-reducing bacteria that use hydrogen. The hydrogen is generated in anaerobic environments by a combination of cathodic protons and anodic electrons, which in turn increases metal corrosion (Hamilton, 1985). Sulphate reducers also produce corrosive metabolites, such as sulphides, that lead to the incorporation of corrosion products, such as iron sulphide within the biofilm matrix.

The corrosion of mild steel in a saline medium is markedly stimulated by abundant deposits of corrosion products of a wide range of chemical compositions, including lepidocrocite (γ-$Fe_2O_3.H_2O$), wustite (FeO) (Edyvean, 1984), goethite (α-$Fe_2.H_2O$) and magnetite (Fe_3O_4) (Videla, 1989).

The inorganic composition of biofilms varies with the chemical composition of the substratum and that of the aqueous phase. However, due to the nature of biofilm and its EPS matrix, any corrosion and/or metabolic products are likely to become trapped and remain in the biofilm for long periods of time.

Extracellular matrix

One of the most notable features of biofilms is their high content (50–90%) of EPS (Characklis & Cooksey, 1983). Understanding the physical and chemical characteristics of the matrix and its relationship to resident organisms is crucial to the understanding of the structure and functioning of biofilms. The terminology for the extracellular material associated with cell aggregates or biofilms varies in the literature, being referred to as slime, capsule, sheath, EPS and glycocalyx. Zobell (1943) suggested the participation of extracellular cementing substances in the adhesion of cells to substrata. The last stage of cell attachment to a surface, 'irreversible' binding, involving specific interactions, is associated with the production of adhesive materials such as exopolysaccharides (Lappin-Scott & Costerton, 1989). Corpe (1970) demonstrated the involvement of acidic polysaccharides in bacterial adhesion, and Fletcher & Floodgate (1973) observed this phenomenon by electron microscopy.

The vast majority of bacterial EPS are polysaccharides. Common sugars such as glucose, galactose, mannose, fructose, rhamnose, N-acetylglucosamine, glucuronic acid, galacturonic acid, mannuronic acid, and guluronic acid, are typical constituents of bacterial polysaccharides (Christensen & Characklis, 1989; Christensen, 1989).

Uhlinger & White (1983) indicated that most polysaccharides and phospholipid accumulation occurrred late in the stationary phase, when the physiological status of the cells showed maximal stress. Dawson et al. (1981) and Marshall (1985) noted that most EPS and enhanced rates of adhesion were seen in starved cells. Investigators have also observed the production of different polysaccharides during exponential growth and in the stationary phase (Uhlinger & White, 1983; Christensen et al., 1985). Wrangstadt and colleagues (1986) induced starvation in exponentially growing cells and observed that soluble, viscous polysaccharide was released, whereas the same polysaccharide was not produced by growing cells. This suggests that starvation triggers the production of a different polymer.

In contrast to these reports regarding rates of EPS production, Lappin-Scott and colleagues (1988a, b, c) and MacLeod et al. (1988) reported less polysaccharide production by starved bacteria than by growing cultures. When grown in rich media, bacteria may produce polymers at a high rate but release these as a slime, rather than retaining the material as a distinct capsule (Christensen & Characklis, 1989). Antibodies made against polymers produced in liquid cultures have been shown to react with biofilm matrices in situ, suggesting that the EPS matrix in biofilms contains some of the same polymers produced in liquid culture by the same organisms (Costerton et al., 1981). Bakke and colleagues (1984) found that the production rate of EPS did not differ quantitatively between biofilms and suspended cultures of Pseudomonas aeruginosa when growth conditions were otherwise the same. In contrast, Turakhia & Characklis (1989) showed that the same organism produced more EPS as biofilm than in suspended culture.

EPS significantly influences the physical properties of the biofilm, including diffusivity, thermal conductivity and rheological properties. EPS, irrespective of charge density or its ionic state, has some of the properties of diffusion barriers, molecular sieves and adsorbents, thus influencing physicochemical processes such as diffusion and fluid frictional resistance. The predominantly polyanionic, highly hydrated nature of EPS also means that it can act as an ion exchange matrix, serving to increase local concentrations of ionic species such as heavy metals, ammonium, potassium etc., while having the opposite effect on anionic groups. It may not have any effect on uncharged potential nutrients, including sugars (Hamilton & Characklis, 1989). However, bacteria are assumed to concentrate and use cationic nutrients such as amines, suggesting that EPS can serve as a nutrient trap, especially under oligotrophic conditions (Costerton *et al.*, 1981). Conversely, the penetration of charged molecules such as some biocides may be at least partly restricted by this phenomenon (Costerton & Lashen, 1984).

Allison (1992) has examined antibiotic penetration of polysaccharide produced by *Ps. aeruginosa* and *Pseudomonas cepacia*. These polysaccharides *together* show a significant increase in viscosity, which is increased even more with Ca^{2+} ions. Some of these polymers are specifically associated with biofilm formation and can significantly reduce the sensitivity of the organisms to a range of antibiotics, including ciprofloxacin. However, Nichols (1989) calculated that adsorption to, or decreased diffusion through, EPS cannot fully explain the antibiotic resistance of biofilm bacteria. Clearly, much work is needed to understand changes to inhibitor sensitivity within a biofilm matrix.

Other roles suggested for the extracellular matrix are as an energy store and site of both intracellular communication and genetic transfer (Characklis & Cooksey, 1983; Sharma *et al.*, 1987).

Characklis and colleagues (1981) reported that the thermal conductivity of a mixed-culture biofilm is similar to that of water, from which it may be concluded that a biofilm provides about 27 times as much resistance to heat transfer as stainless steel of equal thickness. Thus, a very thin biofilm can significantly restrict heat transfer through a stainless steel tube (Christensen & Characklis, 1989).

Rheological measurements conducted with a Weissenburg rheogoniometer on an *in situ* mixed-population biofilm (Characklis, 1980) indicate that the biofilm is viscoelastic. Thus, the viscous properties of the biofilm contribute to increased fluid frictional resistance in flow conduits (Christensen & Characklis, 1989).

Finally, the extracellular matrix may contain particulate materials such as clays, organic debris, lysed cells and precipitated minerals. The composition of different biofilms may be dominated by different components. For example, biofilms on artificial implants or in human hosts tend to be dominated by organic polymers and cell debris, including host-derived material. On corrodable metals in seawater the biofilm may contain large amounts of precipitated metals. Biofilms appear to vary dynamically, and their extracellular matrix composition can clearly change with time.

Trophic state

The nutrient status of the water phase is critically important to biofilm physiology and development and, moreover, the nutrients present in such systems are heterogeneously distributed in both time and space. As a result, bacteria may be exposed alternatively to feast and famine conditions, with the latter predominating. Oligotrophic bacteria are uniquely suited to low-nutrient environments without competitors, and may readily be isolated from such habitats on immersed glass slides (Poindexter, 1981). The genus *Caulobacter* adapts morphologically by synthesizing prosthecae in response to nutrient depletion. These bacteria actually benefit from being attached to the nutritionally enriched substratum, as well as being stationary in a flowing stream of water from which they can scavenge a continuous supply of nutrients. The slowest-growing organisms, however, rarely become a significant component of mature biofilms, where the dominant organisms are mostly the faster-growing copiotrophic species. Copiotrophic bacteria confronted with typical oligotrophic nutrient conditions are faced with a subminimal supply of nutrients. Under these conditions such organisms frequently adopt specific starvation responses (Novitsky & Morita, 1976, 1978). The formation of mini-cells is one of the first manifestations of adaptive starvation survival processes (Morita, 1982). Increasing surface to volume ratio may in this way provide such cells with an advantage under conditions of nutrient limitation (Novitsky & Morita, 1977). Dawson *et al.* (1981) showed that starvation of *Vibrio* DW1 led to an increase in cell numbers, a drop in cell volume and an increase in adhesiveness. Marshall (1988) also concluded that copiotrophic bacteria respond to starvation conditions by becoming smaller and generally more adhesive. Some workers, however, found that once these cells were attached they grew back to their normal size and began to multiply (Kjelleberg *et al.*, 1982; Marshall, 1988). Brown *et al.* (1977) reported that the attachment of cells from a river inoculum on to aluminium foil, within a chemostat vessel, was slight under nitrogen limitation with excess glucose, despite evidence of synthesis of polysaccharide material. Under carbon limitation, on the other hand, extensive mixed populations developed on the surface.

Biofilm cellular morphology

Growth within a biofilm can influence the morphology of certain microbial species. Several investigators (Picologlou *et al.*, 1980; McCoy & Costerton, 1982; Trulear & Characklis, 1982) noted the predominance of filamentous bacteria in biofilm, leading to a relatively low bacterial density. Other changes include the expression and/or exposure of unique surface antigens and extensive exopolysaccharide production to form glycocalices within which the organisms develop (Costerton *et al.*, 1987). Changes in the O-antigen subunits of lipo-polysaccharide may also occur in Gram-negative organisms (Cochrane *et al.*, 1988a, b), and individual vegetative bacteria can show remarkable plasticity of the cell envelope (Brown *et al.*, 1991). Adsorption to surfaces may be sufficient to induce morphological changes. For example, marine *Vibrio* species show a change from a single polar sheathed flagellum to the production of numerous lateral flagella when located on a surface (Stal *et al.*, 1989).

Heterogeneity

In common with most microbial ecosystems, biofilms are generally hetero-geneous in time, in space and usually in genotype.

Temporal heterogeneity

The physical, chemical and biological dynamics of biofilm development normally follow an ordered temporal sequence. Biofilm development involves phases of attachment, growth and polysaccharide production, maturation and the import of other components, usually followed by detachment as the whole film sloughs from its substratum surface.

Starting with a perfectly clean surface, an organic monolayer adsorbs within minutes of exposure and changes the properties of the substratum. It has been demonstrated that a range of materials with diverse surface properties are rapidly conditioned by adsorbing organics, when exposed to natural waters with low organic concentrations. These organic molecules frequently appear to be polysaccharides or glycoproteins. Such adsorbed layers are termed *conditioning films* because they alter the surface properties of the substratum (Characklis & Cooksey, 1983). Presumably, conditioning films have a part to play in modifying the extent of bacterial adhesion to immersed surfaces in natural habitats, since some macromolecules inhibit bacterial adhesion, whereas others have little effect (Fletcher, 1976; Fletcher & Marshall, 1982).

Although the precise mechanisms for adsorption of bacteria to surfaces are still the subject of considerable debate, it is generally accepted that whereas the bacterial cell itself does not make direct contact with the substratum, adsorption is mediated by extracellular structures capable of overcoming electrostatic repulsion effects. The extracellular structures involved in polymer bridging vary with different bacterial types, and include pili, fimbriae and EPS (Characklis *et al.*, 1989).

Besides the surface properties of the substratum, the microbial colonization of a freshly immersed substratum will depend on the flow rate, the trophic state of the bulk water and the type and physiological state of the bacteria adsorbing to the surface. Growth and reproduction of the primary colonizing bacteria may modify the surface characteristics of the substratum, rendering it suitable for subsequent colonization by secondary organisms. In flowing systems, the accumulation of bacteria on the surface is a balance between processes of adsorption and growth on the one hand *versus* detachment on the other (Escher, 1986; Lawrence & Caldwell, 1987). As shear rate increase, fewer and fewer cells can remain attached (Fowler, 1988).

Other primary colonizing bacteria remain fixed to the surface but do not appear to grow or even to lyse. Many biofilm bacteria, however, can develop (Bott & Brock, 1970; Kjelleberg *et al.*, 1982; Escher, 1986; Power & Marshall, 1988). Characklis & Cooksey (1983) point out that, often, copiotrophic species such as *Pseudomonas* attach and grow as primary colonizers, followed by various oligotrophs. Marshall *et al.* (1971) and Corpe (1973) noted that substrata immersed in seawater were initially colonized by rod-shaped bacteria, followed by other organisms such as *Caulobacter, Hyphomicrobium* and *Saprospira*. Scanning electron microscopy has indicated an increased complexity in the microbial community of a biofilm with time; moreover, species composition appears to vary with the substratum type (Mack *et al.*, 1975; Marszalek *et al.*, 1979; Dempsey, 1981). Wimpenny and colleagues (1983) summarized the succession of organisms and the climax community associated with dental plaque formation: first were aerobic species capable of growth on salivary glycoproteins, followed by facultative and anaerobic organisms that utilize the metabolic products of the primary colonizers and produce organic acids. In the case of biofilms that cause corrosion in oil pipelines, the hydrocarbon-degrading microorganisms create the anaerobic conditions and produce the nutrients necessary for growth of the sulphate-reducing bacteria, which are the principal causative organisms of anaerobic corrosion (Hamilton, 1985).

The maturing biofilm incorporates not only secondary colonizing micro-organisms but also other inanimate material, including clay colloids and other minerals as well as organic detritus. The structure of the polysaccharide matrix is not, as many assume, homogeneous: rather it seems to have numerous pores and acts to some extent like a net or strainer, with a flow of water through at least part of the structure.

After growth, the biofilm also undergoes phases of detachment. Detachment is defined as the transfer of biomass from the biofilm to the bulk liquid phase and occurs in two forms, erosion and sloughing (Characklis *et al.*, 1989). Erosion is the continuous removal of small particles from the biofilm at the biofilm/liquid interface, at least partially due to the flow of the aqueous phase. Sloughing is the sporadic detachment of large fragments of the biofilm, resulting from changing conditions within the biofilm. Environmental factors may influence erosion rates. Turakhia & Characklis (1989) have observed that free calcium in the bulk water decreases the erosion rate in a *Ps. aeruginosa* biofilm. Sloughing is frequently noted at high substrate loadings and laminar flow or low shear stresses. Howell & Atkinson (1976) demonstrated that high organic loadings in trickling filters were the driving force for oxygen depletion in the lower depths of the biofilm, which in turn resulted in sloughing. Jansen & Kristensen (1980) have observed the formation of N_2 bubbles under denitrifying biofilms, and consider that sloughing may also result from gas production.

Spatial heterogeneity

The dimensions of a biofilm, the rates of biochemical processes within it and the diffusivity of relevant solute molecules all contribute to differentiation of the structure in space. Spatial differentiation generates a range of habitats providing niches for different physiological types of bacteria.

Concentrations of oxygen, hydrogen sulphide, nitrous oxide and hydrogen ions (pH) can be measured within biofilms by microsensors with tip diameters down to a few micrometres. The use of such sensors has revealed steep concentration gradients, not only in the biofilm itself but also in the aqueous phase immediately above it. This is especially true in films exposed to light in terms of diurnal variations in local chemistry (Revsbech, 1989). In thick biofilms, the upper layers may be aerobic whereas the lower layers adjacent to the substratum are anaerobic because of the respiratory processes of the cells above them (Schaftel, 1982).

With time, a level of organization may develop in which cells of different species form consortia with integrated metabolic processes. Thus, sulphate-reducing bacteria are found in what appear to be aerobic environments as the result of the activities of oxygen-scavenging aerobes.

Differences in community distribution have also been observed by Ritz (1969) when examining sections of dental plaque obtained by cryostat sectioning. The aerobic *Neisseria* were found to be most abundant in young plaque and in the upper layers of mature plaque. Streptococci, which are largely inert to oxygen, were also most abundant in younger plaque and seemed thereafter to be randomly distributed within the plaque. *Veillonella* (anaerobes) were limited to the inner two-thirds of the plaque.

Biofilm thickness also seems to be influenced by species diversity. For example, Christensen & Characklis (1989) reported that a pure culture of *Ps. aeruginosa* biofilm rarely exceeded a thickness of 50 μm in several laboratory studies (Trulear, 1983; Bakke, 1986), whereas a mixed-culture biofilm under virtually the same experimental conditions often reached a thickness greater than 120 μm (Characklis, 1980). However, our results, with a pure culture of *Ps. aeruginosa* in a CDFF, indicate that biofilm thicknesses up to 300 μm can easily be achieved (Kinniment & Wimpenny, 1992). Other factors, including the prevailing physicochemical environment as well as substratum composition, surface texture etc., can all influence biofilm thickness. Bakke (1986) has reported variations in biofilm thickness over the width of a rectangular duct under laminar flow conditions. The variations in thickness found were also a function of film 'age'.

Biofilm activity is influenced by film thickness. Sanders (1966) observed that substrate uptake rate increases in parallel with growth up to a critical film thickness. Tomlinson & Snaddon (1966) suggested that the actively metabolizing part of the film was the aerobic region. Atkinson & Fowler (1974) proposed that an ideal film thickness is equal to the penetration depth or depths of the substrate and/or electron acceptor, since both solute uptake and growth increase until this critical thickness is reached.

Biofilms are not necessarily uniform in density over the entire substratum. Surface cracks and microroughness may contribute to biofilm 'patchiness', as do chemical characteristics of the surface. The resulting patchiness may lead to differentiated microenvironments. Siebel & Characklis (1991) investigated the adsorption of *Klebsiella pneumoniae* on to polycarbonate, and observed an irregular distribution of aggregates of cells forming colonies. When *Ps. aeruginosa* and *K. pneumoniae* were mixed, *Ps. aeruginosa* uniformly colonized the substratum in the area between *K. pneumoniae* aggregates.

Biofilm density can vary with depth through the biofilm. Christensen & Characklis (1989) reported that Watanabe (unpublished data) used a 'microslicer' to separate the layers in a biofilm over 500 μm thick, and noted an increase in density with depth. However, our investigations with *Ps. aeruginosa* biofilms approximately 300 μm thick did not confirm this finding. The use of transmission electron microscopy, examination of the viable count profiles of freeze-sectioned cryoprotected biofilm (unpublished data) and adenylate concentrations through the biofilm (Kinniment & Wimpenny, 1992) suggest the presence of a denser region of viable cells just below the surface of the biofilm.

Biofilm Physiology

Metabolic activity

Although it is generally considered that cells are at an advantage when attached to a surface, the experimental evidence for this is inconclusive (Hamilton, 1987). Over a wide range of organisms and experimental systems, activity assays have given conflicting results.

The influence of solid surfaces on bacterial activity appears to be complex and, particularly in natural environments, may be difficult to assess. First, there is the problem of finding a relevant parameter to measure which will provide a valid indication of metabolic activity. Fletcher (1985) reported some parameters that have been studied, including changes in cell size (Kjelleberg *et al.*, 1982), respiration rate measured by CO_2 production (Bright & Fletcher, 1983a; Kefford *et al.*, 1983), oxygen uptake rate (Jannasch & Pritchard, 1972) or electron transport system activity (Bright & Fletcher, 1983b), substrate uptake rates (Fletcher, 1979; Bright & Fletcher, 1983a, b; Bakke *et al.*, 1984) or substrate breakdown (Estermann *et al.*, 1959), product formation (Hattori & Hattori, 1963) and heat production (Gordon *et al.*, 1983). Although many of these processes may be closely related, change in one type of activity is not always accompanied by corresponding changes in another; care should therefore be taken to specify the type of activity being considered (Fletcher, 1985).

Most measurements deal directly or indirectly with the potential for, and efficiency of, substrate utilization. Many studies that illustrated an increase in activity when growth was associated with surfaces did so only when low-nutrient media were used (Zobell, 1943; Jannasch & Pritchard, 1972; Marshall, 1976; Ladd *et al.*, 1979). Under these conditions surfaces may provide an advantage by assisting the capture and/or uptake of scarce nutrients. There may be no such stimulation by solid surfaces at higher nutrient levels. The size of the nutrient molecule may also be important. Thus, low molecular weight compounds can be transported directly into the cell, whereas macromolecules need to be broken down by extracellular enzymes before they can be assimilated (Fletcher, 1984, 1985). Zobell (1943) could find no evidence that surface activity was enhanced with low molecular weight substrates such as glucose, glycerol or lactate. However, sodium caseinate, lignoprotein and emulsified chitin were active in this respect. Other workers (Hattori & Furusaka, 1959) observed surface effects with low molecular weight substrates. For example, rates of succinate oxidation differed between free-living cells and those attached to an anion exchange resin.

The composition of the solid surface can affect bacterial activity. Thus, the number of attached cells and their distribution on the surface can depend on the surface composition (Murray & van den Berg, 1981; Shimp & Pfaender, 1982). The substratum can also apparently modify activity, due to the physicochemical conditions and forces present at the solid/liquid interface (Bright & Fletcher, 1983b).

Hamilton (1987) reviewed a range of investigations which suggested that, in the majority of cases, attached bacteria demonstrated an increase in metabolic activity, but usually under low-nutrient conditions. Atkinson & Fowler (1974) showed that growth of *Escherichia coli* was improved after surface adsorption, but only at nutrient (glucose) concentrations less than

25 ppm. Jeffrey & Paul (1986) considered a number of studies, the consensus of which confirmed an increase in metabolic activities for surface-associated bacteria at low or zero nutrient concentrations. Their findings indicated that free-living cells were more active than those attached in the presence of added nutrients, whereas the opposite was true for cells metabolizing endogenous reserves.

Other examples of surfaces stimulating metabolic activity were increases in the growth rate and yield of *E. coli* and of the growth and sporulation rates of *Bacillus subtilis* after adsorption (Hattori & Hattori, 1981), stimulation of the rate of ethanol degradation by starved cultures of sulphate-reducing bacteria attached to clay particles (Laanbroek & Geerligs, 1983) and a 20−25% increase in the nitrite oxidation rates of a *Nitrobacter* species after attachment to glass or anion exchange resin (Keen, 1984). Attached bacteria can also apparently express certain extracellular enzyme activities with higher V_{max} and lower K_m than possessed by their planktonic brethren (Hoppe, 1984).

Whereas attachment can be accompanied by increased activity, other investigations have produced different results. *Ps. aeruginosa* grown in steady-state biofilms at various glucose loading and reactor dilution rates, had yield and rate coefficients that were the same as those in suspended cultures, *provided* that the film was sufficiently thin to ensure negligible diffusional resistance to substrate penetration and uptake (Bakke *et al.*, 1984). Bright and Fletcher (1983a, b) studied the effects of adsorption of a marine *Pseudomonas* species to a range of plastic surfaces and found that the nature of the effect depended upon the substratum, the amino acid substrate and its concentration, and the parameter being measured. Assimilation of the amino acid by surface-associated cells was generally greater than, and respiration less than, that shown by free-living bacteria. Measuring heat production by microcalorimetry, Gordon and colleagues (1983) found no increase in the activity of a marine bacterium towards glucose or glutamic acid, after adsorption to hydroxyapatite. In some cases, the reverse was the case. *Nitrosomonas europaea* also showed a fall in activity, together with a reduction in specific growth rate, after attachment to glass slides (Powell, 1985).

Copiotrophic bacteria have been seen significantly to reduce their endogenous respiration rates (Characklis *et al.*, 1989), supporting the concept of dormancy proposed by Stevenson (1978) to account for the prolonged survival of copiotrophs in oligotrophic environments. It seems that, under oligotrophic conditions, copiotrophic bacteria suspended in the aqueous phase are starving, whereas those adsorbed to surfaces may be able to grow and reproduce, especially if the adsorbed nutrient supply is continually replenished, as is the case in open continuous-flow systems (Marshall, 1988; Power & Marshall, 1988).

Biofilm activity and heterogeneity are critically dependent on its thickness.

Although monolayer films show virtually no restrictions on activity due to diffusion, the reverse is true of many thicker biofilms. The latter show bio-chemical, physiological and species differentiation in the presence of steep solute gradients.

Biofilm energetics — adenylate distribution

One measure of the energy status of living cells is to determine concentrations of the major adenylates, ATP, ADP and AMP (adenosine triphosphate, adenosine diphosphate and adenosine monophosphate). It should be remem-bered that ATP turnover in a growing cell is extremely rapid. It was suggested by Chapman & Atkinson (1977) that the quotient ATP + 0.5ADP/ATP + ADP + AMP — the *adenylate energy charge* (EC_A) — gives a valid measure of the energy status of living cells. It has generally been considered that a value for the EC_A greater than $0.75 - 0.8$ units is required for growth of bacterial cells (Chapman & Atkinson, 1977; Knowles, 1977; Karl, 1980). Although most of this work has been carried out in suspended cell systems in the laboratory, adenylates and EC_A values have been used as ecological tools and, more recently, they have been applied to biofilm systems.

Whole films of oral bacteria showed low steady-state EC_A values (Wimpenny *et al.*, 1989; Kemp, 1979). Using a perchloric acid extraction method, Wimpenny and coworkers showed that high EC_A values could be induced in young film exposed to high glucose concentrations. Similarly, Scourfield (1990), using a defined community of oral bacteria, could pulse low EC_A biofilm with sucrose and induce a sharp rise in EC_A values.

Recent studies in our laboratory have detected gradients of metabolic activity, expressed in terms of adenylate distribution, across a pure culture bacterial biofilm (Kinniment & Wimpenny, 1992). *Ps. aeruginosa* biofilms about 300 μm thick were produced in the CDFF. All the procedures used in this work and some of the results are described in detail below.

Measurement of the EC_A of cryostat sections, from the base to the surface of the biofilm, showed a rise of approximately 0.2 units. However, EC_A levels were generally low through the biofilm: highest values were from 0.4 to 0.5, or occasionally 0.6. These low values were substantiated with whole, unfrozen film which was extracted immediately and had not undergone the whole process of cryostat sectioning. These gave average values of approximately $0.35 - 0.45$ units.

The biofilms described by Kinniment & Wimpenny (1992) and by Wimpenny *et al.* (1989) appeared to be growing, albeit slowly. One inter-pretation of the EC_A values noted above is based on spatial heterogeneity. It is possible that only a thin layer at the surface of the biofilm was growing, a larger proportion was viable but in a 'resting' state, and cells at the very base of the biofilm were dead. Another, less likely, explanation, for which we have no direct evidence, is that nucleotides, and especially AMP, were excreted into the surrounding medium as part of a homeostatic mechanism compensating for low rates of ATP formation under starvation conditions (Chapman *et al.*, 1971; Davis & White, 1980).

The presence of 'active' layers in deep biofilms has been observed or suggested in the literature by several workers (Kornegay & Andrews, 1968; Atkinson & Fowler, 1974; LaMotta, 1976). LaMotta (1976) measured ATP levels in a biological film from a wastewater treatment plant. He observed that the ATP content increased proportionately, up to a limiting film thickness of about 320 μm. Measurements of adenylates through the *Pseudomonas* biofilms (Kinniment & Wimpenny, 1992) indicate the presence of an 'active' layer towards the surface of the biofilm, since total adenylates formed a peak just below the interface. These findings were again supported by differences in microbial biomass and in cell density, observed in electron microscope studies, and more recently by using cryosectioning techniques to determine the number of viable cells through the film profile (see below). The results showed a rise in viable counts from the base to the surface of the biofilm of roughly $1-2$ orders of magnitude.

Finally, we might conclude that because low numbers of viable cells were still present near the base of the film, low EC_A values mean that cells within the film structure may be able to survive and possibly grow at lower EC_A levels than are normally assumed for freely suspended cells.

Constant-Depth Biofilm: Rationalization for the Development of a Model Growth System

A major problem when investigating microbial growth from natural environments is the high variability of such systems. For instance, as has been emphasized above, microbial film usually goes through an irregular life-cycle, finally leading to large masses of biofilm removal, often due to the destabilizing effects of cell death and lysis and anaerobic fermentation reactions at the base of the film. This process is akin to growth of a bacterial batch culture, but is much less predictable. There is, therefore, a case for devising a biofilm growth system in which film formation is controlled and reproducible, and is at the same time easy to sample under precisely controlled conditions of nutrient and gas supply. There exist a number of laboratory systems that go some way to satisfying these demands. These include the Robbins device, the rototorque and the CDFF.

Biofilm fermenters

The Robbins device

The Robbins device (McCoy *et al*., 1981; see Chapter 1) has been constructed in a range of materials from admiralty brass to perspex. The common design feature is a recirculating tubing section through which a microbial culture is circulated via a reservoir and a pump. The walls of the tube are fitted with sampling studs whose inner surface is fitted flush with the inner wall of the tube section. Film grows both on the inner wall of the tube and on the studs. The latter may be removed and replaced quite simply during the operation of the system. The film thickness cannot, however, be controlled accurately and changes in the flow rate of circulating medium can lead to removal of biofilm surface layers by liquid shear momentum transfer. One theoretical problem with the Robbins device is that, at slow flow rates, there is likely to be a longitudinal nutrient gradient which could lead to position-dependent changes in biofilm properties. The Robbins device is an excellent model system for a range of biofilms that form in flow systems. These can range from water conduits to indwelling catheters.

The rototorque

The rototorque reactor was developed from the rotating annular reactor devised by Kornegay & Andrews (1967) of the National Science Foundation's Engineering Research Center for Microbial Interfacial Process Engineering at Montana State University, Bozeman, Montana, USA. It consists of two cylinders constructed from perspex (Plexiglas). The inner cylinder rotates at a constant rate. A growth medium plus inoculum passes between the rotor and stator. The latter is fitted with a number of removable glass slide sections on which the biofilm grows. An online torque monitor is used to monitor drag forces on the inner cylinder. Biofilm formation is established under controlled conditions, especially of shear rate.

Constant-depth film fermenters

The earliest CDFF was devised by Atkinson & Fowler (1974). Two models were constructed, each of which had some form of scraper bar to remove excess biofilm from the surface of the substratum. The first system consisted of a roughened glass plate, and the second consisted of a thin steel sheet template from which zones were cut to provide recessed regions in which the biofilm proliferated.

More recently, Wimpenny and his colleagues have developed an enclosed film fermenter capable of growing biofilm in removable pans. In the first model, film pans were located in a fixed plate over which a rotating scraper bar moved. In later versions the scraper blade was fixed and rotated the disc in which the pans were located (Coombe *et al.*, 1984; Peters & Wimpenny, 1988, 1989). The most recent model is described below.

The main design philosophy behind the development of the CDFF is as follows. Any system capable of growing microbes under steady-state conditions has a considerable advantage over closed systems. A steady-state or continuous culture system reaches a dynamic equilibrium, where organisms can proliferate at a relatively constant rate and are to all intents and purposes identical from point to point in time. The chemostat, first described by Monod (1950) and by Novick & Szilard (1950), is a classic example of such a system and microbiological history has proved its value over the last four decades. The most compelling reason for choosing steady-state systems is that their behaviour remains constant for prolonged periods. It is possible against this background, therefore, to perturb one parameter at a time and unequivocally to observe the responses of the system.

More recently, continuous culture systems have evolved to include multi-vessel systems such as the multistage chemostat, and bidirectionally linked systems such as the gradostat and the Herbert device (for review, see Wimpenny, 1988). These systems incorporate some elements of the spatial heterogeneity that characterizes natural microbial ecosystems. There is a logical reason, then, to make the jump from these liquid gradient systems to a steady-state growth model that more closely resembles a natural system. In the CDFF, film develops in a recessed zone whose depth can be selected by the experimenter. The film grows on the surface of a film plug located in a removable film pan. The latter rotates beneath a scraper blade which removes any excess film that forms. It can be seen that two continuous processes are proceeding. First, solute transport takes place across the film surface, leading to an input of nutrients including, where relevant, oxygen. In addition, metabolic products, including carbon dioxide, will diffuse out of the cellular array. Secondly, growth takes place. Cells will grow in a zone dependent on the penetration of nutrients and, for aerobic film, oxygen. The net effect of the growth will be to generate an excess of biomass which will then project above the film-pan surface. This excess is removed from the film by the scraper bar. The biofilm, once it fills the recess, will soon enter what is apparently a steady-state. We have often used the term *quasi steady-state* to define this system. This caveat is necessary since our criteria for 'steady-state' is based only on protein levels, viable counts and, in some cases, dry weight and carbohydrate estimations.

Other goals included the ability to run film growth in an enclosed sterilizable environment, where nutrient supply and gaseous regimen could be completely controlled. Also, the fermenter had to produce a large number of film samples, enabling us to check the statistical reproducibility of our experiments.

The Cardiff constant-depth film fermenter

The CDFF is illustrated in Fig. 4.1. It consists of a borosilicate glass tubing section with top and bottom plates fabricated from stainless steel. A drive shaft enters the fermenter via seals and a bearing assembly, and connects to a rotating steel or polytetrafluoroethylene (PTFE) turntable in which 15 removable film pans are located. Each of the pans is fitted with six removable film plugs. These are recessed with a custom-designed tool to a measured distance, generally 300 μm. Biofilm forms on the surface of the plugs in the space between them and the bottom of a stationary spring-loaded PTFE scraper blade located above the turntable. The fermenter can be sterilized by autoclaving, and is operated under aseptic conditions at constant temperature, rotation speed and gas composition.

Nutrient medium enters the fermenter through a port and is fed to the top of the turntable, where it is distributed over the film plugs by the scraper blade. In operation the fermenter is generally inoculated with an overnight culture of the organism or organisms required. This inoculum is recycled for a period of up to 24 h for organisms to attach to the PTFE surface of the plugs (Fig. 4.2(a)). The system is then connected to an aspirator containing fresh medium, which is pumped into the fermenter via a peristaltic pump at a selected constant rate (Fig. 4.2(b)).

To sample the CDFF the drive motor is switched off so that the pan required is located immediately beneath a sampling port. The latter is flame-sterilized and opened to allow the removal of a complete film pan, which is replaced with a sterile film pan. This operation is carried out with a stainless steel tool which screws into a threaded hole at the centre of the pan.

Film geometry

The biofilm measures approximately 4.75 mm in diameter and is set to 0.3 mm deep. This gives a volume of $5.32 \, mm^3$ (say, $5 \, mm^3$) or $5 \times 10^9 \, \mu m^3$. If we suppose that a microbe has a volume of $1 \, \mu m^3$, then each film plug should contain up to 5×10^9 bacteria. Assuming full pans and a water content in living cells of 80% w/v, we may also suppose that the maximum amount of dry weight present will be about 1 mg. If dry bacterial biomass is about 50% protein, we should expect no more than 500 μg protein per film. In practice, values around half this are found, which is not unreasonable considering that no allowance is made in these calculations for loose packing or interstitial polysaccharide.

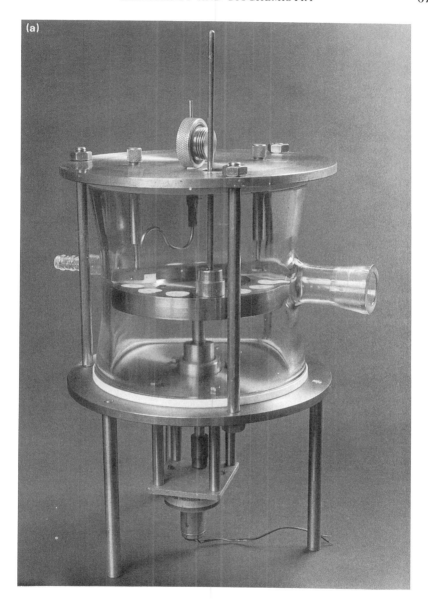

Fig. 4.1. The Cardiff constant-depth film fermenter. (a) The assembled fermenter.

(b)

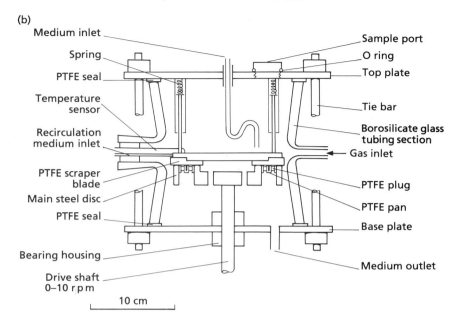

Medium inlet

Spring

PTFE seal

Temperature sensor

Recirculation medium inlet

PTFE scraper blade

Main steel disc

PTFE seal

Bearing housing

Drive shaft 0–10 r.p.m

Sample port

O ring

Top plate

Tie bar

Borosilicate glass tubing section

Gas inlet

PTFE plug

PTFE pan

Base plate

Medium outlet

10 cm

(c)

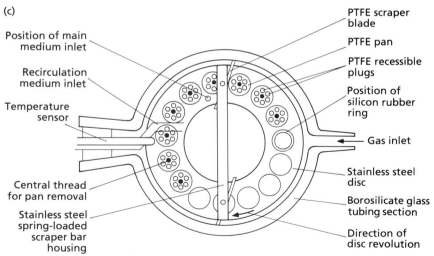

Position of main medium inlet

Recirculation medium inlet

Temperature sensor

Central thread for pan removal

Stainless steel spring-loaded scraper bar housing

PTFE scraper blade

PTFE pan

PTFE recessible plugs

Position of silicon rubber ring

Gas inlet

Stainless steel disc

Borosilicate glass tubing section

Direction of disc revolution

FIG. 4.1. (*continued*) (b) Vertical section through the CDFF. (c) Horizontal section showing details of the film disc and pans.

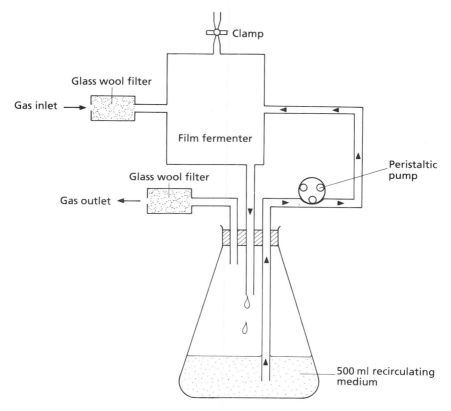

F<small>IG</small>. 4.2. The complete CDFF growth system. (a) During initial recycling of inoculum (adapted from Peters, 1988).

Use of the CDFF to Investigate the Physiology of Biofilm

Systems investigated with the CDFF

The CDFF has been used in Cardiff to investigate the following systems:
1   dental plaque biofilm (Coombe *et al.*, 1984; Scourfield, 1990);
2   a river-water community (Peters & Wimpenny, 1988);
3   growth of a *Pseudomonas* biofilm derived from a metalworking fluid environment (Kinniment & Wimpenny, 1990);
4   growth of organisms derived from contaminated indwelling catheters (Liese Ganderton, unpublished).

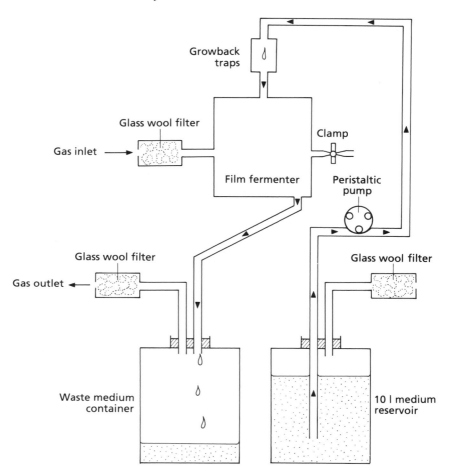

FIG. 4.2. (*continued*) (b) The complete CDFF growth system during the main growth phase (adapted from Peters, 1988).

Growth dynamics of biofilm

There are two relatively simple methods to determine the growth rate of the biofilm itself. One is to determine the accumulation of protein, and the second is to measure the number of colony-forming units in the film. Protein determination was by the standard Folin method (Lowry *et al.*, 1951) modified by an alkaline pretreatment step (boiling for 5 min with 1 mol/l NaOH) to solubilize protein from intact cells. The design of the film pan means that up to six film samples may be removed simultaneously, allowing a statistical assessment of film reproducibility. Often four plugs were used for protein assay, and the remaining two to determine viable count. Figure 4.3(a) shows four replicate growth curves, determined by protein accumulation for *Ps. aeruginosa* biofilm grown using a simplified simulated metalworking fluid medium.

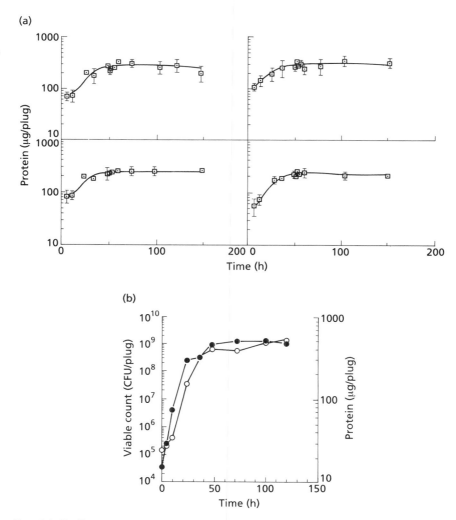

FIG. 4.3. Biofilm growth. (a) Growth of a metalworking fluid isolate (*Pseudomonas aeruginosa*) expressed as protein accumulation showing reproducibility of four separate experiments. Each data point shows the standard deviation of four measurements. (b) Viable count (●) and protein (○) growth curves of a *Pseudomonas aeruginosa* biofilm growing on a synthetic urine medium at 37°C.

Viable count determinations were carried out by various methods. For the metalworking fluid system using *Ps. aeruginosa*, two film plugs were removed with flame-sterilized forceps and the film was then dispersed by vortex-mixing samples for 90 s in 10 ml of a sterile deflocculant solution. This consisted of 0.01% w/v Cirrasol (ICI Organics, Blackley, UK) in 0.01% w/v sodium pyrophosphate (Gayford & Richards, 1970), with about 250 2.5−3.5 mm diameter glass beads. Samples were then serially diluted and plated on to growth media. For biofilm derived from contaminated catheter isolates, samples were dispersed with an ultrasonic cleaning bath for 5 min in 10 ml nutrient broth followed by a further 2 min of vortex mixing. Samples were then treated as in the previous method. Figure 4.3(b) shows growth curves for *Ps. aeruginosa* biofilm grown using a synthetic urine, based on viable count determinations and protein accumulation.

Biofilm formation appears to be exponential, but the actual rates involved depend on the organism and its nutrient environment. It must also be stressed that concentration of nutrients may mean that, in some experiments, growth quickly becomes nutrient-limited.

Other methods of measuring biofilm growth

It is possible to determine biofilm accumulation on the pans by measuring dry weight or a specific cellular component. As well as protein production, carbohydrate accumulation has also been determined. Figure 4.4 shows dry weight and carbohydrate accumulation for a dental plaque system.

Scanning electron microscopy (SEM) and transmission electron microscopy (TEM) of constant-depth biofilm

Biofilm produced in the CDFF has been examined by SEM and TEM by standard techniques for preparation of the material.

The surface of a steady-state *Ps. aeruginosa* biofilm grown on amine−carboxylate medium appears to be densely packed with regular normal rod-shaped bacteria when examined using SEM (Fig. 4.5(a)). Colonization of the surface of the film plug can also be followed as a function of time by SEM. This was done by Scourfield (1990) for the Bowden dental plaque community, a selected group of bacterial species associated with dental plaque (Fig. 4.5(b)). After only 4 h incubation, individual cells can be seen on the PTFE surface, many in the process of dividing. By 24 h, microcolony formation was advanced. After 72 h, the pan surface was completely covered. Steady-state film is also shown. Examination by TEM of the same film and the above *Ps. aeruginosa* film shows that healthy cells are present in the upper two-thirds (about 200 μm) of the film (Fig. 4.6). Below this, and clearly visible in Fig. 4.6(a)

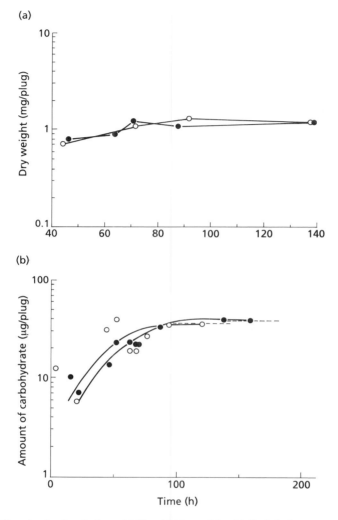

FIG. 4.4. Growth of a dental plaque biofilm. (a) Dry weight. (b) Carbohydrate.

the majority of cells appear to be lysed. It is remarkable that there is a sharp division between the two zones, suggesting that in steady-state biofilm nutrients diffuse downwards to a reproducible position below which cells are starved and/or anaerobic.

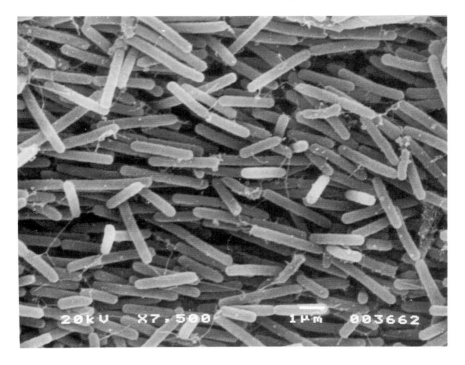

FIG. 4.5. (a) Scanning electron micrograph of the surface of a *Pseudomonas aeruginosa* biofilm grown on amine–carboxylate medium in the CDFF.

Solute gradients in biofilm

Of key importance to biofilm behaviour is diffusion into and out of the structure of key solutes including oxygen, sulphide, and also of acidic or alkaline agents leading to pH changes. Revsbech (1989) has reviewed the application of a range of microelectrodes, including those for oxygen, pH, sulphide and nitrous oxide. In addition, fibre-optic probes have been used to measure light penetration into biofilms containing photosynthetic species.

Kinniment (unpublished) has used an oxygen microelectrode to measure oxygen profiles in the *Ps. aeruginosa* biofilm grown on an amine–carboxylate medium in the CDFF. Figure 4.7 shows one example suggesting that oxygen penetrates about 150 μm into the film.

Cryosectioning biofilm

The intact biofilm is a 'pill' 300 μm thick. It is of the utmost interest to determine the spatial heterogeneity of such a structure. The great majority of work on

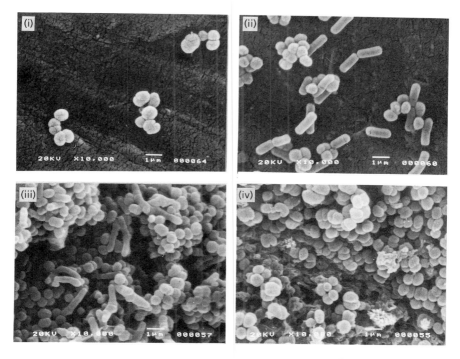

FIG. 4.5. (*continued*) (b) Colonization of the film plug by a group of dental plaque bacteria over different time periods using scanning electron microscopy: (i) 4 h; (ii) 24 h; (iii) 72 h; (iv) steady-state conditions (>100 h).

biofilms has endeavoured to draw conclusions as to its structure and function by examination of the whole film or even of material scraped from the substratum surface. We therefore spent some time developing methods for sectioning biofilm, in particular to examine the distribution of viability and of the adenylates ATP, ADP and AMP. The most successful methods are described below (but see Kinniment & Wimpenny, 1992).

The main aim of all the methods attempted was to transfer the intact film from the film pan while still attached to the film plug, to the cryostat sectioning stub where it would be frozen and then sectioned. The system is illustrated in Fig. 4.8. Layers of sterile 2% w/v Oxoid agar (5 mm) were poured into standard Petri dishes. Agar discs were removed with a sterile No. 15 cork borer, inverted and placed on the surface of the remaining agar in the plate. Circles of lens tissue were cut and one of these was placed on the upper surface of the agar disc. A few microlitres of water were then used to moisten

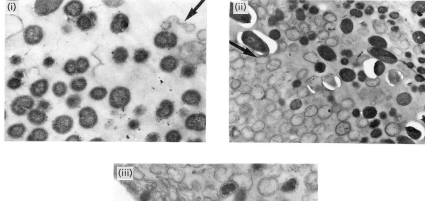

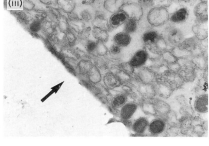

FIG. 4.6. Changes in cell morphology as a function of depth in transmission electron micrographs. (a) The dental plaque community biofilm. Arrows point to: (i) surface of the biofilm; (ii) a region in the lower two-thirds of the biofilm, where a sharp boundary between living and apparently dead cells is indicated; (iii) base of the biofilm (magnification ×7400).

the lens tissue. The next procedure was to locate the agar disc plus lens tissue on a cryostat sample holder. With forceps, a sample-holder was inverted and placed on top of the plug plus tissue. The whole plate was then turned upside down and the sample-holder plus agar disc were carefully separated from the agar in the Petri dish and placed into a polystyrene container.

The next step was to remove a film plug from its pan, invert it and clamp it in a pair of forceps, which were then secured to a micromanipulator arm. The latter was located above the agar disc in the polystyrene box and the film plug was lowered to just above the agar surface. A small drop of water was placed on the agar and the film plug was lowered so that the film surface was just submerged in the drop of water. Liquid nitrogen was poured carefully into the container to the depth of the agar disc, and the whole array allowed to freeze. Once frozen, the forceps were removed from the micromanipulator and the plug was pulled away from the frozen film.

A variant of this method used a sample-holder which had been cooled to −37°C before freezing the agar disc to it. The film and plug, which had

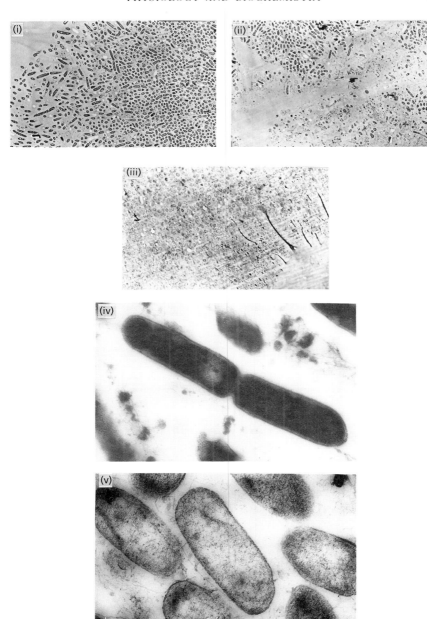

FIG. 4.6. (*continued*) (b) *Pseudomonas aeruginosa* biofilm grown on amine−carboxylate medium: (i) the surface; (ii) near the visible interface between viable and apparently lysing cells; (iii) the film base (magnification ×1300); (iv) 'viable' cell from near the surface; (v) empty, 'dead' cell from near the film base (magnification ×19 000).

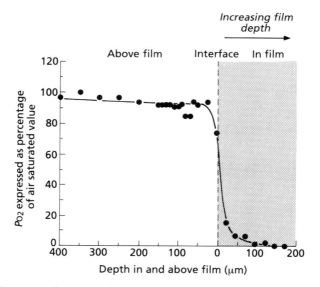

FIG. 4.7. Oxygen profile above and across a *Pseudomonas aeruginosa* biofilm.

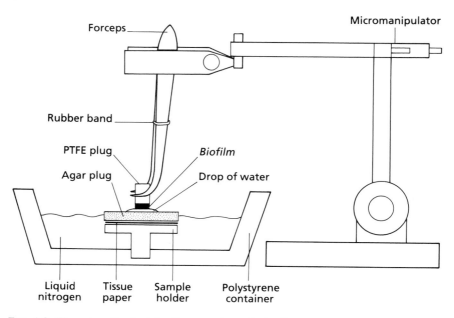

FIG. 4.8. The system for obtaining frozen sections of a biofilm (see text).

already been frozen in liquid nitrogen, were then attached to the agar with a small drop of water. As before, once frozen, the plug was easily removed from the film. The sample-holder plus film was then located on the cryostat microtome arm and 12 μm thick sections were cut.

Distribution of viability across a biofilm profile

A major problem came to light when attempting to perform viable counts on sectioned material. This was that freezing, sectioning, thawing and dispersal of the cells led to a high and variable degree of cell death. Often cell counts fell by four orders of magnitude, therefore it was necessary to reduce this to obtain anything like meaningful results from the work. We finally decided to use dextran (molecular weight 60–90 000) as a cryoprotectant (Ashwood-Smith & Warby, 1971), at a concentration of 25% w/v. The film, while still on its plug in the film pan, was allowed to stand in the aerated dextran solution in nutrient medium at 30°C for 2 h. Freezing and thawing these preparations in liquid nitrogen indicated good protection, with a *maximum* drop in viability of up to fivefold. There was the possibility that the dextran would not penetrate through the biofilm, but separate experiments where dextran was determined in biofilm sections indicated that this was not a problem. Indeed, higher levels of dextran were found near the base of the film than near its surface. This is probably due to the fact that the dextran was located in intercellular space, which would be lower where viable cells were packed closely a little below the surface. On the other hand, the presence of many dead and lysed cells near the film base would allow more dextran to accumulate there.

The distribution of viable cells across the 300 μm film is interesting. In a number of experiments (Fig. 4.9 gives two examples), viability is highest a little below the surface of the film and falls off to about two orders of magnitude less near the film base.

Changes in viability must be related to other evidence concerning film heterogeneity. Viability results correspond to TEMs (see Fig. 4.6(b)), which suggest that there are many dead and lysed cells near the base of the film, whereas cell density is highest at a point a little below the film surface.

Adenylates and the energy status of biofilm

Another important experimental approach is to examine the adenylate status of biofilm. In earlier experiments (Wimpenny *et al.*, 1989; Scourfield, 1990) biofilms of oral bacteria were grown in the CDFF. Adenylates were extracted in perchloric acid and determined with standard firefly lantern assays in a luminometer. EC_A values were determined from the data at different points in the development of a biofilm. In addition, glucose was added to the basic amino acid-containing medium at concentrations from 0 to 100 mg/l. In most cases the average EC_A value for the whole biofilm was between about 0.25 and 0.45. However, at the earliest times and with the highest amount of glucose, these values rose to between approximately 0.6 to just under 1 (Fig. 4.10). These results are interesting, and suggest that the biofilm as a whole may be energy-limited, especially where glucose concentrations were low, but also as the film thickened and diffusion barriers became important. These data fit well with TEMs of the films grown on the same medium containing 10 mg/l glucose. Here it is easy to see a sharp transition between what appear to be healthy cells and the largely empty walls of dead and lysed cells. In a later experiment with the Bowden community, Scourfield (1990) examined changes in adenylates in a biofilm as a function of time, when the biofilm was exposed to a sudden increase in sucrose to emulate what might happen to a natural plaque on being 'fed'! Figure 4.11 clearly shows the rapid rise in EC_A for the film as a whole, and demonstrates what happens when the original glucose concentration is reestablished.

Such global figures are of only limited value, and the distribution of adenylates across the film profile must be determined. This has recently been done (Kinniment & Wimpenny, 1992). Methods used for extracting and assaying adenylates are summarized below, but are described fully in the

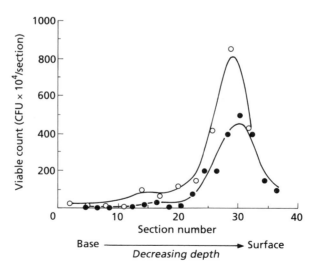

FIG. 4.9. Distribution of CFUs across a film profile of *Pseudomonas aeruginosa* grown on amine–carboxylate medium at 30°C. Film sectioned as described in the text: (●) sections taken in duplicate; (○) sections taken in triplicate.

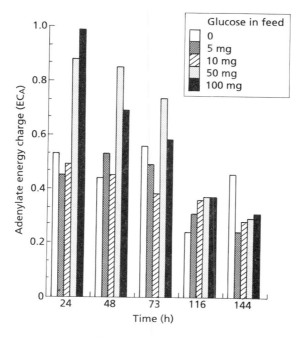

FIG. 4.10. Adenylate energy charge values as a function of film age and of glucose concentration in dental plaque biofilms.

original paper. Sets of two or three sections, obtained as described above, were transferred to 0.5 ml of 2.3 mol/l perchloric acid containing 6.7 mmol/l ethylenediamine tetra-acetic acid (EDTA). The samples were agitated on ice for 15 min and then centrifuged. The supernatant was neutralized with 2 mol/l KOH−0.5 mol/l triethanolamine buffer. The supernatant was separated from the precipitated potassium perchlorate by centrifuging after standing for 5 min and then stored at −70°C before assay.

Standard methods were used to determine ATP, ADP and AMP (Lundin & Thore, 1975a, b) with firefly lantern extract and an LKB 1251 luminometer.

The results of these experiments were interesting. Total adenylates peak in a region a little below the film surface (Fig. 4.12(a)), which again appears to occur where cell density is highest when measured by viability distribution and TEM. The distribution of adenylates shows a trend where AMP falls from the base to the surface of the film while ATP rises, and ADP, though variable, tends to be maintained at about the same level across this region. The data shown were subject to a three-point moving average, since the errors involved in freezing, sectioning and assaying the nucleotides generate a

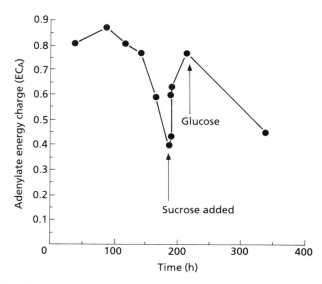

FIG. 4.11. The effects of pulsing 1000 mg/l sucrose on a biofilm of the Bowden community grown on a medium containing 10 mg/l glucose. The EC_A value, high at first, fell as the film aged. The sucrose pulse restored the EC_A values to a high figure.

noisy graph. EC_A values (Fig. 4.12(b)) show a trend which is quite compatible with the earlier dental plaque work: that is, values are low at the base of the film, ranging between 0.25 and 0.35, rising to 0.4–0.5, occasionally 0.6, at the surface of the film. These data must be seen in conjunction with viability, which could range over two orders of magnitude across the same span. The exact status of adenine nucleotides near the base of the film is uncertain. Are they associated only with live cells or are some dispersed in the film matrix outside lysed cells, or bound to other cellular components?

Biofilms and biocides

The response to antimicrobials of microbes in biofilms has been investigated widely in recent years (for a recent review see Nichols, 1991). The virtually unanimous conclusion is that microbes located within the biofilm matrix are more resistant, often by several orders of magnitude, to these agents than are suspended planktonic cultures of the same organism. Such findings apply not only to bona fide antibiotics, for example a wide range of aminoglycosides and β-lactams, but also to a range of other antimicrobial agents, including chlorhexidine and chlorine.

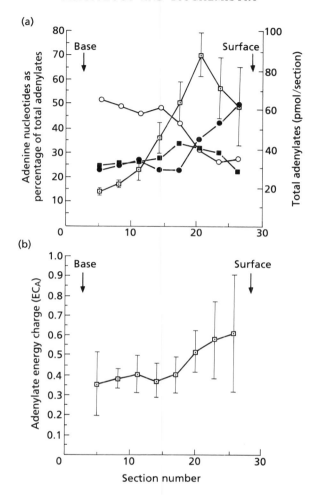

FIG. 4.12. (a) Proportions of adenylates and the concentrations of total adenylates in cryostat sections across a *Pseudomonas aeruginosa* biofilm. (●) % ATP; (■) % ADP; (○) % AMP; (▣) total adenylates. (b) EC_A values as a function of depth across a *Pseudomonas aeruginosa* biofilm. Error bars take into account the fact that the data were transformed into the three-point moving average of the original data to reduce noise.

Kinniment and Wimpenny (unpublished) have investigated the effects of several different biocides commonly used to inhibit growth in commercial metalworking fluids on biofilms of *Ps. aeruginosa*.

Standard biofilms were produced in the CDFF with the amine−carboxylate medium. After the film reached steady state, formaldehyde was added to the influent medium at 400 ppm. Planktonic cultures in the early stationary phase were treated similarly with formaldehyde. Counts and total protein per film were estimated at different times after adding the biocide. It is clear that viable cells in the film are much less susceptible to formaldehyde than are the planktonic cells. In fact, these data suggest that suspended cells may be as much as 50 times more sensitive to the formaldehyde (Fig. 4.13). The effects of formaldehyde at this concentration on the appearance of the biofilm by SEM is shown in Fig. 4.14. Formaldehyde seems to cause extensive aggregation of cells, compared with controls. Other biocides, including chlorocresol and an analytical standard of Kathon 886 MW (Rohm & Haas, Valbonne, France) containing approximately a 12% mixture of the active ingredients: 5-chloro-2-methyl-4-isothiazolin-3-one and 2-methyl-4-isothiazolin-3-one, were also more active against suspended cultures in shaken flasks than in biofilms but apparently not to the same extent. The important message is that virtually all biocides and antibiotics have been through a rigorous selective procedure which involves activity in suspended culture! What is clearly needed is a selection procedure based on a simple biofilm test.

In this chapter we have reviewed some applications of the CDFF which we regard as providing a useful tool for investigating biofilm physiology. The 300 μm film used can be considered as a gradient system where growth rates vary from μ_{max} to 0, depending on position (Fig. 4.15). At the surface the specific growth rate will be highest and may be near μ_{max}; below this point, growth becomes more and more severely limited as a function of depth. At some point — probably between a half and two-thirds of the depth of the biofilm — growth rates fall to zero. Below this point, cells finally die and lyse. Lysed cells may then yield some nutrients, which will diffuse upwards and contribute to further growth. As the film grows there is a net 'extrusion' of cells from the film pan and these are removed by the scraper bar. These cells are, therefore, predominantly the more rapidly growing surface bacteria, but may also include some nutrient-limited organisms from lower levels.

Future Challenges

It is only comparatively recently that microbial biofilms have been perceived to be ubiquitous and possibly the favoured growth habit for most bacteria. There remain many important avenues to explore before we can say that we clearly understand these systems.

One area yielding exciting results at present concerns the spatial organization of biofilm. A clear determinant is oxygen. For instance, the aerobic *Neisseria* are located in the upper parts of a dental plaque system, whereas anaerobic *Veillonella* are found near the base (Ritz, 1969). More specific examples of

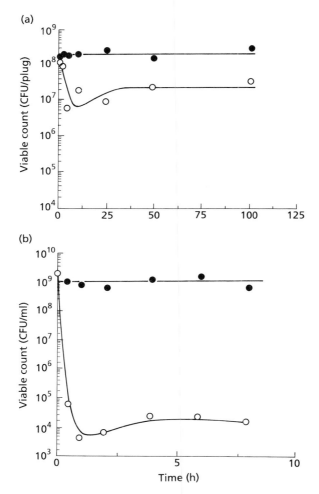

FIG. 4.13. The effects of 400 ppm formaldehyde on (a) biofilm, (b) planktonic cultures of *Pseudomonas aeruginosa.*. (●) Control; (○) treated. Formaldehyde was added at time zero on the graphs.

spatial organization are most commonly seen in dental plaque. These include rosettes and corncob formations. The former may contain *Streptococcus sanguis* surrounded by *Bacterionema loeschii* (Kolenbrander & Anderson, 1988), whereas the latter may be formed either by *Bacterionema matruchotii* surrounded by streptococci (Mouton *et al.*, 1977) or *Fusobacterium nucleatum* and associated *Strep. sanguis*. Coaggregation, first described by Gibbons & Nygaard (1970), is

FIG. 4.14. The effects of 400 ppm formaldehyde on the appearance of a *Pseudomonas aeruginosa* biofilm grown in the CDFF. Pictures were taken after (a, b) 107 h untreated growth, and (c, d) 57 h treatment, where formaldehyde was added after 50 h growth.

a phenomenon based on 'recognition' of one species by another. A simple assay for coaggregation was developed by Kolenbrander & Anderson (1988). Pure cultures of each test organism were mixed under carefully controlled conditions. If coaggregation occurred, the mixed organisms flocculated. Using this simple assay with a wide range of oral bacteria, Kolenbrander's group has generated elegant structural maps of a plaque, illustrating the spatial order of this ecosystem in terms of coaggregation patterns.

From a totally different perspective Bryers and his colleagues (Banks & Bryers, 1990; Bryers & Banks, 1990) have investigated interactions between binary combinations of biofilm organisms grown on a substratum surface. Thus, an autotroph plus a heterotroph were grown together under autotrophic and heterotrophic conditions respectively, and the ratio of each organism was determined after selective ^{14}C labelling. Ratios were determined in 20-μm

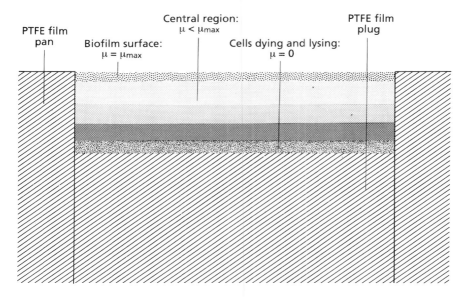

FIG. 4.15. An idealized model of biofilm formation in the CDFF.

thick fixed sections, so that distributions of each species were effectively mapped across the film profile. Under nitrifying conditions the nitrifier predominated at every level in the biofilm. Under heterotrophic conditions, both species were present in approximately equal proportions at the base of the film; however, the heterotroph overgrew the film in the surface layers of older biofilm. In a competition between a *Hyphomicrobium* sp. and a *Pseudomonas putida*, the latter overgrew the former either when the two were inoculated together or when the pseudomonad was added to a pre-existing *Hyphomicrobium* biofilm. These experiments were convincing in that the experimental results fitted a mathematical model of the two systems.

It would, of course, be nice to 'see' the distribution of bacteria in a biofilm. The advent of confocal laser microscopy (CLM) makes this a practical proposition, especially for relatively thin biofilms (see Chapter 5). CLM is capable of generating optical sections through a three-dimensional structure, and data at each level are stored electronically as a single frame. A series of frames is then used to reconstruct a three-dimensional image. It is possible to produce stereoscopic pairs from such images. These are conventionally red – green images, and the use of appropriate spectacles allows the viewer to examine a true three-dimensional object. Such images have been produced by Caldwell and his colleagues (Lawrence *et al.*, 1991).

These approaches, together with information derived from cryosectioning, conventional electron microscopy and the use of microsensors, promises a complete dissection of a number of biofilms before very long.

There remain major problems ahead. Some of these are listed below.

1   What is the relationship between a substratum surface and colonization?
2   What is the role of the conditioning film?
3   What part is played by a surface in concentrating solute molecules?
4   What is the mechanism of bacterial attachment?
5   How is exopolymer production regulated?
6   How do conditions change within a biofilm so that secondary colonization can take place?
7   How do species within a biofilm compete for space and for nutrients?
8   How important is cooperation between species in biofilms?
9   How is solute distribution regulated through the biofilm?
10   How is biofilm destabilized and finally removed from the substratum surface?
11   How do cells in a biofilm adapt to survive profoundly higher concentrations of inhibitory compounds than can be tolerated by suspended cultures of the same organism?

Acknowledgements

The authors gratefully acknowledge support from the Institute of Petroleum (S.L. Kinniment) and from the Science and Engineering Research Coucil for a studentship (M.A. Scourfield). We are also grateful to T.P. Robinson for help with oxygen electrode measurements and to L. Ganderton for allowing us to use some of her data on the growth of catheter biofilm organisms.

References

ALLISON, D.G. (1992) Polysaccharide interactions in bacterial biofilms. In Melo, L.F., Bott, T.R., Fletcher, M. & Capdeville, B. (eds.) *Biofilms: Science and Technology*, pp. 371–376. Kluwer Academic Publishers, Dordrecht.

ASHWOOD-SMITH, M.J. & WARBY, C. (1971) Studies on the molecular weight and cryoprotective properties of polyvinylpyrrolidone and dextran with bacteria and erythrocytes. *Cryobiology*, 8, 453–464.

ATKINSON, B. & FOWLER, H.W. (1974) The significance of microbial film fermenters. *Advances in Biochemical Engineering*, 3, 224–277.

BAKKE, R. (1986) *Biofilm Detachment*. PhD Thesis, Montana State University, Bozeman, MT.

BAKKE, R., TRULEAR, M.G., ROBINSON, J.A. & CHARACKLIS, W.G. (1984) Activity of *Pseudomonas aeruginosa* in biofilms: steady state. *Biotechnology and Bioengineering*, 26, 1418–1424.

BANKS, J.D. & BRYERS, M.K. (1990) Cryptic growth within a binary microbial culture. *Applied Microbiology and Biotechnology*, **33**, 596–601.

BOTT, T.L. & BROCK, T.D. (1970) Growth and metabolism of periphytic bacteria: methodology. *Limnology and Oceanography*, **15**, 333–342.

BRIGHT, J.J. & FLETCHER, M. (1983a) Amino acid assimilation and respiration by attached and free-living populations of a marine *Pseudomonas* sp. *Microbial Ecology*, **9**, 215–226.

BRIGHT, J.J. & FLETCHER, M. (1983b) Amino acid assimilation and electron transport activity in attached and free-living marine bacteria. *Applied and Environmental Microbiology*, **45**, 818–825.

BROWN, C.M., ELLWOOD, D.C. & HUNTER, J.R. (1977) Growth of bacteria at surfaces: influence of nutrient limitation. *FEMS Microbiology Letters*, **1**, 163–166.

BROWN, M.R.W., COSTERTON, J.W. & GILBERT, P. (1991) Extrapolating to bacterial life outside the test tube. *Journal of Antimicrobial Chemotherapy*, **27**, 565–567.

BRYERS, J.D. & BANKS, M.K. (1990) Assessment of biofilm ecodynamics. In de Bont, J.A.M., Visser, J., Mattiasson, B. & Tramper, J. (eds.) *Physiology of Immobilized Cells*, pp. 49–62. Elsevier, Amsterdam.

CHAPMAN, A.G. & ATKINSON, D.W. (1977) Adenine nucleotide concentrations and turnover rates. Their correlation with biological activity in bacteria and yeasts. *Advances in Microbial Physiology*, **15**, 253–306.

CHAPMAN, A.G., FALL, L. & ATKINSON, D.E. (1971) Adenylate energy charge in *Escherichia coli* during growth and starvation. *Journal of Bacteriology*, **108**, 1072–1086.

CHARACKLIS, W.G. (1980) *Biofilm Development and Destruction*. Final report, EPRI CS-1554, project RP902-1. Electric Power Research Institute, Palo Alto.

CHARACKLIS, W.G. & COOKSEY, K.E. (1983) Biofilms and microbial fouling. *Advances in Applied Microbiology*, **29**, 93–138.

CHARACKLIS, W.G., MCFETERS, G.A. & MARSHALL, K.C. (1989) Physiological ecology in biofilm systems. In Characklis, W.G. & Marshall, K.C. (eds.) *Biofilms*, pp. 341–393. John Wiley & Sons, New York.

CHARACKLIS, W.G., NIMMONS, M.J. & PICOLOGLOU, B.F. (1981) Influence of fouling biofilms on heat transfer. *Journal of Heat Transfer Engineering*, **3**, 23–37.

CHRISTENSEN, B.E. (1989) The role of extracellular polysaccharides in biofilms. *Journal of Biotechnology*, **10**, 181–202.

CHRISTENSEN, B.E. & CHARACKLIS, W.G. (1989) Physical and chemical properties of biofilms. In Characklis W.G. and Marshall K.C. (eds.) *Biofilms*, pp. 93–130. Wiley, New York.

CHRISTENSEN, B.E., KJOSBAKKEN, J. & SMIDSRØD, O. (1985) Partial chemical and physical characterization of two extracellular polysaccharides produced by marine, periphytic *Pseudomonas* sp. strain NCMB 2021. *Applied and Environmental Microbiology*, **50**, 837–845.

COCHRANE, M.G., BROWN, M.R.W., ANWAR, H., WELLER, P.H., LAM, K. & LAMBERT, P.A. (1988a) Antibody response to *Pseudomonas aeruginosa* surface protein antigens in a rat model of lung infection. *Journal of Medical Microbiology*, **27**, 255–261.

COCHRANE, M.G., BROWN, M.R.W. & WELLER, P.H. (1988b) Lipopolysaccharide antigens produced by *Pseudomonas aeruginosa* from cystic fibrosis lung infection. *Federation of European Microbiology Societies Letters*, **50**, 241–245.

COOMBE, R.A., TATEVOSSIAN, A. & WIMPENNY, J.W.T. (1984) Factors affecting the growth of thin film bacterial films *in vitro*. In ten Cate, J.M., Leach, S.A. & Arends, J. (eds.) *Bacterial Adhesion and Preventive Dentistry*, pp. 193–205. IRL Press, Oxford.

CORPE, W.A. (1970) An acid polysaccharide produced by a primary film-forming marine bacterium. *Developments in Industrial Microbiology*, **11**, 402–412.

CORPE, W.A. (1973) Microfouling: the role of primary film forming bacteria. *Proceedings of the 3rd International Congress on Marine Corrosion Fouling*, Evanston, pp. 598–609.

COSTERTON, J.W. & LASHEN, E.S. (1984) The inherent biocide resistance of corrosion-causing biofilm bacteria. *Materials Performance*, **23**, 13–16.

COSTERTON, J.W., IRVIN, R.T. & CHENG, K.J. (1981) The bacterial glycocalyx in nature and disease. *Annual Review of Microbiology*, **35**, 399–424.

COSTERTON, J.W., CHENG, K.J., GEESEY, K.G., LADD, P.I., NICKEL, J.C., DASGUPTA, M. & TOMAS, J.M. (1987) Bacterial biofilms in nature and disease. *Annual Review of Microbiology*, **41**, 435–464.

DAVIS, W.M. & WHITE, D.C. (1980) Fluorometric determination of adenosine nucleotide derivatives as measures of the microfouling, detrital and sedimentary microbial biomass and physiological status. *Applied and Environmental Microbiology*, **40**, 539–548.

DAWSON, M.P., HUMPHREY, B.A. & MARSHALL, K.C. (1981) Adhesion, a tactic in the survival strategy of a marine vibrio during starvation. *Current Microbiology*, **6**, 195–198.

DEMPSEY, M.J. (1981) Marine bacterial fouling: a scanning electron microscope study. *Marine Microbiology*, **61**, 305–316.

EDYVEAN, R.G.J. (1984) Interactions between microfouling and the calcareous deposit formed on cathodically protected steel in seawater. *Proceedings of the 6th International Congress on Marine Corrosion and Fouling*, Athens, pp. 469–483.

ESCHER, A.R. (1986) *Bacterial Colonization of a Smooth Surface: An Analysis with Image Analyzer.* PhD Thesis, Montana State University, Bozeman, MT.

ESTERMANN, E.F., PETERSON, G.H. & MCLAREN, A.D. (1959) Digestion of clay-protein, lignin-protein and silica-protein complexes by enzymes and bacteria. *Soil Science Society of America Proceedings*, **23**, 31–36.

FLETCHER, M. (1976) The effects of proteins on bacterial attachment to polystyrene. *Journal of General Microbiology*, **94**, 400–404.

FLETCHER, M. (1979) A microautoradiographic study of the activity of attached and free-living bacteria. *Archives of Microbiology*, **122**, 271–274.

FLETCHER, M. (1984) Comparative physiology of attached and free-living bacteria. In Marshall, K.C. (ed.) *Microbial Adhesion and Aggregation*, pp. 223–232. Springer-Verlag, New York.

FLETCHER, M. (1985) Effect of solid surfaces on the activity of attached bacteria. In Savage, D.C. & Fletcher, M. (eds.) *Bacterial Adhesion*, pp. 339–362. Plenum Press, New York.

FLETCHER, M. & FLOODGATE, G.D. (1973) An electron-microscopic demonstration of an acidic polysaccharide involved in the adhesion of a marine bacterium to solid surfaces. *Journal of General Microbiology*, **74**, 325–334.

FLETCHER, M. & MARSHALL, K.C. (1982) Bubble contact angle method for evaluating substratum interfacial characteristics and its relevance to bacterial attachment. *Applied and Environmental Microbiology*, **44**, 184–192.

FOWLER, H.W. (1988) Microbial adhesion to surfaces. In Wimpenny, J.W.T. (ed.) *Handbook of Laboratory Model Systems for Microbial Ecosystems*, Vol. 2, pp. 139–153. CRC Press, Boca Raton.

GAYFORD, C.G. & RICHARDS, J.P. (1970) Isolation and enumeration of aerobic heterotrophic bacteria in activated sludge. *Journal of Applied Bacteriology*, **33**, 342–350.

GIBBONS, R.J. & NYGAARD, M. (1970) Interbacterial aggregation of plaque bacteria. *Archives of Oral Biology*, **15**, 1397–1400.

GORDON, A.S., GERCHAKOV, S.M. & MILLERO, F.J. (1983) Effects of inorganic particles on metabolism by a periphytic marine bacterium. *Applied and Environmental Microbiology*, **45**, 411–417.

HAMILTON, W.A. (1985) Sulphate-reducing bacteria and anaerobic corrosion. *Annual Review of Microbiology*, **39**, 195–217.

HAMILTON, W.A. (1987) Biofilms: microbial interactions and metabolic activities. *Symposium of the Society for General Microbiology*, **41**, 361–385.

HAMILTON, W.A. & CHARACKLIS, W.G. (1989) Relative activities of cells in suspension and in biofilms. In Characklis, W.G. & Wilderer, P.A. (eds.) *Structure and Function of Biofilms*, pp. 199–219. Wiley, New York.

HATTORI, R. & FURUSAKA, C. (1959) Chemical activities of *Escherichia coli* adsorbed on a resin, Dowex-1. *Nature* (Suppl.), **184**, 1566–1567.

HATTORI, R. & HATTORI, T. (1963) Effect of a liquid–solid interface on the life of microorganisms. *Ecological Reviews*, **16**, 64–70.

HATTORI, R. & HATTORI, T. (1981) Growth rate and molar growth yield of *Escherichia coli* adsorbed on an anion-exchange resin. *Journal of General and Applied Microbiology*, **27**, 287–298.

HOPPE, H.G. (1984) Attachment of bacteria: advantage or disadvantage for survival in the aquatic environment. In Marshall, K.C. (ed.) *Microbial Adhesion and Aggregation*, pp. 283–301. Springer-Verlag, New York.

HOWELL, J.A. & ATKINSON, B. (1976) Sloughing of microbial film in trickling filters. *Water Research*, **10**, 307–315.

JANNASCH, H.W. & PRITCHARD, P.H. (1972) The role of inert particulate matter in the activity of aquatic microorganisms. *Memorie dell'Istituto Italiano di Idrobiologia*, **29**, (Suppl.), 289–308.

JANSEN, J. & KRISTENSEN, G.H. (1980) *Fixed Film Kinetics: Denitrification in Fixed Films*. Report 80-59, Department of Sanitary Engineering, Technical University of Denmark.

JEFFREY, W.H. & PAUL, J.H. (1986) Activity of an attached and free-living *Vibrio* sp. as measured by thymidine incorporation, *p*-iodonitrotetrazolium reduction and ATP/DNA ratios. *Applied and Environmental Microbiology*, **51**, 150–156.

KARL, D.M. (1980) Cellular nucleotide measurements and applications in microbial ecology. *Microbial Reviews*, **44**, 739–796.

KEFFORD, B., KJELLEBERG, S. & MARSHALL, K.C. (1983) Bacterial scavenging: utilization of fatty acids localized at a solid/liquid interface. *Archives of Microbiology*, **133**, 257–260.

KEEN, G.A. (1984) *Nitrification in Continuous Culture: The Effect of pH and Surfaces*. PhD Thesis, University of Aberdeen.

KEMP, C.W. (1979) Adenylate energy charge: a method for the determination of viable cell mass in dental plaque samples. *Journal of Dental Research*, **68(D)**, 2192–2197.

KINNIMENT, S.L. & WIMPENNY, J.W.T. (1990) Biofilms and biocides. *International Biodeterioration*, **26**, 181–194.

KINNIMENT, S.L. & WIMPENNY, J.W.T. (1992) Measurements of the distribution of adenylate concentrations and adenylate energy charge across *Pseudomonas aeruginosa* biofilms. *Applied and Environmental Microbiology*, **58**, 1629–1635.

KJELLEBERG, S., HUMPHREY, B.A. & MARSHALL, K.C. (1982) The effects of interfaces on small starved marine bacteria. *Applied and Environmental Microbiology*, **43**, 1166–1172.

KNOWLES, C.J. (1977) Microbial metabolic regulation by adenine nucleotide pools. *Symposium of the Society of General Microbiology*, **27**, 241–283.

KOLENBRANDER, P.E. & ANDERSON, R.N. (1988) Intergeneric rosettes: sequestered surface recognition among human periodontal bacteria. *Applied and Environmental Microbiology*, **54**, 1046–1050.

KORNEGAY, B.H. & ANDREWS, J.F. (1967) *Characteristics and Kinetics of Fixed-Film Biological Reactors*. Final report, grant WP-01181, Washington, D.C.: Federal Water Pollution Control Administration.

KORNEGAY, B.H. & ANDREWS, J.F. (1968) Kinetics of fixed-film biological reactors. *Journal of Water Pollution Control Federation*, **40**, R460–R468.

LAANBROEK, H.J. & GEERLIGS, H.J. (1983) Influence of clay particles (illite) on substrate utilization by sulphate-reducing bacteria. *Archives of Microbiology*, **134**, 161–163.

LADD, T.I., COSTERTON, J.W. & GEESEY, G.G. (1979) Determination of the heterotrophic activity of epilithic microbial populations. In Costerton, J.W. & Colwell, R.R. (eds.) *Native*

Aquatic Bacteria: Enumeration, Activity and Ecology, pp. 180–195. American Society for Testing and Materials, Philadelphia.

LaMOTTA, E.J. (1976) Kinetics of growth and substrate uptake in a biological film system. *Applied and Environmental Microbiology*, **31**, 286–293.

LAPPIN-SCOTT, H.M. & COSTERTON, J.W. (1989) Bacterial biofilms and surface fouling. *Biofouling*, **1**, 323–342.

LAPPIN-SCOTT, H.M., CUSACK, F. & COSTERTON, J.W. (1988a) Nutrient resuscitation and growth of starved cells in sandstone cores: a novel approach to enhanced oil recovery. *Applied and Environmental Microbiology*, **54**, 1373–1382.

LAPPIN-SCOTT, H.M., CUSACK, F., MACLEOD, F.A. & COSTERTON, J.W. (1998b) Starvation and nutrient resuscitation of *Klebsiella pneumoniae* isolated from oilwell waters. *Journal of Bacteriology*, **64**, 541–549.

LAPPIN-SCOTT, H.M., CUSACK, F., MACLEOD, F.A. & COSTERTON, J.W. (1988c) Starvation and survival of *Klebsiella pneumoniae* In Whittenbury, R., Gould, G.W., Banks, J.G. & Board, R.G. (eds.) *Homeostatic Mechanisms in Micro-organisms*, 253. FEMS Symposium **44**.

LAWRENCE, J.R. & CALDWELL, D.E. (1987) Behavior of bacterial stream populations within the hydrodynamic boundary layers of surface microenvironments. *Microbial Ecology*, **14**, 15–27.

LAWRENCE, J.R., KORBER, D.R., HOYLE, B.D., COSTERTON, J.W. & CALDWELL, D.E. (1991) Optical sectioning of bacterial biofilms. *Journal of Bacteriology*, **173**, 6558–6567.

LOWRY, O.H., ROSEBROUGH, N.J., FARR, A.L. & RANDALL, R.J. (1951) Protein measurement with the folin phenol reagent. *Journal of Biological Chemistry*, **193**, 265–275.

LUNDIN, A. & THORE, A. (1975a) Analytical information obtainable by evaluation of the time course of firefly bioluminescence in the assay of ATP. *Analytical Biochemistry*, **66**, 47–63.

LUNDIN, A. & THORE, A. (1975b) Comparison of methods for extraction of bacterial adenine nucleotides determined by firefly assay. *Applied Microbiology*, **30**, 713–721.

McCOY, W.F. & COSTERTON, J.W. (1982) Fouling biofilm development in tubular flow systems. *Developments in Industrial Microbiology*, **23**, 441.

McCOY, W.F., BRYERS, J.D., ROBBINS, J. & COSTERTON, J.W. (1981) Observations of biofouling film formation. *Canadian Journal of Microbiology*, **18**, 910–927.

MACLEOD, F.A., LAPPIN-SCOTT, H.M. & COSTERTON, J.W. (1988) Plugging of a model rock system by using starved bacteria. *Applied and Environmental Microbiology*, **53**, 1365–1372.

MACK, W.N., MACK, J.P. & ACKERSON, A.O. (1975) Microbial film development in a trickling filter. *Microbial Ecology*, **2**, 215–226.

MARSHALL, K.C. (1976) *Interfaces in Microbial Ecology*. Harvard University Press, Cambridge, MA.

MARSHALL, K.C. (1985) Bacterial adhesion in oligotrophic habitats. *Microbiological Sciences*, **2**, 321–326.

MARSHALL, K.C. (1988) Adhesion and growth of bacteria at surfaces in oligotrophic habitats. *Canadian Journal of Microbiology*, **34**, 503–506.

MARSHALL, K.C., STOUT, R. & MITCHELL, R. (1971) Selective sorption of bacteria from seawater. *Canadian Journal of Microbiology*, **17**, 1413–1416.

MARSZALEK, D.S., GERCHAKOV, S.M. & UDEY, L.R. (1979) Influence of substrate composition on marine microfouling. *Applied and Environmental Microbiology*, **38**, 987–995.

MONOD, J. (1950) La technique de culture continué. Théorie et applications. *Annales de l'Institut Pasteur*, **79**, 390–409.

MORITA, R.Y. (1982) Starvation-survival of heterotrophs in the marine environment. *Advances in Microbial Ecology*, **6**, 171–198.

MOUTON, C., REYNOLDS, H. & GENCO, R.J. (1977) Combined micromanipulation, culture and immunofluorescent techniques for isolation of the coccai organisms comprising the 'corn-cob' configuration of human dental plaque. *Journal de Biologie Buccale*, **5**, 321–332.

MURRAY, W.D. & VAN DEN BERG, L. (1981) Effect of support material on the development of fixed films converting acetic acid to methane. *Journal of Applied Bacteriology*, **51**, 257–265.

NICHOLS, W.W. (1989) Susceptibility of biofilms to toxic compounds. In Wilderer, P.A. & Characklis, W.G. (eds.) *Structure and Function of Biofilms*, pp. 321–332. Wiley-Interscience, New York.

NICHOLS, W.W. (1991) Biofilms, antibiotics and penetration. *Reviews in Medical Microbiology*, **2**, 177–181.

NOVICK, A. & SZILARD, L. (1950) Description of the chemostat. *Science*, **112**, 715–716.

NOVITSKY, J.A. & MORITA, R.Y. (1976) Morphological characterisation of small cells resulting in nutrient starvation of a psychrophilic marine Vibrio. *Applied and Environmental Microbiology*, **32**, 617–622.

NOVITSKY, J.A. & MORITA, R.Y. (1977) Survival of a psychrophilic marine Vibrio under long-term nutrient starvation. *Applied and Environmental Microbiology*, **33**, 635–641.

NOVITSKY, J.A. & MORITA, R.Y. (1978) Possible strategy for the survival of marine bacteria under starvation conditions. *Marine Biology*, **48**, 289–295.

PETERS, A.C. (1988) *A Constant Depth Laboratory Model Film Fermenter*. PhD Thesis, University College Cardiff, Cardiff.

PETERS, A.C. & WIMPENNY, J.W.T. (1988) A constant depth laboratory model film fermenter. *Biotechnology and Bioengineering*, **32**, 263–270.

PETERS, A.C. & WIMPENNY, J.W.T. (1989) A constant depth laboratory model film fermenter. In Wimpenny, J.W.T. (ed.) *Handbook of Laboratory Model Systems for Microbial Ecosystems Research*, pp. 175–195. CRC Press, Boca Raton.

PICOLOGLOU, B.F., ZELVER, N. & CHARACKLIS, W.G. (1980) Biofilm growth and hydraulic performance. *Journal of Hydraulic Division, American Society of Civil Engineers*, **106**, 733–746.

POINDEXTER, J.S. (1981) Oligotrophy: fast and famine existence. *Advances in Microbial Ecology*, **5**, 63–89.

POWELL, S.J. (1985) *Inhibition of Nitrosomonas europaea by Nitrapyrin: The Role of Surfaces*. PhD Thesis, University of Aberdeen.

POWER, K. & MARSHALL, K.C. (1988) Cellular growth and reproduction of marine bacteria on surface-bound substrate. *Biofouling*, **1**, 163–174.

REVSBECH, N.P. (1989) Microsensors: spatial gradients in biofilms. In Characklis, W.G. & Wilderer, P.A. (eds.) *Structure and Function of Biofilms*, pp. 129–144. Wiley, New York.

RITZ, H.L. (1969) Fluorescent antibody staining of neisseria, streptococcus and veillonella in frozen sections of human dental plaque. *Archive of Oral Biology*, **14**, 1073–1083.

SANDERS, W.M. III. (1966) Oxygen utilization by slime organisms in continuous culture [in stream surveys and self purification analyses]. *Air and Water Pollution*, **10**, 253.

SCHAFTEL, S.O. (1982) *Processes of Aerobic/Anaerobic Biofilm Development*. MSc Thesis, Montana State University, Bozeman, MT.

SCOURFIELD, M.A. (1990) *An Investigation into the Structure and Function of Model Dental Plaque Communities Using a Laboratory Film Fermenter*. PhD Thesis, University of Wales College of Cardiff, Cardiff.

SHARMA, A.P., BATTARSBY, N.S. & STEWART, D.J. (1987) Techniques for the evaluation of biocide activity against sulphate-reducing bacteria. In Board, R.G., Allwood, M.C. & Banks, J.G. (eds.) *Preservatives in the Food, Pharmaceutical, and Environmental Industries*, pp. 165–175. Blackwell Scientific Publications Oxford.

SHIMP, R.J. & PFAENDER, F.K. (1982) Effects of surface area and flow rate on marine bacterial growth in activated carbon columns. *Applied and Environmental Microbiology*, **44**, 471–477.

SIEBEL, M.A. & CHARACKLIS, W.G. (1991) Observations of binary population biofilms. *Biotechnology and Bioengineering*, **37**, 778–789.

STAL, L.J., BOCK, E., BOUWER, E.J., DOUGLAS, L.J., GUTNICK, D.L., HECKMANN, K.D., HIRSCH, P., KOLBEL-BOELKE, J.M., MARSHALL, K.C., PROSSER, J.I., SCHUTT, C. & WATANABE,

Y. (1989) Group report. Cellular physiology and interactions of biofilm organisms. In Characklis, W.G. & Wilderer, P.A. (eds.) *Structure and Function of Biofilms*, pp. 269–286. Wiley, New York.

STEVENSON, L.H. (1978) A case for bacterial dormancy in aquatic systems. *Microbial Ecology*, **4**, 127–133.

TOMLINSON, T.G. & SNADDON, D.H.M. (1966) Biological oxidation of sewage by films of microorganisms. *Air and Water Pollution*, **10**, 865–881.

TRULEAR, M.G. (1983) *Cellular Reproduction and Extracellular Polymer Formation in the Development of Biofilms*. MSc Thesis, Montana State University, Bozeman, MT.

TRULEAR, M.G. & CHARACKLIS, W.G. (1982) Dynamics of biofilm processes. *Journal of Water Pollution Control Federation*, **54**, 1288–1301.

TURAKHIA, M.H. (1986) *The Influence of Calcium on Biofilm Processes*. PhD Thesis, Montana State University, Bozeman, MT.

TURAKHIA, M.H. & CHARACKLIS, W.G. (1989) Activity of *Pseudomonas aeruginosa* in biofilms: effect of calcium. *Biotechnology and Bioengineering*, **33**, 406–414.

UHLINGER, D.J. & WHITE, D.C. (1983) Relationship between the physiological status and the formation of extracellular polysaccharide glycocalyx in *Pseudomonas atlantica*. *Applied and Environmental Microbiology*, **45**, 64–70.

VIDELA, H.A. (1989) Metal dissolution/redox in biofilms. In Characklis, W.G. & Wilderer, P.A. (eds.) *Structure and Function of Biofilms*, pp. 301–320. Wiley, New York.

WIMPENNY, J.W.T. (1988) Bidirectionally linked continuous culture: the gradostat. In Wimpenny, J.W.T. (ed.) *Handbook of Laboratory Model Systems for Microbial Ecosystems*, Vol. 1, pp. 73–98. CRC Press, Boca Raton.

WIMPENNY, J.W.T., LOVITT, R.W. & COOMBS, J.P. (1983) Laboratory model systems for the investigation of spatially and temporally organised microbial ecosystems. In Slater, J.H., Whittenbury, R. & Wimpenny, J.W.T. (eds.) *Microbes in their Natural Environments*, pp. 423–462. Cambridge University Press, Cambridge.

WIMPENNY, J.W.T., PETERS, A. & SCOURFIELD, M. (1989) Modeling spatial gradients. In Characklis, W.G. & Wilderer, P.A. (eds.) *Structure and Function of Biofilms*, pp. 111–127. Wiley, New York.

WRANGSTAD, M., CONWAY, P.L. & KJELLEBERG, S. (1986) The production and release of an extracellular polysaccharide during starvation of a marine *Pseudomonas* sp. and the effect thereof on adhesion. *Archives of Microbiology*, **145**, 220–227.

ZOBELL, C.E. (1943) The effect of solid surfaces upon bacterial activity. *Journal of Bacteriology*, **46**, 39–56.

5: Confocal Laser Scanning Microscopy of Adherent Microorganisms, Biofilms and Surfaces

S.P. GORMAN, W.M. MAWHINNEY AND C.G. ADAIR

School of Pharmacy, Medical Biology Centre, The Queen's University of Belfast, Belfast BT9 7BL, UK

The development of immunofluorescence with fluorochrome-coupled antibodies as highly specific stains for measuring intracellular parameters has contributed to epifluorescence becoming one of the most commonly used microscopic techniques in biological research (Tsien *et al.*, 1985; Paradiso *et al.*, 1987). However, with conventional fluorescence microscopy, the illumination of the entire field of view of the specimen with intense light at the excitatory wavelength excites fluorescence emissions throughout the whole depth of the specimen, resulting in out-of-focus blur. This is partially corrected by the use of flattened cell cultures, or by restricting sample thickness to about 10 μm. Unfortunately, these practices distort the views obtained, and the cutting of sections introduces technical difficulties and complicates the interpretation of three-dimensional structures. These problems may be partially alleviated by the use of computer techniques (Agard & Sedat, 1983), although results are not immediate.

With conventional microscopy the whole area of interest is imaged either on to a screen or directly on to the retina of the eye. This, although suitable in many cases, does not permit subsequent electronic processing or easy adaptation to take advantage of resolution-enhancement apparatus. Electron microscopy requires specimen preparation involving dehydration, which may cause disruptive shrinkage, and slicing, which dictates that three-dimensional images can be obtained only by lengthy reconstruction from hundreds of serial sections. Confocal laser scanning microscopy (CLSM) forms a bridge between light and electron microscopy, affording penetrative views of specimens. The absence of 'flare' and the excellent contrast obtained by the non-invasive imaging of structures allow subsequent reconstruction of three-dimensional (3D) images. Although CLSM has been applied extensively to biological

Microbial Biofilms:
Formation and Control

materials (Shotton, 1989), few studies have been made of its potential in microbiology (Lawrence et al., 1991). The purpose of this chapter is to show the potential of CLSM in the investigation of microbial biofilm, biomaterial surface microrugosity and the invasive potential of microorganisms.

Confocal Laser Scanning Microscopy

In scanning optical microscopy (SOM), an image is constructed by moving (scanning) a diffraction-limited spot of light relative to the specimen. The information so gathered can then easily be collated for computer enhancement and analysis. SOM is therefore the optical equivalent of scanning electron microscopy (SEM). With a scanning optical microscope, a scanning point source illuminates a small region of the object while a scanning point detector detects light only from that area. In the confocal laser scanning microscope the basic scanning optical arrangement remains unchanged, except for the source of illumination, which employs a scanning laser beam. Currently, most commercially available apparatus is fitted with low-power air-cooled argon lasers. These can emit a variety of wavelengths, of which the two strongest lines are at 488 nm — the excitation maximum of fluorescein — and at 514 nm, capable of causing emissions from rhodamine and Texas red. More expensive lasers capable of emitting in the ultraviolet region (tunable dye lasers), are available. Laser systems provide high-intensity illumination, giving good system sensitivity and improved fluorescence resolution.

Confocal microscopy is particularly significant in biomedical studies. It overcomes the inherent limitation of classic light microscopy, namely the out-of-focus blurring or fluorescence flare. It is often a source of frustration, when using conventional epifluorescence, that structures can be seen but not captured on film. By removing this background out-of-focus haze, the confocal system can allow photographic imaging of these structures. It has the added advantage that preparations normally regarded as overstained or deemed to have unacceptable levels of background staining can be successfully viewed.

By the use of point scanning, rather than full-field illumination, optical sectioning is possible. Instead of scanning in the x,y plane it is possible to scan in the x,z plane, while retaining all the advantages of confocal microscopy. Thus, images of optical sections parallel to the optical axis of the microscope are generated. Current systems permit optical sectioning down to submicrometre levels, and these sections may then be digitally enhanced to provide a three-dimensional, non-invasive image of subsurface organelles. Hence, cell interiors within living tissue can be investigated without the artefacts introduced by preparing specimens for SEM, and functional cellular architecture (the structural relationship between various cell organelles) may be elucidated. Much of this architecture occurs in the range $0.1-1\,\mu m$, which cannot be adequately resolved by conventional light microscopes.

CLSM demonstration of the invasive potential of microorganisms

The application of CLSM to investigation of the invasive potential of bacteria has recently been reported. Goldner *et al.* (1991) employed the stereoscopic dimension of the microscope to observe the interaction of *Bacteroides fragilis* with tissue cell monolayers in a series of consecutive penetrative images. Cultures were grown as required under anaerobic conditions (95% H_2, 5% CO_2) in tryptone glucose yeast (Oxoid) medium to which were added different concentrations of cysteine. The organisms were conditioned by culturing in the medium at a cysteine concentration of 0.5 g/l, and then prepared for testing by culturing at a concentration of 0.05 g/l cysteine, which was found to be more favourable for penetration of tissue cells. The bacteria were harvested at midexponential phase (*ca.* 5−7 h, OD_{660} 0.4−0.5), centrifuged, washed and resuspended in Hank's balanced salt solution to a level of 10^8 cfu/ml. HeLa cell monolayers were grown on 16 mm diameter well coverslips and rinsed three times with 37°C Hank's solution. Prewarmed (37°C, 0.1 ml) bacterial suspension was added to the monolayers on each coverslip, where the proportion of bacteria to tissue cells was in the order of 10 bacteria/cell. The coverslips were placed in a humidified incubator with a 5% CO_2 atmosphere at 37°C for periods of 12 h and 24 h before staining with acridine orange (7 mg in 50 ml Hank's solution) at 37°C for 1 min. After rinsing with Hank's solution to remove excess dye, further quenching of extracellular fluorescence was carried out by staining with crystal violet (50 mg in 50 ml physiological saline, 0.15 mol/l sodium chloride). The coverslips, protected against drying with one or two drops of physiological phosphate-buffered saline (PBS), were then mounted on glass slides, with the monolayer side in contact with the slide, and examined by CLSM. From successive observations in the interior of the cell, a series of images pointed to a comparatively enhanced penetration after 24 h, compared with 12 h of contact.

CLSM of microbial biofilm on peritoneal catheters

Patients on continuous ambulatory peritoneal dialysis (CAPD) frequently suffer from recurrent episodes of peritonitis and their catheter may have to be surgically removed. Examination of such catheters by SEM often reveals the presence of a sessile microbial biofilm, which may result in resistance of the causative organism to standard treatment protocols, suggesting a possible cause for relapsing peritonitis (Evans & Holmes, 1987; Mawhinney *et al.*, 1991a). In the majority of catheter sections examined, however, only erythrocytes and 'inflammatory cells' are observed on the catheter surface, although the presence of bacteria on the catheters is indicated by laboratory microbiological procedures and may be confirmed by transmission electron microscopy (TEM) (Mawhinney *et al.*, 1990). CLSM of the catheter surface was employed to facilitate direct observation of fluorescent dye-stained microbial biofilm underlying the occluding erythrocyte and inflammatory cell layer (Mawhinney *et al.*, 1991b).

Examination of peritoneal catheter surface topography by CLSM

Surface topography, including the degree of surface roughness (microrugosity), plays an important role in initial bacterial adherence. Implants with porous surfaces are much more susceptible to infection after implantation because microorganisms tend to become sequestered in these cavities and avoid host defences (Dinnen, 1977). Muller *et al.* (1984) found that prosthetic silicone implants, which were relatively smooth, tended to develop cracks and fissures with extended wear, thereby creating anchorage sites for bacteria. Grooves or roughened surfaces on CAPD catheters, in particular, provide niches for easier bacterial adhesion and subsequent biofilm formation (Fessia *et al.*, 1988). Commercially available catheters may vary in their susceptibility to biofilm formation in respect of the degree of inherent surface microrugosity. CLSM offers optimum clarity and resolution of fine surface detail, and such data can be electronically manipulated to create three-dimensional and topographical images.

Method

CLSM of microbial biofilm on peritoneal catheters

Peritoneal catheters

Tenckhoff peritoneal catheters were obtained by surgical removal from CAPD patients with recurrent peritonitis and from patients without clinical infection at the time of renal transplantation. Catheters were transferred aseptically into sterile bags for immediate examination. Several 1-cm sections were cut from the cuff and intraperitoneal regions. These were split longitudinally and one portion was placed in 5% (w/v) glutaraldehyde in cacodylate buffer (0.1 mol/l, pH 7.4) for subsequent microscopic examination; the other portion was retained for microbiological identification of adherent bacteria.

Biofilm isolation and microorganism identification

Biofilm attached to the catheter surface was dislodged into sterile spent dialysate by a combination of gentle scraping, vortexing (30 s) and ultra-sonication (30 s) in a 150 W ultrasonic bath operating at a nominal frequency of 50 Hz. These procedures have been shown to be harmless to the infecting microorganisms (Gorman, 1991). Disrupted biofilm was examined for growth on Columbia blood agar (Oxoid) and isolated microorganisms were typed according to a scheme similar to that suggested by Ludlam *et al.*, (1989) which involves assignment of an antibiogram and a biotype.

Scanning and transmission electron microscopy

For SEM, glutaraldehyde-fixed catheter portions were washed in water for 2 h and then dehydrated through a 30, 50, 75, 90 and 100% ethanol series, allowing 30 min in the lower concentrations and progressing to 60 min in the higher concentrations. They were then dried at the critical point of carbon dioxide and coated with gold/palladium. Prepared samples were scanned with a Jeol 35 CF microscope.

TEM required a 12-h continuous wash cycle, in water, of the glutaraldehyde-fixed catheter portions. These were post-fixed in 1% w/v osmium tetroxide for 1 h, followed by a 30 min wash in water. They were then dehydrated through an ethanol series of 10, 20, 40% (30 min each), followed by 60, 80 and 100% (60−120 min each) and embedded in Epon or TAAB epoxy resin (TAAB Laboratories, Reading, Berkshire) for 36−48 h. Samples were cut into ultrathin sections, which were then stained for 15 min in uranyl acetate followed by 5 min in lead citrate before examination at 10 kV in a Phillips 400 transmission electron microscope.

Confocal laser scanning microscopy

Glutaraldehyde-fixed catheter portions were attached to glass microscope slides with epoxy resin and stained with acridine orange (0.5% w/v in sodium citrate buffer, pH 6.6, 0.1 mol/l) for 3 min. Samples were examined at a wavelength of 488 nm by CLSM with long working-distance lenses (Biorad Lasersharp MRC 500).

Examination of peritoneal catheter surface topography by CLSM

Peritoneal catheter collection and preparation

Over an 18-month study period, 29 peritoneal Tenckhoff catheters were gathered aseptically after surgical removal because of recurrent peritonitis, catheter blockage, patient unsuitability for CAPD, or from patients without clinical infection after renal transplantation. Six catheters with a dwell-time of more than 12 months were selected for examination and three unused, sterile catheters were used as controls.

Sections (1 cm) were cut from the cuff and intraperitoneal regions and split longitudinally. One portion was retained for microbiological identification of adherent bacteria, while the other portion was washed clean (as determined by microscopy) by mild, intermittent (3×20 min) ultrasonication in sterile water. Control samples were treated similarly by ultrasonication before microscopic examination. Samples were then air-dired and mounted, and unstained sections examined directly by reflective CLSM (Leica UK) or sputter-coated for examination by SEM.

The image-processing software of the CLSM included simulated fluorescence processing (shadow imaging), topographical imaging with measurement line and topographical image with 3D plot. The latter allowed measurement of the ratio of the 3D surface area to the area of the projection of the sur-

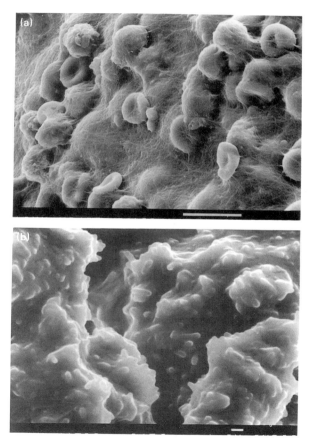

FIG. 5.1. SEMs of peritoneal catheters from patients with peritonitis. (a) Gross deposit of biofilm overlaid with inflammatory cells and erythrocytes (bar = 10 μm). (b) An unobstructed view of a *Pseudomonas aeruginosa* biofilm (bar = 1 μm).

face on to a plane to give a measure of the surface roughness. In all cases measurements were taken in three areas deemed typical for the catheter (Gorman *et al.*, 1991).

Discussion

Biofilm examination

Examination by SEM of the surface of an infected Tenckhoff catheter, after a dwell time of 18 months in a patient prone to recurrent peritonitis, revealed a typical mass of 'inflammatory cells' with occasional erythrocytes (Fig. 5.1(a)). Frequently, when catheter sections taken from patients with peritonitis were examined by SEM, identifiable areas of bacterial infection were not observed. In these cases extensive deposits of 'inflammatory cells' and erythrocytes were present on the catheter. However, as *Staphylococcus epidermidis* or *Staphylococcus aureus* was invariably recovered from culture of the disrupted deposit, bacterial contamination may be presumed to underlie these occluding cells. This is in

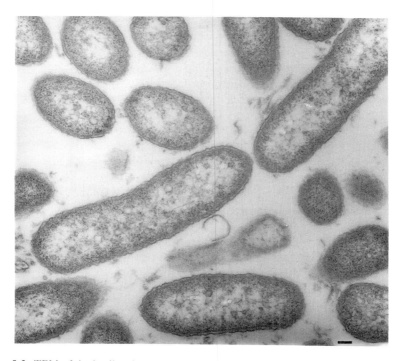

FIG. 5.2. TEM of the biofilm shown in Fig. 5.1(b), consisting of *Pseudomonas aeruginosa* (bar = 0.1 µm).

contrast to the clear SEM view (Fig. 5.1(b)) of a microbial biofilm on a similar catheter surface. The microorganism isolated from this biofilm was *Pseudomonas aeruginosa*, and the rod-shaped form of the bacterium may clearly be observed enmeshed in the glycocalyx or biofilm exudate. Examination of the *Pseudomonas* biofilm by TEM shows the size and distribution of the bacteria within the biofilm (Fig. 5.2).

Catheter surface deposits were examined by CLSM to a depth of 22 μm to gain a penetrative view of the biofilm with a fluorescent dye to highlight entrained bacteria. This facilitated focusing of the specimen below the super-ficial 'inflammatory cells' and erythrocytes to demonstrate the presence of an underlying bacterial biofilm. The biofilm is shown as a fluorescent band circling the rim of a catheter pore, and has become detached from the left side of the pore (Fig. 5.3(a)).

A higher magnification (Fig. 5.3(b), scale bar 25 μm) view of this biofilm, taken by penetrating the catheter pore to a depth of 80 μm, shows discrete entities of $1-2$ μm diameter clearly visible within the biofilm (arrows). High-intensity fluorescence is visible towards the centre, indicating gross bacterial presence in this area. The distribution of bacteria within the hydrated matrix of the biofilm is generally shown to be more dilute than in the TEM observations. This may reflect the effects of the dehydration process during the preparation of specimens for TEM. The polysaccharide matrix of bacterial biofilms is extensively hydrated, and may be dehydrated to 1% of its volume during preparation for electron microscopy (Sutherland, 1977). Therefore, the extent of the condensed residue observed through such preparative tech-niques underestimates the true nature of the glycocalyx matrix and biofilm microorganisms.

Additionally, sputter−cryo techniques have revealed that biofilm is often coated with a dense layer of glycocalyx, which may obscure the underlying microbial population (Richards & Turner, 1984). Howgrave-Graham & Wallis (1991) have recently described the difficulties in obtaining SEM and TEM images of glycocalyx and biofilm populations *in situ*. Neither glutaraldehyde, osmium tetroxide nor formaldehyde adequately fixed the glycocalyx, such that dehydration with acetone, ethanol or 2,2-dimethoxypropane caused its dissolution. In contrast, sputter−cryo (speciments frozen to −180°C, gold−palladium coated and examined in SEM at −180°C), freeze-dried and air-dried granules showed a dense mat of glycocalyx, which obscured the microbial population within. The use of CLSM, as in this study, offers further insight into the nature of occluded biofilm.

Surface examination

All catheters removed from patients in this study were shown by SEM and microbiological techniques to have attached viable microbial biofilm. SEMs of

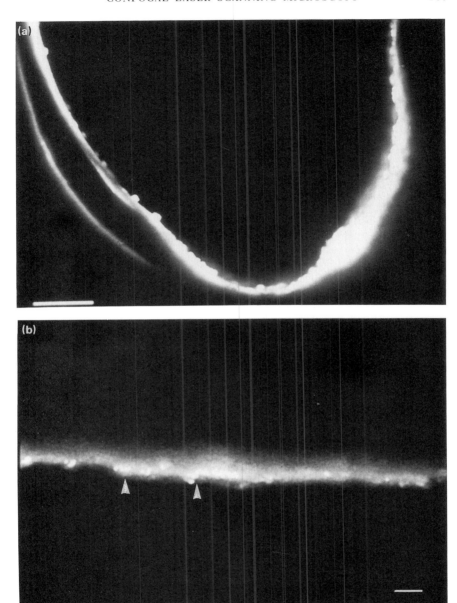

FIG. 5.3. CLSM of the deposit within a catheter pore. (a) The deposit, consisting of high levels of *Staphylococcus epidermidis* from culture, is detached from the left side of the pore (bar = 50 μm). (b) A higher magnification view (bar = 25 μm) of the deposit within the pore shows clusters of cocci (arrows).

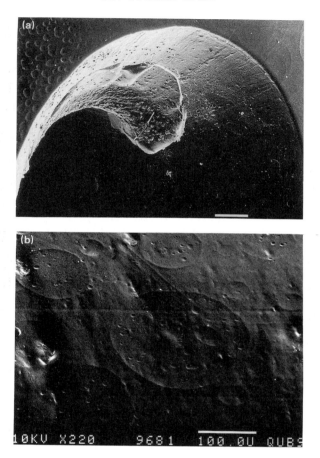

FIG. 5.4. SEM showing (a) scoring and pitting of the outer surface and pore of a CAPD catheter (bar = 100 μm), and (b) defects in the catheter surface (bar = 100 μm).

control catheters showed deep scoring of the outer surface and particularly within the catheter pores (Fig. 5.4(a)). Numerous other surface defects were apparent on the catheter outer surface including depressions or pits ranging in diameter from 100 μm to <10 μm (Fig. 5.4(b)).

The application of CLSM-simulated fluorescence processing, or shadow-imaging, clearly showed the typical large grooves and ridges apparent in those catheters *in situ* for 18 months, and the topographical image of the same area with measurement line allowed estimation of the peak−trough heights of the surface (Fig. 5.5). Peak−trough height measurements for such catheters ranged up to 7.2 μm. Examination of the control catheters by CLSM showed

FIG. 5.5. CLSM view of the surface of a catheter removed from a patient following 18 months' continuous ambulatory peritoneal dialysis. Topographical imaging mode was employed with application of a measurement line to quantify the surface microrugosity.

a less convoluted surface with more regular aberrations. Further information may be gained by examination of the 3D contour map of the catheter surface (Fig. 5.6). 3D imaging allowed comparison of the area of the catheter surface with the area of the original surface projected on to a plane. The contour maps provided values by this application in the range 4.5−5 for long-dwelling catheters and 3.5−4 for control catheters ($p < 0.0005$; unpaired t-test), which shows the detrimental effects of time on the surface microrugosity of indwelling CAPD catheters.

These investigations indicate that the manufacturing process causes pitting and deep scoring of the CAPD catheter surface and pores. Substantial changes in the surface topography also appear in long-term indwelling catheters. It is not clear, however, whether these channels are a product of the host environment or are the result of the activities of the attached microbial biofilm. However, they may offer a 'safe-haven' for microorganisms, that protects them from the action of antibiotics and from the shear forces exerted by active dialysis in CAPD.

CLSM holds great promise in many research areas, including, as demon-

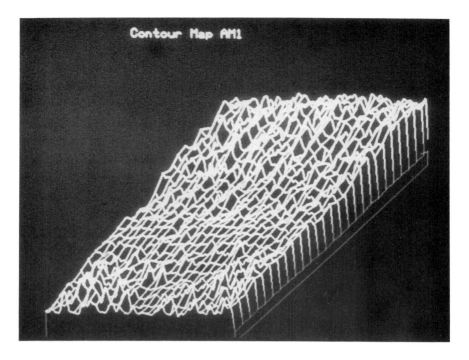

Fig. 5.6. Contour map representation of the CAPD catheter surface area described in Fig. 5.5, obtained by CLSM.

strated here, the examination of biological samples the thickness of which does not permit the use of conventional microscopy. Further improvements in computational capabilities and the advent of faster scanning laser systems should allow further development of confocal microscopy as a major biological tool.

Acknowledgements

We are grateful for the financial support of the Northern Ireland Kidney Research Fund, EHSSB and the Pharmaceutical Society (NI). The assistance of Mathew Issouckis, Leica UK Ltd., the Staff of the Renal Unit (BCH) and QUB Electron Microscopy Unit is gratefully acknowledged.

References

AGARD, D.A. & SEDAT, J.W. (1983) Three dimensional architecture of a polythene nucleus. *Nature (London)*, **302**, 676–681.

DINNEN, P. (1977) The effect of suture material in the development of vascular infection. *Surgery*, **91**, 61–63.

EVANS, R.C. & HOLMES, C.J. (1987) Effect of vancomycin hydrochloride on *Staphylococcus epidermidis* biofilm associated with silicone elastomer. *Antimicrobial Agents and Chemotherapy*, **31**, 889–894.

FESSIA, S.L., AMIRANA, O. & CARR, K.L. (1988) Biofilm formation on commercially available plastic tubing. *Advances in Peritoneal Dialysis*, **4**, 253–256.

GOLDNER, M., COQUIS-RONDON, M. & CARLIER, J.P. (1991) Demonstration by confocal laser scanning microscopy of invasive potential with *Bacteroides fragilis*. *Microbiologica*, **14**, 71–75.

GORMAN, S.P. (1991) Microbial adherence and biofilm production. In Denyer, S.P. & Hugo, W.B. (eds.) *Mechanisms of Action of Chemical Biocides*, pp. 271–295. Society for Applied Bacteriology Technical Series No. 27. Blackwell Scientific Publications, Oxford.

GORMAN, S.P., MAWHINNEY, W.M. & ADAIR, C.G. (1991) The influence of catheter surface microrugosity on the development and persistence of recurrent peritonitis in CAPD. *Pharmacotherapy*, **11**, 267.

HOWGRAVE-GRAHAM, A.R. & WALLIS, F.M. (1991) Preparation techniques for the electron microscopy of granular sludge from an anaerobic digester. *Letters in Applied Microbiology*, **13**, 87–89.

LAWRENCE, J.R., KORBER, D.R., HOYLE, B.D., COSTERTON, J.W. & CALDWELL, D.E. (1991) Optical sectioning of microbial biofilms. *Journal of Bacteriology*, **173**, 6558–6567.

LUDLAM, H.A., NOBLE, W.C., MARPLES, R.R. & PHILIPS, I. (1989) The evaluation of a typing scheme for coagulase-negative staphylococci suitable for epidemiological studies. *Journal of Medical Microbiology*, **30**, 161–165.

MAWHINNEY, W.M., ADAIR, C.G. & GORMAN, S.P. (1990) Confocal laser scanning microscopic investigation of microbial biofilm associated with CAPD catheters. *Pharmacotherapy*, **10**, 241.

MAWHINNEY, W.M., ADAIR, C.G. & GORMAN, S.P. (1991a) Development and treatment of peritonitis in continuous ambulatory peritoneal dialysis. *International Journal of Pharmacy Practice*, **1**, 10–18.

MAWHINNEY, W.M., ADAIR, C.G. & GORMAN, S.P. (1991b) Examination of microbial biofilm on peritoneal catheters by electron and confocal laser scanning microscopy. *Proceedings of the 10th Pharmaceutical Technology Conference*, Bologna, Italy, **2**, pp. 652–660. Solid Dosage Research Unit, Liverpool.

MULLER, W.A., COHN, L.H. & SCHOEN, F.J. (1984) Infection within a degenerated Starr–Edwards silicone rubber poppet in the aortic valve position. *American Journal of Cardiology*, **54**, 1146.

PARADISO, A.M. & TSIEN, R.Y. & MACHEN, T.E. (1987) Digital image processing of intracellular pH in gastric oxyntic and chief cells. *Nature (London)*, **325**, 447–450.

RICHARDS, S.R. & TURNER, R.J. (1984) A comparative study of techniques for the examination of biofilms by scanning electron microscopy. *Water Research*, **18**, 767–773.

SHOTTON, D.M. (1989) Confocal scanning optical microscopy and its applications for biological specimens. *Journal of Cell Science*, **94**, 175–206.

SUTHERLAND, I.W. (1977) *Surface Carbohydrates of the Prokaryotic Cell*. Academic Press. London.

TSIEN, R.Y., RINK, T.J. & POENIE, M. (1985) Measurement of cytosolic free Ca^{2+} in individual small cells using fluorescence microscopy with dual excitation wavelengths. *Cell Calcium*, **6**, 145–157.

6: Attachment in Disease

S. PATRICK[1] AND M.J. LARKIN[2]

[1]*Department of Microbiology and Immunobiology, The Queen's University of Belfast, Grosvenor Road, Belfast BT12 6BN; and* [2]*School of Biology and Biochemistry, The Queen's University of Belfast, Belfast BT9 5AG, UK*

The attachment of pathogenic bacteria to mammalian cells is a pivotal step in the colonization of the host. Attachment is mediated by the interaction of ligands on the bacterial surface and receptor molecules on the mammalian cell. Bacteria have evolved with a range of attachment mechanisms, from the non-specific in terms of cell type and cell surface receptor, to the highly specific. Virtually all of the known structures on the surface of bacteria are involved in some way in attachment to target cells, either by promoting or interfering with attachment. For example, bacterial surface structures may impede attachment and uptake by phagocytic cells. On the other hand, attachment may be a prelude to the phagocytosis of the bacterium by a normally non-phagocytic host cell.

The structures involved are fimbriae (sometimes referred to as pili), flagella, lipopolysaccharide (LPS), extracellular polysaccharides in the form of capsules or glycocalyx, lipoteichoic acids and outer membrane proteins (OMP). Fimbriae, first named by Duguid *et al.* (1955) after the Latin for 'thread' or 'fibre', are generally accepted as very important in bacterial attachment to host cells. Brinton (1959) named similar structures pili after the Latin for 'hair-like' and both names are still in use. Expression of such adhesive structures, or adhesins, by the bacteria may be regulated by environmental changes associated with the host cells or the host cells themselves. In the case of *Salmonella* spp., an initial attachment event triggers the biosynthesis of further adhesin molecules (Finlay *et al.*, 1989).

The interaction of the bacterial adhesin with the host cell can also alter the functions of the host cell. These may include the induction of plasmin formation, rearrangement of the cytoskeleton to induce phagocytosis of the

Microbial Biofilms:
Formation and Control

bacterium and, possibly, cytokine release (Hoepelman & Tuomanen, 1992).

Studies of attachment in disease must consider the characterization of the relevant surface structures on the bacteria, the interaction between the bacterium and the target cell surface and the characteristics of the target cell surface. Such studies necessarily involve a wide range of biochemical and microbiological techniques too numerous to detail here, and the reader is directed to Hancock & Poxton (1988) for a full account. This chapter will deal with techniques and approaches adopted in our work on the virulence of the obligate anaerobe *Bacteroides fragilis*, in relation to the identification of surface structures by microscopy and their antigenicity. This will provide an insight into the likely problems and factors to be considered when approaching the study of a pathogen for the first time.

Characterization of Surface Structures by Microscopy

Before embarking on any study of bacterial attachment it is essential to define the surface structures expressed on the surface of the organism, and to ensure that the population to be studied is homogeneous with respect to expression of the surface structure of interest. The antigenic heterogeneity of the structures should also be examined because the parts of the bacterial molecule involved in attachment may not be expressed by all of the bacteria within a population. For example, Type 1 fimbriae, involved in attachment by *Escherichia coli*, may be subject to both antigenic and phase variation. In the latter there is a switch from expression to non-expression of the fimbriae at a rate of $10^{-2}-10^{-3}/$ cell/generation as the result of inversion of a 314 base-pair segment of DNA (Tennent *et al.*, 1990). Considerable heterogeneity has been shown with respect to surface structure/antigen expression in *B. fragilis* (Patrick & Lutton, 1990). Subpopulations of bacteria, separated by density gradient centrifugation from a single strain, have different haemagglutinating properties, which suggests that attachment studies should be approached with caution (Patrick *et al.*, 1988). A number of simple precautions should, therefore, be taken. First, the bacteria should be grown in a defined medium to facilitate the control of available nutrients. For example, nutritional conditions can alter the size of capsules expressed by *E. coli* (Sutherland, 1977). Secondly, the phase of growth of the culture should be known. Stationary-phase cultures should be avoided because such populations are not homogeneous for any metabolic parameter, and some of the population will be either dead, dying or utilizing nutrients released from dead cells. Ideally, bacteria in continuous culture should be used to ensure their metabolic uniformity. Alternatively, cultures should be inoculated from a standard inoculum, stored in batches at $-70°C$ or in liquid nitrogen, and always harvested at a particular phase of growth, such as the late exponential phase. Thirdly, variation in the expression of surface structures with growth phase should also be determined. Finally, bacteria grown *in vivo*, either in an experimental model (e.g. Day *et al.*, 1980; Patrick *et al.*, 1984), or taken directly from clinical specimens, must be examined for the expression of surface structures implicated in attachment by the *in vitro* studies.

The following methods and techniques are regarded as generally appropriate for any bacterial population to be studied.

Light microscopy

Negative staining with India ink in a wet mount and observation by light microscopy should be undertaken to determine whether capsules are present on the bacterium of interest. This is an essential first step which lacks the drawback of dry preparations, where shrinkage of both the negative stain and the bacterial cell may give false-positive results. Once the presence of capsules has been established, dry staining methods can then be compared. If the latter give the same results, they can be used in further studies. Dry staining methods are more useful if time is limited and a number of samples are to be examined, because prepared slides can be stored in the dark at room temperature and examined later.

Wet India ink method (Cruickshank, 1965)

A large loopful of India ink is mixed on a clean glass slide with a small quantity of broth culture or bacteria taken from a colony. A coverslip is then placed on the suspension and pressed down hard between sheets of blotting paper. The preparation is viewed by phase-contrast microscopy at ×1000 magnification. Encapsulated bacterial cells should be seen surrounded by clear areas corresponding to the capsule, with a background of India ink particles.

Dry eosin/carbol fuchsin method (Cruickshank, 1965)

Carbol fuchsin staining solution (Ziehl–Nielsen's carbol fuchsin diluted 1:5) has the following composition: basic fuchsin (5 g), absolute ethanol (40 ml), phenol (5% w/v in distilled water; 500 ml). The fuchsin is dissolved in the alcohol and is added to the phenol solution. For capsule staining this solution is further diluted 1:5 in distilled water and can be stored in the dark at room temperature. Eosin staining solution consists of four parts eosin solution (10% w/v water-soluble yellowish or bluish erythrosin in distilled water) and one part serum (human, rabbit, sheep or calf) heated at 56°C for 30 min. A crystal of thymol may be added as a preservative. After standing at room temperature for a few days, the mixture is centrifuged and the supernatant is retained. It is stable at room temperature for at least a year.

Staining method. To stain the bacterial cells mix one drop of broth culture or bacterial suspension in one drop of carbol fuchsin on a clean glass slide and allow to stand for 30 s. Then add one drop of eosin solution and leave for 1 min. Spread a film of the suspension along the slide with a second clean slide, as in preparing a blood smear, allow to dry at room temperature, and examine by light microscopy at ×1000 magnification. The bacteria, particularly those without capsules, will be much clearer if bright-field phase-contrast microscopy is used. This gives the bacteria a greenish tinge against the pink background of the eosin. The matt background of the eosin stain can be enhanced by adding a loopful of Percoll (20% in saline; Pharmacia Biosystems (UK) Ltd, Milton Keynes, UK) to the bacterial suspension on the slide. Figures 6.1(a, b and c) illustrate bacteria stained using the eosin carbol fuchsin technique.

Loops should be carefully flamed and allowed to cool before dipping into the staining solution. For both staining techniques control slides should be periodically prepared without bacterial culture, to ensure that the stains have not become contaminated.

Electron microscopy

Electron microscopy is an essential tool which will reveal capsules not visible by light microscopy, and also whether fimbriae are present on the bacterial surface.

Ultrathin sectioning of embedded bacteria

This method can be used for determining the presence of capsular material, and a variety of different methods for embedding, fixation and staining is available. The method chosen is usually determined by trial and error, because different types of bacterial polysaccharide capsule may have different properties. Ruthenium red staining after osmium tetroxide and glutaraldehyde fixation is useful because ruthenium red stains negatively charged polysaccharides. For successful staining, commercial ruthenium red should first be purified to remove contaminating ruthenium purple. The use of specific antibody to stabilize capsules has the disadvantages that structural detail of the capsular material is lost, and it precludes subsequent immunolabelling to identify particular structures. If immunolabelling of ultrathin sections is to be carried out, osmium tetroxide fixation and ruthenium red stain may destroy the antigenicity of the polysaccharide epitopes. For this reason glutaraldehyde/

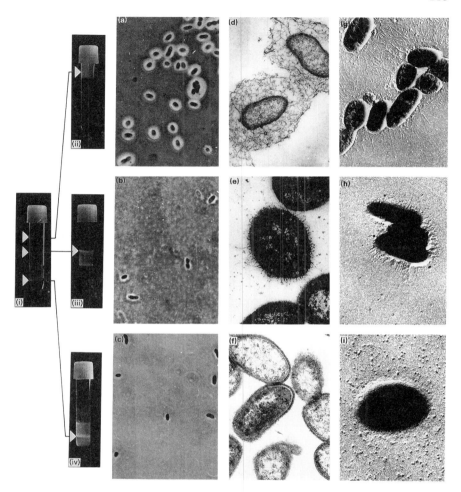

FIG. 6.1. Appearance of *Bacteroides fragilis* enriched by Percoll density gradient centrifugation.
(i) Unenriched broth; after overnight subculture (ii) from 0–20% interface layer (large capsule
population); (iii) 20–40% interface layer (small capsule population); (iv) 60–80% interface layer
(electron-dense population). Micrographs of large capsule population cells (a, d, g); small capsule
population cells (b, e, h); non-capsulate (electron-dense layer) population cells (c, f, i). Eosin/
carbol fuchsin-negative stain (a, b, c); ultrathin section electron microscopy (d, e, f); platinum–
gold shadowing electron microscopy (g, h, i).

paraformaldehyde fixation is used, but this has the disadvantage that capsules are no longer visible. If facilities are available for cryofixation and freeze substitution embedding, the appearance of capsular material may be retained and immunogold labelling may still be successful. The use of Lowicryl K4M resin and dimethyl formamide as a dehydrating agent may also allow visualization of capsular material with concomitant immunogold labelling (Bayer, 1990).

Purification of ruthenium red (after Luft, 1971)
1   Place 0.25 g of commercially available ruthenium red in a mortar with a few drops of 0.5 mol/l ammonia solution and grind with a pestle. Transfer the suspension to a V-shaped quick-fit test tube. Adjust the final volume to 10 ml.
2   Incubate the solution at 60°C for 30 min in a water bath, with frequent vigorous stirring with a glass Pasteur pipette.
3   Cool the test tube under tapwater and centrifuge in a bench centrifuge at 1500 g for 5 min.
4   Carefully withdraw the supernatant fluid and place in a clean 50-ml beaker.
5   Allow the supernatant to evaporate in a desiccator containing 100 g each of anhydrous $CaSO_4$, NH_4CO_3 and $NaOH$.
6   Store in the desiccator in the dark until ready to use.

Fixation and staining procedure
1   Wash bacteria in cacodylate buffer (0.1 mol/l, pH 6.8) or add a small quantity of bacteria directly to a large volume of fixative/stain.
2   Incubate in fixative/stain of glutaraldehyde (2.5% v/v), ruthenium red (1 mg/ml) in cacodylate buffer (0.1 mol/l) for 1 h at 4°C in the dark.
3   Wash three times in buffer. The bacterial suspension will form a granular pellet. Do not break this up but try to keep it as granular lumps throughout the further procedures by stirring gently with the end of a Pasteur pipette, rather than by sucking up and down. This greatly reduces the loss of material during the procedure.
4   Resuspend in osmium tetroxide (1% w/v), ruthenium red (1 mg/ml) in cacodylate buffer, as before. Incubate for 3 h at room temperature in the dark. Wash twice in buffer.
5   Dehydrate in graded ethanol by centrifuging and suspending sequentially in 30, 50, 75, 95 and 100% ethanol. Repeat the 100% ethanol treatment. Filter the 100% ethanol through sodium sulphate on Whatman No. 1 filter paper immediately before use to ensure that it is completely dry. Ensure that the samples are not in any alcohol solution for longer than 15 min, including the centrifugation period, though this limits the number of samples that can be processed at any time.
6   Infiltrate and embed in the resin of choice — LR White is one of the

simplest to use and is compatible with immunogold labelling of ultrathin sections. After the final dehydration step, the pelleted bacteria are suspended in resin−dried 100% ethanol (1:1) for 1 h and then left overnight in 100% resin. The resin should be changed two or three times during the next day. The samples should always be in containers left open to the air as the resin, if enclosed, may begin to polymerize prematurely. Finally, place the sample, in resin, into a gelatine or Beem capsule (Agar Scientific Ltd, Cambridge, UK) and gently centrifuge to pellet the sample. A paper label can be inserted into the resin before polymerization by incubation at 60°C overnight.

7   Ultrathin sections can then be cut, stained — for example with uranyl acetate — and viewed by electron microscopy.

Negative staining of whole cells for electron microscopy

This is useful to determine the presence of fimbrial structures and flagella. Glow-discharging of formavar/carbon-coated grids, within about 30 min of use, will improve the retention of material on the grid. Grids are dipped into the bacterial suspension in distilled water, allowed to drain on filter paper and then dipped into a drop of stain, such as methylamine tungstate 2% (w/v) in distilled water, and again allowed to drain. Immunogold labelling may be carried out before negative staining. Negative stain may also reveal condensed capsular material as a result of the dehydration of the polysaccharides, which generally contain 99% water. The interpretation of the appearance of this condensed material is frequently difficult. The drying that occurs as a result of placing the bacterium on the grid and into the microscope vacuum, as opposed to the controlled dehydration after fixation and staining for thin section, may not always give this material a consistent appearance.

Platinum−gold shadowing of whole cells

This is similar to negative staining in that whole cells are placed on an electron microscopy grid and shadowed at an angle (e.g. 20°) with platinum and gold in a high-vacuum coating unit such as a Balzer BAE 120 (Lutton et al., 1989). Again, interpretation of the structures observed may be difficult because dried capsular material may have a fibrillar appearance (see Figs. 6.1(g) and (h)). Definition of the structures associated with the surface of the bacterium should ideally involve an electron microscopic comparison of whole cells and thin sections, as well as light microscopy. Figure 6.1 compares three populations as seen by light microscopy ((a), (b) and (c)), electron microscopy of ultrathin sections ((d), (e) and (f)) and platinum−gold shadowed whole bacteria ((g), (h) and (i)). The main points to note are that (i) the large and small capsules after platinum−gold shadowing are fibrillar in appearance; (ii) the bacteria, which are apparently non-capsulate by light microscopy, have a marginal electron-dense layer in ultrathin sections; this layer is termed electron-dense because it is visible by electron microscopy in both the presence and absence of ruthenium red, and suggests that it may have non-polysaccharide components (Patrick *et al.*, 1986); and (iii) the small round objects visible after platinum−gold shadowing in both the electron-dense layer population and within the 'fibrils' of the large capsule are extracellular vesicles that bud from the outer membrane (Lutton *et al.*, 1991). If these are present in a bacterial population, their possible involvement in or interference with attachment of the bacteria to host cells should be taken into account.

Immunological Characterization of the Surface Structures

Definition of the antigenicity of the observed surface structures facilitates the study of the antigenic variation that may be evident within strains. Ideally, monoclonal antibodies (MAbs) specific for particular surface structures should be prepared. An excellent account of the production of MAbs is given by Goding (1986). Immunofluorescence microscopy and flow cytometry can then be used to monitor antigenic variation. Immunogold electron microscopy will assist in locating the epitope within the bacterium. Methods for immunogold labelling are provided by Palak & Varndell (1984) and Beesley (1989).

Immunofluorescence microscopy

Preparation and coating of slides

Multiwell slides (Flow Laboratories Ltd, Rickmansworth, UK) are soaked overnight in a weak solution of Decon (approx. 0.1% v/v), thoroughly rinsed in tapwater and then in distilled water, and dried. A slightly turbid suspension (30 µl) of bacteria diluted in phosphate-buffered saline (PBS; (g/l) NaCl, 8.00; K_2HPO_4, 1.21; KH_2PO_4, 0.34; pH 7.3) is placed in each well. Optimum concentrations of bacteria should be determined empirically, so that an even distribution of a single layer of bacteria is seen on microscopy. Slides are dried at 37°C in a fan incubator and fixed by placing them in 100% methanol at −20°C for 10 min. The slides are then dried at room temperature and wrapped with a sachet of silica gel. They may be stored at −70°C for many months, depending on the stability of the antigens.

Two-step immunolabelling procedure

This is based on Johnson *et al.* (1978). On removal from storage at −70°C, the slide is allowed to come to room temperature. Primary antibody (30 µl), diluted in PBS or neat hybridoma supernatant, is applied to the well. The slide is then incubated at 37°C in a humidified plastic box. From this stage onwards the wells must *not* be allowed to dry out.

The primary antibody is then washed off with PBS with a wash bottle. Care must be taken to ensure that antibody from one well does not run over into another. The PBS is therefore directed to the centre of the slide, so that antibody is washed over the outer edge of the slide.

The slide is then placed in a bath (e.g. a glass staining jar) of PBS and mixed with a magnetic stirrer for at least 30 min. The back of the slide should face the magnetic 'flea', in case it damages the bacterial film. The back of the slide is then dried. The spaces between the wells are also gently dried with a strip of filter paper; the wells themselves must not be dried.

The appropriate dilution of secondary antibody (e.g. antimouse IgG fluorescein isothiocyanate (FITC) conjugate; 30 µl) is then applied to the well and the incubation and washing steps are repeated. Finally, the slide is covered with a mounting fluid, such as 10% (w/v) glycerol in PBS, containing an antibleaching agent (e.g. photobleaching retardant; Syva Corporation, USA) and covered with a large coverslip. If this is sealed with nailvarnish, the slides can be stored for a few weeks at 4°C in the dark.

Dual labelling can be carried out by adding an extra step. After applying the primary antibody, such as a murine MAb, and washing, a rabbit polyclonal antiserum, known to label all the bacteria within a population, may be added. After washing, the bacteria are incubated with two conjugates together, such as FITC-conjugated antirabbit immunoglobulin and tetramethylrhodamine β-isothiocyanate (TRITC)-conjugated antimouse immunoglobulin. The slide is then mounted as above. The bacteria can be examined by fluorescence microscopy with the appropriate filters. The double label allows the detection of non-homogeneous labelling of a bacterial population by a MAb. Double exposure of colour film with the same field of view and with filters suitable for FITC and TRITC will give pictures of a single double-labelled field. A simpler method is to observe the same field by phase-contrast microscopy and fluorescence microscopy.

Preparation of Homogeneous Populations

Once structural or antigenic heterogeneity has been established for a bacterial species, consideration must be given to methods for producing homogeneous populations, before attachment is studied. Two methods are outlined below.

Physical size and density gradient centrifugation

Density gradient centrifugation can be used to separate bacteria that express different surface structures. Discontinuous density gradients, also referred to as step gradients, can be prepared with sucrose or other materials, such as Percoll (Pharmacia Biosystems (UK) Ltd, Milton Keynes, UK). The latter has the advantage that it is supplied sterile and, when diluted, can be made isotonic with physiological saline at a physiological pH. The bacterial populations can be separated on a bench centrifuge, rather than by ultracentrifugation. In this way, bacteria suspended in Percoll retain their viability. Bacteria harvested from the gradients can be used either directly or, if expression of the surface structures is relatively stable, they can be subcultured before assays of attachment.

Isopycnic density centrifugation has been used to separate *E. coli* possessing colonization factor I (Giesa *et al.*, 1982). Continuous gradients were prepared in an ultracentrifuge, and the bacterial populations were then applied to the gradients and centrifuged on a bench centrifuge. Discontinuous gradients may be prepared as detailed below, and have been used to separate *B. fragilis* with different sizes of capsules (Patrick & Reid, 1983; it should be noted that photomicrographs b and c in this publication are transposed). Since each bacterium may vary, empirical studies should be carried out with different percentage dilutions of Percoll and different centrifugation times. A list of publications in which Percoll has been used may be obtained from Pharmacia at the above address.

Preparation of Percoll

If bacteria are to be cultured after separation, the following manipulations must be conducted under aseptic conditions. Percoll is supplied sterile and should be dispensed aseptically in 90-ml volumes and stored at 4°C in sterile bottles sealed with tape.

Percoll is made up in 0.15 mol/l sodium chloride and this will be called '100% Percoll'. A small volume is removed and the amount of 1 mol/l HCl required to bring the pH to approximately 7, based on the number of drops from a standard 20 µl dropper, is noted. The Percoll should be well mixed as each drop of HCl is added, and will become more cloudy in appearance. The required volume of HCl per millilitre of the 100% Percoll is calculated, sterile HCl is added aseptically, a small volume is removed and the pH determined. The required volume of HCl per 100 ml Percoll should remain constant within a single batch of Percoll, but this may vary between batches and should always be rechecked.

Preparation of gradients

The 100% Percoll (pH 7) can be further diluted with 0.15 mol/l NaCl to give 80%, 60%, 40% and 20% (v/v) suspensions. These are then placed into

11 ml/16 mm diameter sterile plastic screw-capped tubes, starting with 80% at the bottom. The other layers, in decreasing concentration, are then added, 2 ml per layer. If only a few tubes are required the gradient steps can be layered carefully by hand with a Pasteur pipette. The initial contact of one layer with another is critical for the integrity of the interface. If the Percoll is added too quickly the layers will mix. The Percoll should be allowed to run gently down the side of the tube on to the surface of the layer below. If large numbers of gradients are required, a peristaltic pump can be used to prepare them. Stiff narrow-gauge tubing should be inserted into the soft tubing passing through the pump. To ensure asepsis, 70% (v/v) industrial methylated spirits in water should be passed through the tubing, followed by sterile saline. After preparation the interfaces should be clearly visible. Less mixing between layers occurs if the Percoll suspensions are used straight from the 4°C refrigerator. If the refrigerator is vibration-free, gradients can be stored overnight at 4°C. The interfaces will be less obvious on the second day, but the system will work adequately.

Separation of bacteria

With a Pasteur pipette *ca.* 2−2.5 ml of a broth culture is carefully layered on to the gradient, which is centrifuged in a bench centrifuge with a swing-out head. The relative centrifugal force and the time of centrifugation for a particular bacterium will vary, and can only be determined empirically: between 2000 and 3000 g for 20−40 min is a good starting point. First, after centrifugation, the culture medium on top of the 20% layer is removed with a clean pipette and bacteria are then removed from each of the interfaces with a Pasteur pipette. Percoll in the 20% layer is then carefully removed, without disturbing the 20−40% interface, and discarded. Bacteria at the 20−40% layer are removed next with a clean pipette, the 40% Percoll is discarded, and so on down the gradient. The use of a clean pipette and discarding of the Percoll between the interfaces minimizes the contamination of one layer with another. Even if a distinct band of bacterial culture is not visible at an interface, it should still be removed, examined by microscopy and subcultured. In the case of *B. fragilis* there may be too few cells with a large capsule to be visible as a band on top of the 20% Percoll. They can, however, be enriched by subculture from the interface (Fig. 6.1(i) to (iv)). Since the enrichment may not be absolute, it is wise to repeat the density gradient centrifugation after culture from the interface layers.

Immunological separation

The use of immunomagnetic beads is a simple method for enriching bacteria that express particular surface epitopes. Immunomagnetic beads can be bought already conjugated to either antimouse or antirabbit immunoglobulin. If MAbs are available for particular epitopes, these can be attached to the immunomagnetic beads and then used to remove bacteria carrying that epitope from a mixed population (Fig. 6.2). Since it is difficult to remove the bacteria from the beads, this technique relies on the subculture of the bacteria and enrichment of bacteria expressing the epitope.

Coating of antimouse IgG-magnetic beads with murine monoclonal antibody

The precise volumes of antibody to be used will depend on its titre. We have used the following procedure with neat supernatant from a hybridoma cell line. Immunomagnetic beads (50 µl; Dynabead M-450, coated with antimouse IgG; Dynal Ltd, New Ferry, Wirral, UK) are pipetted into a 1-ml glass test tube. The tube is clipped to a plastic-coated magnet called a magnetic particle concentrator (MPC; Dynal Ltd) for at least 1 min. While the test tube is still clipped to the MPC, the fluid is removed with a pipette and 5 ml sterile PBS

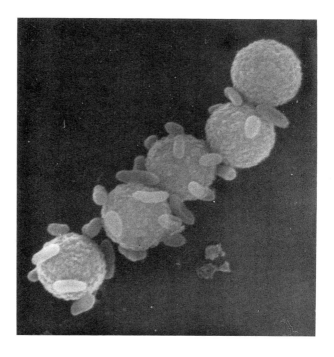

FIG. 6.2. *Bacteroides fragilis* attached to antimouse IgG-coated immunomagnetic beads via an anti-*B. fragilis* polysaccharide mouse monoclonal antibody.

is added. The test tube is removed from the influence of the magnet and the contents are vortexed. The tube is then replaced in the MPC and the PBS removed and replaced. This process is repeated twice. These washing steps remove the sodium azide in which the beads are supplied. The beads are finally made up to 50 µl in PBS and again vortexed. Monoclonal antibody supernatant (1 ml) is added and the beads are incubated at room temperature for 2−24 h with very gentle rocking. The beads are then washed again four times in PBS as above, with gentle rocking for 30 min each time. They are finally suspended in 1 ml PBS, and can be stored at 4°C for up to 2 weeks.

Selective enrichment of bacteria

Bacteria are suspended at a concentration of $10^7 - 10^8$/ml in a defined minimal medium, suitable for the culture of the bacteria, containing 0.02% (w/v) Tween 20. The optimal concentration of bacteria should be determined for different bacteria and antibodies. Beads (0.5 ml), precoated with specific MAb, are added to 10 ml of bacterial suspension and the mixture is gently shaken by hand for up to 2 min, placed in the MPC and washed three times with broth. The 2-min incubation time is critical and it should not be exceeded. The beads are then added to fresh broth and a non-relevant MAb, such as one specific for a viral protein, should be used as a control to monitor non-specific binding. The enriched cultures can then be monitored by fluorescence microscopy or flow cytometry (Lutton *et al.*, 1991) to determine whether there has been enrichment for a particular epitope.

Study of Attachment

Populations with clearly defined surface structures/antigens can be used in studies of attachment. The most commonly used methods involve adherence of bacteria to erythrocytes (haemagglutination), epithelial cells or cells in tissue culture. These methods are often preliminary to determination of the specific attachment mechanism involved. Other, more non-specific, methods of assessing the ability to attach involve determination of the hydrophobicity of the bacterial surface (see Chapter 2).

Surface hydrophobicity

Hydrophobic interactions are considered to play a major role in the association of bacteria with phagocytes and epithelial cells. In our studies (unpublished data) we have used two of the most common methods for determining the surface hydrophobicity of homogeneous populations of *B. fragilis* separated by Percoll density gradient centrifugation.

Adherence to hydrocarbons

This method is based on the principle that, after brief mixing, hydrophobic bacteria will bind to hydrophobic solvents such as xylene or cyclohexane (Rosenberg *et al.*, 1980). Bacteria are centrifuged and washed twice in PUM buffer (g/l, $K_2HPO_4 \cdot 3H_2O$, 22.2; KH_2PO_4, 7.26; $MgSO_4 \cdot 7H_2O$, 0.2; urea, 1.8; pH 7.1). The bacteria are then resuspended in PUM buffer and the $OD_{550\,nm}$ is determined. The bacterial suspension (1 ml) is then added to a series of 10 mm diameter glass test tubes. Volumes of the hydrocarbon, in a range from 50 to 500 µl, are added to each tube. The tubes are incubated at 30°C for 10 min and are then uniformly shaken for 2 min. The hydrocarbon phase is allowed to separate completely before the lower aqueous phase is carefully removed with a Pasteur pipette. A PUM buffer blank is included as a control. The change in $OD_{550\,nm}$ is then determined for each volume of added hydrocarbon, and attachment is expressed as the percentage decrease, which represents the proportion of bacteria excluded from the aqueous phase.

Hydrophobic interaction chromatography

This determines the proportion of a population of bacteria retained on a hydrophobic gel column, and we have modified the methods of Ismaeel *et al.* (1987) and Mozes & Rouxhet (1987). Columns are constructed from Pasteur pipettes plugged with glass wool, loaded with octyl sepharose CL-4B (Pharmacia Biosystems (UK) Ltd, Milton Keynes, UK) to a bed height of 30 mm and equilibrated with an equilibrating solution (ES) of 4 mol/l NaCl in 0.05 mol/l citrate buffer (pH 4.7). Bacteria are washed and resuspended in ES, 100 µl of the bacterial suspension is diluted in 3 ml of ES and the $OD_{555\,nm}$ determined (OD_0, which may be adjusted to 1). The original bacterial suspension (100 µl) is also applied to the column, followed by 3 ml of ES, which is collected and the $OD_{550\,nm}$ determined (OD_1). Distilled water (3 ml), adjusted to the same pH, is then applied to the column, collected and the $OD_{550\,nm}$ also determined (OD_2). The proportion of bacteria retained by the column at high ionic strength R_F is given by $100\,(OD_0 - OD_1)/OD_0$. The proportion of bacteria retained on the column at low ionic strength R_L is given by $100(OD_0 - OD_2)/OD_0$.

A simpler alternative method is to add 5 ml bacterial suspension in distilled water or quarter-strength Ringer's solution to a column and determine the decrease in $OD_{550\,nm}$.

Haemagglutination

Haemagglutination is one of the simplest methods for the determination of the adherence of bacteria to animal cells. An excellent reference for haemagglutination methods is Old (1985). Two methods can be used, either the static settling or the rocked-tile method. Both techniques are very simple but each has its advantages and disadvantages. It is recommended that both are used to determine haemagglutination, and that a known positive strain is included in each test as a control.

Preparation of erythrocytes and bacterial suspension

Whole blood, treated with either heparin or citrate to prevent coagulation, should be centrifuged at $200\,g$ for 10 min and the cells washed gently three times in PBS. The cells are then made up in PBS to 2 or 3% packed-cell volume. The erythrocytes can be stored at 4°C for up to 7 days. Blood from as wide a range of species as possible should be used.

Bacteria are centrifuged, washed and resuspended in PBS. A range of different concentrations of bacteria should be used to determine their optimal concentration. It is useful to start at about 1×10^9 bacteria/ml and proceed with doubling dilutions.

Rocked-tile method

Equal volumes of the erythrocyte and bacterial suspension in PBS are mixed on a chilled tile by rocking. Haemagglutination is judged 'by eye' as the formation of granular clumps of erythrocytes within 5 min (Fig. 6.3). We have found that this test can easily be carried out on a microscope slide and haemagglutination viewed as clumping under a low-power objective. To determine whether sugars inhibit haemagglutination, an equal volume of a 2% (w/v) solution of the sugar of interest is added to the erythrocytes before the bacteria.

Static settling method

Equal volumes of erythrocyte and bacterial suspension in PBS are added to round-bottomed wells. The large-size WHO plates are useful because they are large enough for the haemagglutination to be obvious, but microtitre plates can also be used. The stationary plates are incubated for a standard time, either at 4°C or at room temperature. In the absence of haemagglutinating bacteria, the erythrocytes sediment to the bottom of the well under gravity, and tumble into the centre of the well to form a discrete button. In the presence of haemagglutinating bacteria this process is retarded, and the erythrocytes remain in suspension for longer and settle to form a lawn over the bottom of the well. This will also eventually collapse into a button in the

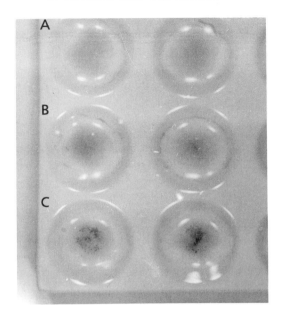

FIG. 6.3. Haemagglutination (HA) of *Bacteroides fragilis* determined by the rocked-tile method: A, capsulate population, HA-negative; B, mixture of capsulate and non-capsulate population, HA-positive; C, non-capsulate (electron-dense layer) population, HA-positive.

centre of the well if the well is left stationary for sufficient time. The extent to which the sedimentation of the erythrocytes is retarded will vary according to the concentration of bacteria and how well they agglutinate the erythrocytes. The incubation time is therefore critical: if the plates are left for too long, positive haemagglutination may be missed. False-positive reactions may occur at very high bacterial concentrations if the bacteria have large capsules that retard the bacterial sedimentation under gravity. The physical size of the bacterium in suspension may prevent the erythrocytes from settling to the centre of the well even after a long incubation period (Patrick *et al.*, 1988). This highlights the need to confirm haemagglutination by both the rocked-tile and the static settling methods.

Where unreproducible haemagglutination results are obtained for a particular bacterial type, this may be due to variation in the proportion of haemagglutinating cells within a single population. Separation of a strain into subpopulations may help to determine whether this is the case. By artificially mixing the subpopulations, the proportion of haemagglutinating bacteria required for a positive result can be determined.

Haemagglutination of bacterial populations grown under different environmental conditions can be compared with the artificially mixed populations. This can be used to determine the selective pressures that favour a higher proportion of haemagglutinating bacteria. To test the artificial mixtures, the haemagglutinating subpopulation (A) is mixed with the non-haemagglutinating subpopulation (B) in the following amounts: (µl) 20A:180B; 40A:160B and so on in 20 µl steps, finishing with 180A:20B. This can conveniently be done directly into the incubation wells with a standard 20 µl dropper pipette. This results in different ratios of the two bacterial types, but the same final bacterial concentration. Erythrocyte suspension (0.2 ml) is then added and the degree of haemagglutination monitored. A gradation in the degree of haemagglutination should be observed (Fig. 6.4).

Adherence to epithelial cells

There is a wide variety of methods for studying the degree to which bacteria adhere to the surface of epithelial cells. The intention here is to illustrate the methods generally adopted in such studies (see Chapter 8), some of which we have used in our studies on *B. fragilis* (unpublished results).

Collection of epithelial cells

Epithelial cells are usually collected from the buccal mucosa of a number of human volunteers by gentle scraping with a blunt and sterile instrument such as a wooden spatula or tongue depresser. They may also be collected from sources such as the urine of healthy females by centrifugation of freshly voided urine. The cells are suspended in PBS. We have also obtained cells from bovine and ovine peritoneal mesothelium by gentle scraping and suspending in PBS.

The cells are usually washed two or three times by centrifugation at 200 g for 10 min and resuspended in PBS. The percentage viability of the cells can be determined by dye-exclusion, for example, with trypan blue (Gorman *et al.*, 1986). The concentration of cells in the suspension is adjusted to 10^5 cells/ml after direct counting with a haemocytometer.

The adherence assay

Bacterial cells are washed twice by centrifugation and are resuspended in PBS to a concentration of 10^8 bacteria/ml. This is usually done by adjusting the $OD_{550 \, nm}$, based on a previous viable count calibration. The bacterial suspension (250 µl) is then mixed with an equal volume of epithelial cells in an Eppendorf centrifuge tube and incubated at 37°C for 1 or 2 h, which is usually sufficient for the bacteria to attach to the cells. For initial tests it is prudent to determine the

F𝙸𝙶. 6.4. Haemagglutination (HA) of *Bacteroides fragilis* capsulate and non-capsulate populations mixed in a series of proportions: A, nine parts non-capsulate (electron-dense layer) : one part capsulate, strongly HA-positive to B, one part non-capsulate : nine parts capsulate, HA-negative; C, control without bacteria.

optimum incubation time for attachment. The cells are then washed by centrifugation at $200\,g$ for 10 min to remove any bacteria not firmly attached. If an Eppendorf centrifuge with a low speed facility is not available, the Eppendorf tubes can be lowered into the tube holder of a bench centrifuge swing-out head with a length of cotton thread. A PBS control without added bacteria should always be included to assess the background level of commensal bacteria adherent to the epithelial cells.

The cells are finally resuspended in 500 µl PBS, smeared onto a microscope slide and fixed with methanol. A variety of stains, from fluorescent acridine orange to Giemsa stain (Sigma Chemical Co (UK) Ltd, Poole, UK), can be used to detect the adherent bacteria. We have used negative phase-contrast with a ×100 oil-immersion objective to observe bacteria without staining. Giemsa stain, which rapidly stains nuclear material in about 45 s, is also satisfactory and can be washed off with methanol. The bacteria stain blue, as does the nucleus of the epithelial cells. Other methods that can be used to determine the attachment of bacterial cells include the use of radioisotope-labelled bacteria (Calderone *et al.*, 1984; Gorman *et al.*, 1987). The adherence assay is carried out essentially as above, except that the bacterial cells are labelled with ^{3}H-leucine. After incubation with the epithelial cells, the mixture is filtered through a membrane filter (8 µm pore size, Millipore (UK) Ltd, Watford, UK), washed several times with PBS, transferred to a tissue solubilizer (e.g. NCS, Amersham International Plc, Amersham, UK) before the addition of scintillant and scintillation counting. The percentage of bacteria attached to the cells is given by:

$$\frac{\text{cpm sample} - \text{cpm background}}{\text{cpm bacterial suspension}} \times 100$$

Data evaluation

Bacteria attached to at least 50 cells and 50 control cells should be counted. The distribution of bacterial attachment to cells may be important in analysing the data, for example when comparing two sets of data. Most published work assumes that the distribution is normal and hence uses a *t*-test, which is parametric and assumes a normal distribution, to compare the mean and standard deviation from two sets of data. In our experiments with *B. fragilis* (unpublished data), however, and in other published work, it is clear that adherent bacteria are not normally distributed. It would therefore be more appropriate to use a non-parametric and distribution-free statistical analysis, such as the Mann−Whitney U-test. However, there can also be problems with this test because it ranks the data and does not work well if cells have the same number of attached bacteria. The suitability of both these tests is reviewed by Rosenstein *et al.* (1985) and Woolfson *et al.* (1987). In the latter the conclusion was that, despite the non-normal distributions, the Student's *t*-test was probably the more useful method (see Chapter 19).

Adherence to other cells

Experiments similar to those with epithelial cells have been carried out with cultured cells. Another approach is to assess the adherence of bacteria to a monolayer of cells cultured on glass cover-slips (Hartley *et al.*, 1978). Here, the cover-slips (6×22 mm) are treated with EDTA (elthylenediamine tetra-acetic acid) and placed into individual glass tubes. They are then seeded with about 10^5 cells (e.g. HeLa cells) and incubated at 37°C for 2 days to produce a confluent monolayer. The coverslips are then washed with two changes in PBS, and the bacterial suspension is added. After further incubation for 30 min, the cells are washed five times in PBS, fixed with methanol, stained with Giemsa and mounted on a microscope slide.

Inhibition of adherence

In order further to characterize the nature of the interaction between the bacteria and the cell surface, substances likely to inhibit attachment specifically can be added to the adherence assay. Such studies often involve the purification of bacterial surface structures involved in adhesion, such as the fimbriae, and modification of the adherence assay. The details of these techniques are beyond the scope of this chapter. A general review of the approaches adopted can be found in Arp (1988).

Addition of sugars

Sugars, such as mannose which inhibits haemagglutination by *E. coli* Type 1 fimbriae (Duguid *et al.*, 1966), may be added to the assay. Haemagglutination that is inhibited by mannose is termed mannose-sensitive, and it is accepted that the receptor on the erythrocytes contains mannose. Mannose-resistant adhesion may involve other sugar receptors on the cells: for example, K99 fimbriae of enterotoxigenic *E. coli* bind to Neu5Gc(α2-3)Gal(β1-4)Glc-(β1-1) ceramide (Smit *et al.*, 1984) and the P fimbriae of *E. coli*, involved in pyelonephritis, bind to Gal(α1-4)Gal-disaccharide moieties associated with the P blood group antigens. The possibilities are extensive. Generally, sugars of interest are added at about 2% (w/v) to the prepared erythrocytes or epithelial cells before adding the bacteria. If it is desired to determine the host cell receptor for a particular bacterial ligand, reference should be made to Karlson & Stromberg (1986). A general discussion of the importance and diversity of oligosaccharide moieties in mammalian systems is provided by Rademacher *et al.* (1988).

Addition of monoclonal antibodies

Since MAbs bind specifically to target epitopes on bacterial surface structures, they can be a powerful tool for the determination and characterization of surface structures associated with adhesion and their interaction with target cells. It is assumed that binding of the MAb to an epitope on the surface structure of interest will interfere with adherence. This approach has been successfully used for *Porphyromonas* (*Bacteroides*) *gingivalis* fimbriae (Isogai *et al.*, 1988). Before adding a MAb in a haemagglutination or adherence assay, it is necessary further to purify the MAb. This precaution should also be adopted when using monospecific polyclonal antibodies. We have found that serum, including fetal calf serum, added to cell culture media, can cause bacterial clumping and this interferes with the adherence assay. Ascites fluid or cell culture supernatant is centrifuged at $20\,000\,g$ for 20 min and the resulting supernatant chromatographed on Sephadex G200 (Pharmacia Biosystems (UK) Ltd) with PBS as the elution buffer, and the immunoglobulin — usually IgG — is collected. The MAb is then added to the erythrocyte or cell suspension, before adding bacteria to the adherence assay. This should be done as twofold serial dilutions of the MAb in PBS, usually starting at about 1 mg/ml.

Conclusions

The interaction between the bacterial surface and the surfaces of mammalian cells remains one of the most important steps in the process of pathogenesis. The interactions are often complex and depend on the species of bacterium and type of target cell. It is clear that an approach adopted for one pathogen may be unsuitable for others. Moreover, bacterial surface structures may be subject to environmentally induced variation: surface structures or antigens expressed during an infection may be different from those in laboratory culture. The definition of bacterial growth conditions is therefore important. In addition, genetically controlled phase variation mechanisms may operate within a population of bacteria in culture. This may result in the heterogeneous expression of adhesion factors within a single population grown under the same environmental conditions. Therefore, the use of a defined homogeneous population of bacteria is also essential. A wide variety of microbiological, immunological, chemical and biochemical techniques has been used to dissect the processes involved in attachment. A concerted attempt to understand these interactions necessarily involves the use of all of these techniques in combination.

Acknowledgements

S.P. acknowledges the support of a project grant from the Medical Research Council of the UK.

References

ARP, L.H. (1988) Bacterial infection of mucosal surfaces: an overview of cellular and molecular mechanisms. In Roth, J.A. (ed.) *Virulence Mechanisms of Bacterial Pathogens*, pp. 3–19. American Society for Microbiology, Washington DC.

BAYER, M.E. (1990) Visualisation of the bacterial polysaccharide capsule. In Jann, K. & Jann, B (eds.) *Bacterial Capsules, Current Topics in Microbiology and Immunology 150*, pp. 129–157. Springer-Verlag, Berlin.

BEESLEY, J.E. (1989) Immunocytochemistry of microbiological organisms: a survey of techniques and applications, In Bullock, R.G. & Petrusz, P. (eds.) *Techniques in Immunocytochemistry, 4*, pp. 67–93. Academic Press, London.

BRINTON, C.C. (1959) Non-flagellar appendages of bacteria. *Nature*, **183**, 782–786.

CALDERONE, R.A., LEHERER, N. & SEGAL, E. (1984) Adherence of *Candida albicans* to buccal and vaginal epithelial cells: ultrastructural observations. *Canadian Journal of Microbiology*, **30**, 1001–1007.

CRUICKSHANK, R. (1965) *Medical Microbiology: A Guide to the Laboratory Diagnosis and Control of Infection*, 11th edn, pp. 657–660. Churchill Livingstone, Edinburgh.

DAY, S.E., VASLI, K.K., RUSSELL, R.J. & ARBUTHNOTT, J.P. (1980) A simple method for the study in vivo of bacterial growth and accompanying host response. *Journal of Infection*, **2**, 39–51.

DUGUID, J.P., ANDERSON, E.S. & CAMPBELL, I. (1966) Fimbriae and adhesive properties in salmonellae. *Journal of Pathology and Bacteriology*, **92**, 107–138.

DUGUID, J.P., SMITH, I.W., DEMPSTER, G. & EDMUNDS, P.N. (1955) Non-flagellar filamentous appendages ('fimbriae') and haemagglutinating activity in *Bacterium coli*. *Journal of Pathology and Bacteriology*, **70**, 335–348.

FINLAY, B.B., FRY, J., EDWIN, P.R. & FALKOW, S. (1989) Passage of *Salmonella* through polarized epithelial cells: role of the host and bacterium. *Journal of Cell Science*, **11** (Suppl.), 99–107.

GIESA, F.R., ZAJAC, I., BARTUS, H.F. & ACTOR, P. (1982) Isopycnic separation of *Escherichia coli* cultures possessing colonisation factor antigen 1. *Journal of Clinical Microbiology*, **15**, 1074–1076.

GODING, J.W. (1986) *Monoclonal Antibodies: Principles and Practices*, 2nd edn, Academic Press, London.

GORMAN, S.P., McCAFFERTY, D.F., WOOLFSON, A.N. & ANDERSON, L. (1986) Decrease in adherence of bacteria and yeasts to human mucosal epithelial cells by noxythiolin in vitro. *Journal of Applied Bacteriology*, **60**, 311–317.

GORMAN, S.P., McCAFFERTY, D.F., WOOLFSON, A.D. & JONES, D.S. (1987) Reduced adherence of micro-organisms to human mucosal epithelial cells following treatment with Taurolin, a novel antimicrobial agent. *Journal of Applied Bacteriology*, **62**, 315–320.

HANCOCK, I.C. & POXTON, I.R. (1988) *Bacterial Cell Surface Techniques*. John Wiley, Chichester.

HARLOW, E. & LANE, D. (1988) *Antibodies: A Laboratory Manual*. Cold Spring Harbor Laboratory, New York.

HARTLEY, C.L., ROBBINS, C.M. & RICHMOND, M.H. (1978) Quantitative assessment of bacterial adhesion to eukaryotic cells of human origin. *Journal of Applied Bacteriology*, **45**, 91–97.

HOEPELMAN, A.I.M. & TUOMANEN, E.I. (1992) Consequences of microbial attachment: directing host cell functions with adhesins. *Infection & Immunity*, **60**, 1729–1733.

ISMAEEL, N., FURR, J., PUGH, W. & RUSSELL, A. (1987) Hydrophobic properties of *Providencia stuartii* and other Gram-negative bacteria measured by hydrophobic interaction chromatography. *Letters in Applied Microbiology*, **5**, 91–95.

ISOGAI, H., ISOGAI, E., YOSHIMURA, F., SUZUKI, T., KAGOTA, W. & TAKANO, K. (1988) Specific inhibition of adherence of an oral strain of *Bacteroides gingivalis* 381 to epithelial cells by monoclonal antibodies against the bacterial fimbriae. *Archives of Oral Biology*, **33**, 479–485.

JOHNSON, G.D., HOLBOROW, E.J. & DOWLING, D.J. (1978) Immunofluorescence and immunoenzyme techniques. In Weir, D.M. (ed.) *Handbook of Experimental Immunology*, 3rd edn, Ch. 2. Blackwell Scientific Publications, Oxford.

KARLSON, K. & STROMBERG, N. (1986) Overlay and solid phase analysis of glycolipid receptors for bacteria and viruses. *Methods in Enzymology*, **137**, 220–232.

LUFT, J.H. (1971) Ruthenium red and violet. 1 Chemistry purification, methods of use for electron microscopy and mechanism of action. *Anatomical Record*, **171**, 347.

LUTTON, D.A. (1991) *Variability of Surface Structure Expression in Bacteroides fragilis*. PhD Thesis, Queen's University, Belfast.

LUTTON, D.A., PATRICK, S., VAN DOORN, J., EMMERSON, M. & CLARKE, G. (1989) Expression of *Bacteroides fragilis* fimbrial antigen in vitro and in vivo. *Biochemical Society Transactions*, **17**, 758–759.

LUTTON, D.A., PATRICK, S., CROCKARD, A.D., STEWART, L.D., LARKIN, M.J., DERMOTT, E. & McNEILL, T.A. (1991) Flow cytometric analysis of within strain variation in polysaccharide expression by *Bacteroides fragilis* by use of murine monoclonal antibodies. *Journal of Medical Microbiology*, **35**, 229–237.

MOZES, N. & ROUXHET, P. (1987) Methods for measuring hydrophobicity of microorganisms. *Journal of Microbiological Methods*, **6**, 99–112.

OLD, D.C. (1985) Haemagglutination methods in the study of *Escherichia coli*. In Sussman, M. (ed.) *The Virulence Determinants of Escherichia coli: Reviews and Methods*, pp. 287–231. Academic Press, London.

PALAK, J.E. & VARNDELL, I.M. (1984) *Immunolabelling for Electron Microscopy*. Elsevier, Amsterdam.

PATRICK, S. & LUTTON, D.A. (1990) *Bacteroides fragilis* surface structure expression in relation to virulence. *Médecine et Maladies Infectieuses*, **20** hors serie, 19–25.

PATRICK, S. & REID, J.H (1983) Separation of capsulate and non-capsulate *Bacteroides fragilis* on a discontinuous density gradient. *Journal of Medical Microbiology*, **16**, 239–241.

PATRICK, S., REID, J.H. & COFFEY, A. (1986) Capsulation of in vitro and in vivo grown *Bacteroides* species. *Journal of General Microbiology*, **132**, 1099–1109.

PATRICK, S., REID, J.H. & LARKIN, M.J. (1984) The growth and survival of capsulate and non-capsulate *Bacteroides fragilis* in vitro and in vivo. *Journal of Medical Microbiology*, **17**, 237–246.

PATRICK, S., COFFEY, A., EMMERSON, A.M. & LARKIN, M.J. (1988) The relationship between cell surface structure expression and haemagglutination in *Bacteroides fragilis. FEMS Microbiology Letters*, **50**, 67–71.

RADEMACHER, T.W., PAREKH, R.B. & DWEK, R.A. (1988) Glycobiology. *Annual Review of Biochemistry*, **57**, 785–838.

ROSENBERG, M., GUTNIK, D. & ROSENBERG, E. (1980) Adherence of bacteria to hydrocarbons: a simple method for measuring cell surface hydrophobicity. *FEMS Microbiology Letters*, **9**, 29–33.

ROSENSTEIN, I.J., GRADY, D., HAMILTON-MILLAR, J.M.T. & BRUMFIT, W. (1985) Relationship between the adhesion of *Escherichia coli* to uroepithelial cells and the pathogenesis of urinary infection: problems in methodology and analysis. *Journal of Medical Microbiology*, **20**, 335–344.

SMIT, H., GAASTRA, W., KAMERLING, J.P., VLIEGENHART, J.F.G. & DE GRAAF, F.K. (1984) Isolation and characterization of the equine erythrocyte receptor for enterotoxigenic *Escherichia coli* K99 fimbrial adhesin. *Infection and Immunity*, **46**, 578–584.

SUTHERLAND, I.W. (1977) Bacterial exopolysaccharides — their nature and production. In Sutherland, I.W. (ed.) *Surface Carbohydrates of the Prokaryotic Cell*, pp. 47–68. Academic Press, London.

TENNENT, J.M., HULTGREN, S., MARKLUND, B.I., FORSMAN, K., GORANSSON, M., UHLIN, B.E. & NORMARK, S. (1990) Genetics of adhesin expression in *Escherichia coli*. In Iglewski, B.H. & Clark, V.L. (eds.) *The Molecular Basis of Pathogenesis*; Vol. XI, *The Bacteria*, pp. 79–110. Academic Press, London.

WOOLFSON, A.D., GORMAN, S.P., MCCAFFERTY, D.F. & JONES, D.S. (1987) On the statistical evaluation of adherence assays. *Journal of Applied Bacteriology*, **63**, 147–151.

7: Medical Device-Associated Adhesion

M.H. WILCOX

Department of Experimental and Clinical Microbiology, University of Sheffield Medical School, Beech Hill Road, Sheffield S10 2RX, UK

The use of implantable devices is now commonplace in many branches of medicine. The major complication associated with such implants is infection, particularly with coagulase-negative staphylococci (CNS) (Phillips, 1989). The major pathogens found on urinary catheters and joint prostheses are, however, *Escherichia coli* and *Staphylococcus aureus*, respectively. In general, Gram-negative bacterial and fungal infection of medical devices is more serious than sepsis with CNS and this often necessitates both implant removal and intensive antimicrobial treatment. However, the striking prevalence of CNS-associated infection makes these organisms the primary pathogens in this area of medicine.

CNS were long considered to be relatively harmless skin commensals and their isolation from clinical specimens was thought to represent contamination. However, the situation has changed dramatically and CNS are now recognized as important pathogens. Their emergance as major pathogens undoubtedly reflects the development of a wide range of medical implants (Table 7.1). The manipulation of catheters, particularly at the time of insertion, but also once *in situ* during sampling, is associated with a high risk of contamination with skin flora such as CNS.

Septicaemia caused by CNS has increased markedly in incidence (Pfaller & Herwaldt, 1988) (Table 7.2). It commonly occurs in immunocompromised patients, who often have indwelling devices, and is associated with significant mortality. Crude mortality figures of 27–78% have been reported, but these include deaths attributable to underlying disease. The excess mortality due to CNS infection is approximately 12%, and the median increased hospital stay is 14 days (Pfaller & Herwaldt, 1988). Treatment of CNS infections is difficult, and antibiotic-resistant strains represent a particular problem. The successful eradication of infection often involves the removal of the indwelling

Microbial Biofilms:
Formation and Control

TABLE 7.1. *Medical devices associated with infection due to CNS*

	Infections caused by CNS (%)
Continuous ambulatory peritoneal	
dialysis catheters (Tenckhoff)	30−50*
Cerebrospinal fluid shunts	35−80
Intravascular lines and cannulae	50−75
Prostheses:	
cardiac valves	20−70
cardiac pacemakers	50
joints	20−40
vascular grafts	10−40
Urinary catheters	5−10

* The devices referred to may be made of a variety of polymers and more than one polymer type may be used in a single device. The infection rates quoted are composite and relate to a variety of devices that may be used for each purpose.

TABLE 7.2. *Bacteraemia due to CNS in a single-centre study, Iowa, USA, 1945−1987 (Pfaller & Herwaldt, 1988)*

	CNS bacteraemia per 10 000 admissions	Bacteraemia due to CNS (%)
1945−55	0.5	1.5
1974−78	1.3	−
1980	5.2	8
1987	38.6	26

device, which may not only involve risk to the patient but is also time-consuming and costly. In addition to septicaemia, peritonitis, endocarditis, mediastinitis, meningitis and loosening of joint prostheses are well recognized complications of device-related infections caused by CNS. Discussion of the many methods for the laboratory diagnosis of infection, associated with the wide array of indwelling devices currently in use, is, however, beyond the scope of this chapter, but several salient points are worth emphasizing and the details of a widely used approach are given below.

Laboratory Diagnosis of Device-Associated Infection

The accurate diagnosis and cause of device-associated infection cannot be determined unless the correct samples are taken. Wherever possible, the device should be submitted to the laboratory. Swabs taken from skin entry sites, for example, may provide useful adjunctive information, but should not be relied upon alone. It is evident that some episodes of intravascular catheter-related sepsis result from microbial colonization of side ports and hub caps. In one study it was noted that 70% of episodes of septicaemia were caused by strains that were identical to hub-cap isolates (Linares *et al.*, 1985). Where it is not practical to send the whole or part of an implant, for example with a long catheter, the sections representing the skin entry site and the tip should be collected in separate, sterile containers.

The frequent presence on the device of only small numbers of firmly adherent microorganisms requires that vigorous vortexing or, more preferably, sonication is used to dislodge pathogens. Additionally, broth-enrichment culture should be used to increase the yield of positive cultures. Blood or other body fluids may be obtained through or in conjunction with the device. Quantitative culture of blood taken, both through an intravascular catheter and peripherally, is used to diagnose implant infection, and catheter-related bacteraemia is characterized by colony counts at least tenfold higher in the former (Weightman *et al.*, 1988). Peritoneal fluid from patients on continuous ambulatory peritoneal dialysis (CAPD) with bacterial peritonitis, may contain only one colony-forming unit (cfu) per millilitre, in spite of a pronounced white-cell response (Rubin *et al.*, 1980). Lytic agents such as Triton X may be used to release organisms held within phagocytes (Gould & Casewell, 1986).

The problem with the broth culture approach mentioned above is that contaminated (colonized) implants are not differentiated from infected ones (Maki, 1981). Maki *et al.* (1977) described a simple semiquantitative culture technique that is now widely used for the diagnosis of intravascular catheter-related infection.

Role-plate technique

Two 5-cm sections from long catheters (20−60 cm) — the skin entry part and the tip — or the whole of shorter catheters, are examined. Each segment is rolled back and forth across the surface of a blood agar plate at least four times. After culture at 37°C, infected catheters are identified by the presence of 15 cfu or more. The method is applicable to catheter-associated infection of whatever cause, but the isolation of even low numbers of coliform or fungal microorganisms may well be of significance.

None of the patients with catheter-associated septicaemia had semiquanti-tative culture results of <15 cfu (Maki *et al.*, 1977). The positive predictive value of the technique is approximately 25%. The figure of 15 cfu is arbitrary, and most infected catheters yield a confluent growth with this method. The main criticism of the technique is its failure to detect microorganisms adherent to the inner surface of the catheter. The method is often, therefore, combined with broth culture. It is also likely that if large numbers of bacteria are present at the skin entry site, these will be transferred to the line tip during removal.

Since CNS were recognized as the most frequent cause of device-related infection, there have been a multitude of publications about the pathogenicity of these organisms. Most of these have focused on the ability of CNS to produce slime and to adhere to and grow on medical devices. Since confusion has arisen because of the interchangeable use of the terms 'slime production' and 'adherence', these phenomena will now be considered.

Slime Production by Coagulase-Negative Staphylococci

CNS slime was probably first referred to by Bayston & Penny (1972) as a 'mucoid substance' associated with the colonization of Spitz−Holter shunts. Ten years later, Christensen *et al.* (1982) described an extracellular poly-saccharide slime material, and noted that slime-producing CNS isolates appeared to be associated with intravenous catheter-related sepsis. Slime is thought to be distinct from the capsule in CNS, and strains may produce one, both or neither of these. It is loosely bound to the cell surface and tends to surround clusters of bacteria, rather than being associated with single cells. Slime is water-soluble and hence is easily removed by washing (Peters *et al.*, 1987).

The production of slime may be enhanced by medium supplementation, for example with substances such as casein hydrolysate, glucose, galactose, lactose and maltose (Christensen *et al.*, 1982; Peters *et al.*, 1987). Tryptone soya broth has frequently been used in slime production/adherence assays (see below). Peters *et al.* (1987) used pmS 110-broth as a medium to promote slime production in CNS.

pmS 110-broth

This consists of peptone (Difco) 10 g and yeast extract (Difco) 2.5 g dissolved in 50 ml of 0.05 mol/l phosphate buffer plus 3% saline. It is dialysed against 950 ml of the same saline phosphate buffer for 48 h at 4°C. After autoclaving the dialysate, 10 g mannitol and 2 g lactose each dissolved in a small volume of dialysate, are sterilized by filtration, and added.

Several chemically defined media have been described for the preparation of CNS slime. Hussain *et al.* (1991a) developed a medium consisting of glucose, 18 amino acids, two purines and six vitamins, which yielded slightly larger quantitites of slime when compared with tryptone soya broth. This approach has helped to establish that components of complex media, and in particular of agar, may become incorporated into slime harvested *in vitro* (Drewry *et al.*, 1990; Hussain *et al.*, 1991b).

Many studies refer to CNS strains as 'slime-producers' or 'non-producers', slime-positive or slime-negative, respectively. However, as slime production is so medium-dependent, such terms are relevant only to the particular experimental conditions employed. Most CNS of the *Staphylococcus epidermidis* species group appear to produce slime, to a greater or lesser extent, when suitable incubation conditions are provided (Peters *et al.*, 1987).

The original method for detecting slime production in CNS consisted of using stains for polysaccharide to identify bacteria adherent to the sides of culture tubes (Christensen *et al.*, 1982).

Tube assay for slime production

CNS strains are cultured in 4 ml tryptone soya broth in either glass or polystyrene tubes, which should be scratch-free. After overnight incubation in air at 37°C, the contents of the tubes are carefully poured off. Normal saline (1 ml) is then added and the tubes are rolled lightly. After discarding the saline, 1 ml of polysaccharide stain such as Congo red or alcian blue is added, and the tubes are rolled again. Slime-positive strains coat the inside of the tubes with a layer that stains appropriately. A ring of stained material at the air/liquid interface is, however, not indicative of slime production. The subjective nature of the interpretation of results can create problems, and often necessitates repetition of the assays.

Despite its widespread use, however, this technique is basically flawed, since it relies on bacteria being able to adhere to the specific material of which the culture vessel is made, so that any extracellular polysaccharide produced may then be detected by staining. Adherence and slime production are not the same, and their relationship will be further discussed below. Furthermore, it is unclear whether polysaccharide-specific stains, such as alcian blue, also detect bacterial cell capsular material, if this is present. The colonial appearance of CNS on solid media is not a reliable guide of slime production (Peters *et al.*, 1987).

Mannose-specific lectins, such as Concanavalin A, which react with sugars on the bacterial cell surface, have been used in a slide agglutination assay as a marker for the presence of slime (Ludwicka *et al.*, 1984). This assumes that mannose is always present in CNS slime. Interestingly, slide agglutination has been shown to remain unaltered after bacteria have been sufficiently washed to remove the water-soluble extracellular slime material (Wilcox *et al.*, 1991a). Hence, lectins may, in fact, bind to sugars present in the cell wall rather than to slime.

Adherence and slime production by coagulase-negative staphylococci

Many attempts to determine whether slime production is a marker of CNS pathogenicity have, in fact, measured bacterial adherence, with associated extracellular polysaccharide formation. Ishak *et al.* (1985), for example, measured CNS 'slime production' by the tube adherence method (see above) after culture for 18−24 h. 'Adherence' to Teflon catheters was also assessed by immersing segments of these into suspensions of CNS for 20 min and then estimating the number of adherent cells by the roll-plate technique. They concluded that adherence of CNS to Teflon catheters was not significantly affected by slime production. However, both assays were actually measuring adherence, the former by cells cultured overnight, and the latter in a non-culture situation over 30 min. Others have shown that CNS slime production, as assessed by reactivity with mannose-specific lectins, may occur in strains that are non-adherent in the tube test.

CNS infections associated with nosocomial septicaemia, cerebrospinal fluid shunts, Tenchkoff catheters or prosthetic devices in general, have been shown to be caused by organisms, usually either blood culture contaminants or skin commensals which are significantly more adherent than controls (Pfaller & Herwaldt, 1988). Also, CNS infections resistant to antimicrobial treatment, that relapse soon after its cessation, and infections characterized by shunt blockage, are associated with organisms that show greater adherence *in vitro*. Kristinsson *et al.* (1986) found that adherent CNS isolated from CAPD patients with peritonitis were associated with a 50% chance of relapse within several days of stopping antibiotic treatment. Conversely, only 17% of peritonitis episodes caused by non-adherent CNS recurred ($p \leq 0.002$). Some workers, however, have failed to find a relationship between adherence and pathogenicity. This is probably best explained on the basis of subjective differences in the interpretation of tube assay results. The following quantitative test, which measures the adherence of CNS to microtitre tray wells, is a definite improvement (Christensen *et al.*, 1985).

Microtitre tray adherence assay

Polystyrene microtitre trays with 96 flat-bottomed wells are used as culture vessels. CNS strains are inoculated in quadruplicate into wells, prepared by diluting overnight cultures, to give an initial inoculum of 10^6 cfu/ml and a final volume of 250 µl. Tryptone soya broth is the usual medium of choice. After incubation at 37°C for 18−24 h, the well contents are carefully aspirated and non-adherent organisms are removed by washing three times with normal saline. The washing procedure should take place with the trays slightly tilted and the saline directed against the side of the well, so that it washes across the bacterial biofilm. Remaining adherent bacteria are fixed in 25% (v/v) formaldehyde and stained with crystal violet that has been centrifuged to remove debris. The plates are washed thoroughly under the tap to remove excess stain, and are then air-dried at room temperature. Bacterial adherence to the bottom of wells is measured spectrophotometrically at 546 nm with an ELISA plate reader. Adherence scores for each strain are calculated as the difference between the mean optical density at 546 nm for the test and control wells. The reproducibility of the assay largely depends on the washing procedure. In experienced hands, the intra- and intertray coefficients of variation should be approximately 10%.

It should be emphasized that this assay technique does not differentiate between adherence and bacterial growth, but provides a composite picture of the two processes. Modifications of the protocol, including alternative media, need to be assessed for their effects on bacterial growth yield before ascribing any differences in adherence to such changes.

Although slime production may be an important virulence factor in CNS infections, many biological properties having been described for slime, it is not now considered to be involved in the initial phase of bacterial adherence. Its role in CNS adherence is probably to consolidate bound bacterial cells, and as such it contributes to the formation of a biofilm. Peters *et al.* (1982) showed that CNS adherent to catheter surfaces after incubation for 12 h were smooth, but later acquired a coat of slime material. It has also been observed that the adherence of washed, as compared with unwashed, slime-producing CNS is the same (Hogt *et al.*, 1986).

An interesting approach to the study of the distinct processes of adherence and slime production by CNS has recently been described by Van Pett *et al.* (1990).

Differential labelling of bacterial cells and extracellular material

Strains are cultured in tryptone soya broth containing either of the two radioactive labels ^{3}H-thymidine and ^{14}C-glucose. After incubation, 1 ml Carnoy's fixative is added for 30 min and is then removed by washing twice with 3 ml normal saline; 1 ml of 0.2 mol/l sodium hydroxide is added and the tube heated for 1 h at 85°C to solubilize the biofilm. The level of radioactivity in the sample is then measured by liquid scintillography. The ratio of ^{3}H, uptake of which represents bacterial adherence, to ^{14}C, uptake of which represents glycocalyx formation, is high for 'non-slime producing' strains and low for 'slime producers'.

Scanning and transmission electron microscopy (SEM and TEM) may also be used to differentiate between adherence and slime production (Wilcox *et al.*, 1991a). Squares (1 cm) of polymer material are suspended by wire beneath the medium surface. After culture, the squares are removed carefully, washed lightly in normal saline and adherence, plus or minus slime production, is then assessed by SEM. Planktonic bacteria are recovered by centrifugation, fixed in glutaraldehyde with postfixation in buffer containing ruthenium red, and slime is observed by TEM (Mayberry-Carson *et al.*, 1984).

Experimental Variables of *in vitro* Adherence

Three important variables that have more often than not been ignored in studies *in vitro* of device-associated bacterial adherence are the type of surface material examined, the influence of atmosphere type during incubation, and the culture medium employed.

Polymer surfaces

A wide variety of materials is used in the manufacture of implantable medical devices. These range from metals, which are crystalline in structure, to the majority of medical polymers such as silicone, which are amorphous. A few polymers have mixed structures, gaining rigidity from crystalline zones and toughness from amorphous regions. Although polymer materials have a smooth macroscopic appearance, examination of their surfaces at high magnification frequently reveals indentations and crevices. Several studies have shown that CNS appear initially to adhere to these areas, possibly because of favourable electrostatic forces at such sites (Cheesborough *et al.*, 1985) (see Chapter 4).

High molecular weight polymers are generally thought to be resistant to bacterial deterioration. However, the use of additives such as plasticizers and stabilizers during the production of some materials provides potential weak points. For example, polyvinylchloride contains low molecular weight plasticizers that are vulnerable to attack by *Pseudomonas aeruginosa* and *Serratia marcescens* (Mears, 1979). Microbes possessing such ability may be at a considerable advantage in nutrient-deprived environments, if they can utilize released substances as a food source. It is interesting to note that eluates derived from a range of intravascular catheter materials have been shown to support the growth of *E. coli* and *Ps. aeruginosa in vitro*, but not of staphylococci (Lopez-Lopez *et al.*, 1991). Furthermore, the eluate from siliconized latex was shown to be toxic for *E. coli*. Franson *et al.* (1986) found that CNS adherent to catheter segments immersed in phosphate-buffered saline, survived for at least 96 h, whereas control bacterial suspensions were non-viable at this time. The introduction of nutrient media at 96 h led to a proliferation of catheter-associated CNS growth. This suggests that adherent CNS may survive in a relatively dormant state until the nutrient supply improves.

Catheter manufacturers generally aim to make biocompatible catheters with a low thrombogenic potential, rather than catheters with antibacterial properties. It has been shown that polyurethane catheters are the most thrombogenic *in vivo*, closely followed by those made from silicone; catheters made from polyurethane coated with hydromer were the least thrombogenic (Borrow & Crowley, 1985). There is almost no information about the relationship between polymer type and the deposition of host proteins that occurs on immersion into a body fluid.

Surprisingly few studies have investigated whether certain polymer materials are preferentially colonized by CNS (Gristina *et al.*, 1987). The degree of adherence to polymers by *S. epidermidis in vitro* is greater than that of *S. aureus*, whereas the latter preferentially binds to metals. This is consistent with the clinical observation that, although transcutaneous metallic fixation screws cross an environment where CNS predominate, they have a predilection for *S. aureus* infections. Hogt *et al.* (1987) examined a range of copolymers of differing hydrophobicities and surface charges, and noted significant differences in the degrees of CNS adherence.

Modified microtitre tray adherence assay

By modifying the microtitre tray adherence method outlined above, it is possible to examine CNS adherence to both silicone rubber and polystyrene, and possibly other polymers, within the same system (Wilcox *et al.*, 1991b). Discs of silicone rubber (Dow Corning, Reading, UK) 5 mm in diameter, cut from sheets with a paper hole punch, are affixed aseptically to the bottom of the microtitre tray wells with silicone grease (BDH, Poole, UK). Care should be taken to ensure that a good seal is achieved, to prevent loss of discs during washing, and that grease does not contaminate the upper surface of the silicone rubber. The method is otherwise as detailed above.

Adherence assays have traditionally employed either glass or polystyrene surfaces, neither of which is clinically relevant. Hickman and Broviac catheters, used for long-term large-vein access, and Tenckhoff catheters, which allow peritoneal dialysis fluid exchange in patients undergoing CAPD, are made of silicone rubber. The above method has clearly shown that CNS adherence to silicone rubber is different from that to polystyrene (Wilcox *et al.*, 1991b). It would seem sensible, therefore, that materials relevant to *in vivo* conditions are used when studying CNS adherence *in vitro*.

Atmosphere type

The adherence of CNS has generally been studied under aerobic conditions. The role of gas tension, and in particular that of carbon dioxide, on CNS adherence does not appear to have been examined until recently. Exopolysaccharide production by *Ps. aeruginosa* is affected by the oxygen concentration, and it has been suggested that this accounts for the refractoriness to antibiotic therapy of *Pseudomonas* endocarditis in the left side of the heart (Bayer *et al.*, 1989). These workers did not, however, examine the effect of CO_2 tensions, despite a marked difference in the P_{CO_2} on the two sides of the circulatory system. The adherence of CNS cultured in air compared to that in air supplemented with 5% CO_2 is markedly different (Wilcox *et al.*, 1991b). With tryptone soya broth as the culture medium, most but not all CNS strains adhere much better in air in the microtitre tray assay. Results obtained with nutrient broth are less uniform (Denyer *et al.*, 1990).

A modest fall in pH (mean 0.24 units in tryptone soya broth) is observed in cultures incubated in a CO_2-enriched atmosphere, compared with incubation in air. Since this variable is inextricably linked to P_{CO_2} in the fluid, alterations in bacterial adherence should be ascribed with caution to only one or other parameter. Significant differences in the cell-surface composition of CNS cultured in these two atmospheres have been demonstrated by X-ray photoelectron spectroscopy and cell-wall protein profiles (Denyer *et al.*, 1990). The precise reasons for the effect of CO_2 on CNS adherence remains unknown.

It should be noted that 5% CO_2 is a physiological concentration, found, for example, in venous blood and in the peritoneal fluid of patients on CAPD (Wilcox *et al.*, 1990). Hence, CNS in device-associated infections will be exposed to such CO_2 levels *in vivo*. Studies of CNS adherence *in vitro* should take this factor into account, if it is desired to apply conditions representative of those found *in vivo*.

Media

The conventional complex and chemically defined culture media used in the study of CNS slime and adherence, have already been outlined. These have usually been selected on the basis of their promotion of either adherent growth or slime production. Such nutrient-rich environments are, however, far removed from those that occur *in vivo*, where pathogens must adapt to the scarce food resources available. Bacterial growth in body fluids has been shown to result in alterations in cell surface structure, for example those associated with the need to obtain iron (Brown & Williams, 1985). In addition, proteins present in such biological fluids are known to be deposited on foreign surfaces and so to influence the adherence of bacteria. The deposition of albumin on polymer surfaces, for example, reduces the subsequent adherence of CNS (Peters *et al.*, 1987).

For these reasons, it is not surprising that the adherent growth observed with CNS in the microtitre tray model is markedly reduced, when sterile used peritoneal dialysis fluid (PUD) is chosen as a culture medium, rather than tryptone soya broth. It is, however, difficult directly to compare the results of this adherence assay with such media, because bacterial growth yield is at least 10-fold less after culture in PUD than in tryptone soya broth (Wilcox *et al.*, 1990). Previous studies that correlate CNS pathogenicity with adherence potential have generally involved bacteria grown in tryptone soya broth, although whether such a relationship is demonstrable in isolates cultured in body fluid appropriate to the conditions *in vivo*, has not been determined.

A further note of caution should be sounded in connection with the use of body fluids. The bicarbonate buffering system in body fluids *in vivo* is linked to the level of dissolved carbon dioxide. Once a fluid is removed from the patient, carbon dioxide tends to diffuse out of solution, resulting in a rise in pH. For example, it has been shown that unless, after collection from the patient, PUD is manipulated in a 5% CO_2 atmosphere, a pH rise of greater than one unit takes place, together with the precipitation of calcium, magnesium, phosphate and proteins (Table 7.3) (Wilcox *et al.*, 1990). The growth of CNS in PUD is markedly inhibited in air compared with air plus 5% CO_2. Hence, when a body fluid is used as the culture medium in order to simulate conditions *in vivo*, care must be taken to prevent alterations of the natural buffering system, to avoid use of a medium unrepresentative or even inhibitory to bacterial growth. Manipulations such as filtration, centrifugation or acidification may significantly alter biological fluids subsequently used as bacterial culture media, so that they no longer represent conditions *in vivo*.

TABLE 7.3. *Biochemical variables in pooled PUD after incubation at 37°C for 24 h in two gaseous environments*

	Air plus 5% CO_2	Air
Na^+	133*	133
K^+	4.3	4.3
Ca^{2+}	1.55	1.18
Mg^{2+}	1.0	0.7
Cl^-	108	108
HCO_3^-	16	16
PO_4^{3-}	1.39	1.13
Urea	24.5	24.5
Glucose	25.7	26.5
Creatinine (mol/l)	864	869
Protein (mg/l)	840	740
pH	7.37	8.6

* Concentrations in mmol/l, except where otherwise indicated.

Conclusions

Particular aspects of the aetiology of medical device-related infection have not been well defined and these have intentionally not been discussed here. For example, it is unclear what mechanisms are responsible for the initial adherence of CNS to polymer surfaces *in vivo*, and the search for specific bacterial adhesins continues. The factors that affect the progression from device colonization to infection are also poorly understood. In particular, the host-derived factors that influence CNS adherence deserve further study. It is evident that the *in vitro* study of medical device-associated infection has been hindered by a lack of a standardized methodology, and by a failure to appreciate the pronounced effect on CNS adherence of the parameters outlined above. It is hoped that future workers will be better equipped to control for such factors *in vitro*, in their search for the mechanisms that occur *in vivo*.

References

BAYER, A.S., O'BRIEN, T., NORMAN, O.C. & NAST, C.C. (1989) Oxygen-dependent differences in exopolysaccharide production and aminoglycoside inhibitory bactericidal interactions with *Pseudomonas aeruginosa* — implications for endocarditis. *Journal of Antimicrobial Chemotherapy*, **23**, 21–35.

BAYSTON, R. & PENNY, S.R. (1972) Excessive production of mucoid substance in *Staphylococcus SIIA*; a possible factor in the colonisation of Holter shunts. *Developmental Medicine and Child Neurology*, **14** (Suppl. 27), 25–28.

BORROW, M. & CROWLEY, J.G. (1985) Evaluation of central venous catheter thrombogenicity. *Acta Anaesthesiology Scandanavia*, **81** (Suppl.), 59–64.

BROWN, M.R.W. & WILLIAMS, P. (1985) The influence of environment on envelope properties affecting survival of bacteria in infections. *Annual Reviews of Microbiology*, **39**, 527–556.

CHEESBOROUGH, J.S., ELLIOT, T.S.J. & FINCH, R.G. (1985) A morphological study of bacterial colonisation of intravenous cannulae. *Journal of Medical Microbiology*, **19**, 149–157.

CHRISTENSEN, G.D., SIMPSON, W.A., BISNO, A.L. & BEACHEY, E.H. (1982) Adherence of slime-producing strains of *Staphylococcus epidermidis* to smooth surfaces. *Infection and Immunity*, **37**, 318–326.

CHRISTENSEN, G.D., SIMPSON, W.A., YOUNGER, J.J., BADOUR, L.M., BARRETT, F.F., MELTON, D.M. & BEACHEY, E.H. (1985) Adherence of coagulase-negative staphylococci to plastic tissue culture plates: a quantitative model for the adherence of staphylococci to medical devices. *Journal of Clinical Microbiology*, **22**, 996–1006.

DENYER, S.P., DAVIES, M.C., EVANS, J.A., FINCH, R.G., SMITH, D.G.E., WILCOX, M.H. & WILLIAMS, P. (1990) Influence of carbon dioxide on the surface characteristics and adherence potential of coagulase-negative staphylococci. *Journal of Clinical Microbiology*, **28**, 1813–1817.

DREWRY, D.T., CALBRAITH, L., WILKINSON, B.J. & WILKINSON, S.G. (1990) Staphylococcal slime: a cautionary tale. *Journal of Clinical Microbiology*, **24**, 559–561.

FRANSON, T.R., SHETH, N.K., MERON, L. & SOHNLE, P.G. (1986) Persistent *in vitro* survival of coagulase-negative staphylococci adherent to intravascular catheters in the absence of conventional nutrients. *Journal of Clinical Microbiology*, **24**, 559–561.

GOULD, I. & CASEWELL, M.W. (1986) Laboratory diagnosis of peritonitis during continuous ambulatory peritoneal dialysis. *Journal of Hospital Infection*, **7**, 155–160.

GRISTINA, A.G., HOBGOOD, C.D. & BARTH, E. (1987) Biomaterial specificity, molecular mechanisms and clinical relevance of *S. epidermidis* and *S. aureus* infections in surgery. In Pulverer, G., Quie, P.G. & Peters, G. (eds.) *Pathogenicity and Clinical Significance of Coagulase-Negative Staphylococci*, pp. 143–157. Gustav Fischer Verlag, Stuttgart.

HOGT, A.H., DANKERT, J. & FEIJEN, J. (1986) Cell surface characteristics of coagulase-negative staphylococci and their adherence to fluorinated polyethylene (propylene). *Infection and Immunity*, **51**, 294–301.

HOGT, A., DANKERT, J. & FEIJEN J. (1987) Adhesion of staphylococci onto biomaterials. In Pulverer, G., Quie, P.G. & Peters, G. (eds.) *Pathogenicity and Clinical Significance of Coagulase-Negative Staphylococci*, pp. 113–131. Gustav Fischer Verlag, Stuttgart.

HUSSAIN, M., HASTINGS, J.G.M. & WHITE, P.J. (1991a) A chemically defined medium for slime production by coagulase-negative staphylococci. *Journal of Medical Microbiology*, **34**, 143–147.

HUSSAIN, M., HASTINGS, J.G.M. & WHITE, P.J. (1991b) Isolation and characterisation of the extracellular slime made by coagulase-negative staphylococci in a chemically defined medium. *Journal of Infectious Diseases*, **163**, 534–541.

ISHAK, M.A., GROSCHEL, D.H.M., MANDELL, G.L. & WENZEL, R.P. (1985) Association of slime with pathogenicity of coagulase-negative staphylococci causing nosocomial septicaemia. *Journal of Clinical Microbiology*, **22**, 1025–1029.

KRISTINSSON, K.G., SPENCER R.C. & BROWN, C.B. (1986) Clinical importance of production of slime by coagulase-negative staphylococci in chronic ambulatory peritoneal dialysis. *Journal of Clinical Pathology*, **39**, 117–118.

LINARES, J., SITGES-SERRA, A., GARAU, J., PERONS, J.L. & MARTIN, R. (1985) Pathogenesis of catheter sepsis: a prospective study with quantitative and semiquantitative cultures of hub and segments. *Journal of Clinical Microbiology*, **21**, 357–360.

LOPEZ-LOPEZ, G., PASCUAL, A. & PEREA, E.J. (1991) Effect of plastic catheter material on bacterial adherence and viability. *Journal of Medical Microbiology*, **34**, 349–353.

LUDWICKA, A., UHLENBRUCK, G., PETERS, G., SENG, P.N., GRAY, E.D., JELJASZEWICZ, J. & PULVERER, G. (1984) Investigation on extracellular slime substance produced by *Staphylococcus epidermidis*. *Zentralblatt für Bakteriologie Hygiene A*, **258**, 256–267.

MAKI, D.G. (1981) Infections associated with intravascular catheters. In Wenzel, R.P. (ed.) *Handbook of Hospital Acquired Infections*, pp. 371–512. CRC Press, Boca Raton.

MAKI, D.G., WEISE, C.E. & SARAFIN, H.W. (1977) A semi-quantitative method for identifying intravenous-catheter-related infection. *New England Journal of Medicine*, **296**, 1305–1309.

MAYBERRY-CARSON, K.J., TOBER-MEYER, B., SMITH, J.K., LAMBE, D.W. & COSTERTON, J.W. (1984) Bacterial adherence and glycocalyx formation in osteomyelitis experimentally induced with *Staphylococcus aureus*. *Infection and Immunity*, **43**, 825–833.

MEARS, D.C. (ed.) (1979) *Materials and Orthopaedic Surgery*. Williams & Wilkins, Baltimore.

PETERS, G., LOCCI, R. & PULVERER, G. (1982) Adherence and growth of coagulase-negative staphylococci on surfaces of intravenous catheters. *Journal of Infectious Diseases*, **146**, 479–482.

PETERS, G., SCHUMACHER-PERDREAU, F., JANSEN, B., BEY, M. & PULVERER, G. (1987) Biology of *S. epidermidis* extracellular slime. In Pulverer, G., Quie, P.G. & Peters, G. (eds.) *Pathogenicity and Clinical Significance of Coagulase-Negative Staphylococci*, pp. 15–31. Gustav Fischer Verlag, Stuttgart.

PFALLER, M.A., & HERWALDT, L.A. (1988) Laboratory, clinical and epidemiological aspects of coagulase-negative staphylococci. *Microbiology Reviews*, **1**, 281–299.

PHILLIPS, I. (ed.) (1989) *Focus on Coagulase-Negative Staphylococci*. Royal Society of Medicine Services, London.

RUBIN, J., ROGERS, W.A., TAYLOR, H.M., DALE EVERETT, E., PROWANT, B.F., FRUTO, L.V. & NOLPH, K.D. (1980) Peritonitis during continuous ambulatory peritoneal dialysis. *Annals of Internal Medicine*, **92**, 7–13.

VAN PETT, K., SCHURMAN, D.J. & LANS SMITH, R..(1990) Quantitation and relative distribution of extracellular matrix in *Staphylococcus epidermidis* biofilm. *Journal of Orthopaedic Research*, 8, 321−327.

WEIGHTMAN, N.C., SIMPSON, E.M., SPELLER, D.C.E., MOTT, M.G. & OAKHILL, A. (1988) Bacteraemia related to indwelling central venous catheters: prevention, diagnosis and treatment. *European Journal of Clinical Microbiology and Infectious Diseases*, 7, 125−129.

WILCOX, M.H., SMITH, D.G.E., EVANS, J.A., DENYER, S.P., FINCH, R.G. & WILLIAMS, P. (1990) Influence of carbon dioxide on growth and antibiotic susceptibility of coagulase-negative staphylococci cultured in human peritoneal dialysate. *Journal of Clinical Microbiology*, 28, 2183−2186.

WILCOX, M.H., HUSSAIN, M., FAULKNER, M.K., WHITE, P.J. & SPENCER, R.C. (1991a) Slime production and adherence by coagulase-negative staphylococci. *Journal of Hospital Infection*, 18, 327−332.

WILCOX, M.H., FINCH, R.G., SMITH, D.G.E., WILLIAMS, P. & DENYER, S.P. (1991b) Effects of carbon dioxide and sub-lethal levels of antibiotics on adherence of coagulase-negative staphylococci to polystyrene and silicone rubber. *Journal of Antimicrobial Chemotherapy*, 27, 577−587.

8: Antimicrobial and Other Methods for Controlling Microbial Adhesion in Infection

S.P. Denyer[1], G.W. Hanlon[1], M.C. Davies[2] and S.P. Gorman[3]

[1]Department of Pharmacy, University of Brighton, Moulsecoomb, Brighton BN2 4GJ; [2]Department of Pharmaceutical Sciences, University of Nottingham, Nottingham NG7 2RD; and [3]School of Pharmacy, Medical Biology Centre, The Queen's University of Belfast, Belfast BT9 7BL, UK

Microbial adherence, directly to host tissue or in association with indwelling medical devices, contributes significantly to the success of many clinical infections and the associated biofilm exhibits marked resistance to antimicrobial therapy (Costerton, 1984). Growing recognition of the interplay between adherence and pathogenicity has led to the examination of possible approaches for its control. This might be achieved by careful selection of antimicrobial agents that limit adhesion to host tissue, the inclusion of antimicrobial compounds in medical devices, or the modification of biomaterial surfaces to hinder adhesion.

Regulation of Adhesion to Epithelial Cells by Antimicrobial Chemotherapy

The potential effect of antimicrobial agents on the adherence of microorganisms to tissue can readily be investigated *in vitro* with exfoliated epithelial cells or cells cultivated in cell culture. Although exfoliated cells are clearly the most representative, donor-related factors such as smoking, chronic obstructive pulmonary disease, diabetes mellitus and drug therapy can significantly alter microbial adherence (Funfstuck *et al.*, 1987) and monolayer cell culture often provides an alternative experimental system giving more reproducible results. Examples of both types of approach are given below, and experimental conditions should be carefully controlled because microbial adherence to epithelial cells may be affected by several factors.

Microbial Biofilms:
Formation and Control

The nature of the microbial challenge employed in the adherence assay in terms of clinical isolate or laboratory strain, growth phase, concentration of organisms, single or multiple challenge, the growth medium in terms of composition, pH and osmolality, and incubation conditions must be determined and defined. When examining the effect of antimicrobial agents on adhesion, it is usual to pre-expose the microorganisms to agents at concentrations representative of those achieved *in vivo*. The antimicrobial agent may then be removed in the supernatant following centrifugation, and the microbial cells resuspended in 0.01 mol/l phosphate-buffered saline (PBS; pH 7.3) before use in the adherence assay. In this way potential toxic effects of antimicrobial agents on the isolated epithelial cells can be avoided, but it should be remembered that any effect of the drug on the epithelial cell surface cannot be modelled under such conditions.

*Preparation and use of exfoliated epithelial cells
in adherence assays*

Standard preparation methods for two cell types are given below, but an alternative technique has been described by Valentin-Weigand *et al.* (1987). Depending on the nature of the study, epithelial cells may be obtained from an individual or collected from a number of volunteers and pooled. Cell viability should be ascertained before they are used for the adherence assay. The percentage of live cells may be estimated by the addition of trypan blue dye (0.01%) to an equal volume of the cell suspension. This is then mixed and incubated at 25°C for 10 min, placed in a counting chamber and examined with the light microscope. Cells are designated viable if they exclude the trypan blue.

Preparation of mucosal epithelial cells

The adherence of bacteria to mucosal epithelial cells has often been studied by incubating washed bacterial suspensions with buccal, pharyngeal or vaginal epithelial cells obtained by gentle swabbing of the appropriate tissue surface. Cells are collected in 0.01 mol/l PBS, pH 7.3, washed three times in PBS by centrifugation (500 g for 20 min) and resuspended in PBS. Alternatively, after vacuum-filtration on to a membrane of approximately 10 μm pore diameter (Millipore Ltd) they are resuspended in PBS. Cells are counted in a haemocytometer and adjusted to the desired concentration, generally 10^5 cells/ml, with PBS. Epithelial cell suspensions should be used within 2 h. A human buccal epithelial cell is shown in Fig. 8.1 with attached blastospore and invasive pseudomycelial forms of *Candida albicans*. Uroepithelial cells can be obtained by the centrifugation at 500 g for 20 min of freshly voided midstream

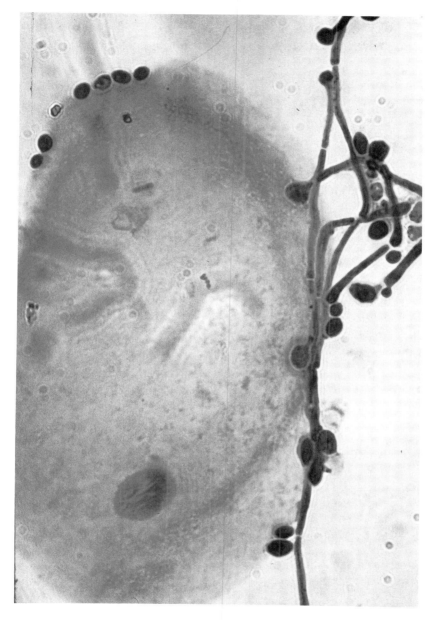

FIG. 8.1. Light micrograph of crystal violet-stained *Candida albicans* blastospores and pseudomycelia adhering to a buccal epithelial cell.

urine from non-bacteriuric individuals collected in a sterile container by the clean-catch procedure.

Preparation of cornified epithelial cells

These are usually obtained from healthy non-atopic volunteers. The skin is cleaned with 70% alcohol and corneocytes are collected, by gentle scraping of the proximal forearm with a Teflon policeman or scapel blade, in PBS, which may contain 0.1% Triton X-100. The cells are vigorously shaken until a predominately single cell suspension is obtained. Cells collected in this manner are essentially free of nucleated keratinocytes and are considered to be a homogeneous population of corneocytes. They are washed three times in PBS to remove adherent microorganisms and resuspended to the desired concentration after counting in a haemocytometer.

Adherence assay

The adherence assay described by Svanborg-Eden (1978) continues to serve as a model for such investigations. Generally, 1-ml volumes each of microbial suspension (*ca*. 10^8 cells/ml) and epithelial cells (*ca*. 10^5 cells/ml) are mixed and incubated with gentle shaking at 37°C for a designated time, usually 1 h. Epithelial cells are recovered and unattached microorganisms removed with two cycles of centrifugation and washing in 3 ml PBS ($150\,g$ for 10 min), 50–100 μl of the epithelial cell/microbial preparation is then fixed to a glass slide by air drying and stained with Giemsa stain. A filtration step is often included in the assay, either to replace or follow the centrifugation/washing cycles to ensure complete removal of unattached microorganisms. If filtration is to follow centrifugation, the unattached microorganisms in the supernatant liquid are discarded and the remaining cell pellet is resuspended in, and washed with, 50 ml PBS on polycarbonate filters (pore diameter *ca*. 12 μm). Filters with epithelial cells and adherent bacteria are then stained. The numbers of bacteria adhering to each of 50 (minimum) epithelial cells are counted by light microscopy at a magnification of ×1000. Large-cell yeasts such as *Candida* spp. adhere in much smaller numbers to epithelial cells than do bacteria, and may readily be counted at a magnification of ×400.

Preparation and use of epithelial cell monolayers in adherence assays

Epithelial cell lines can be obtained from commercial cell culture laboratories and maintained by standard methods (Freshney, 1986). Unlike studies with exfoliated cells, cell culture requires strict asepsis to avoid contaminating microbial growth. Aseptic practices are frequently supported by the inclusion of antibiotics in the cell culture growth medium. It is essential to remove these antibiotics from the final assay mixture in order to eliminate their possible effect on microbial adhesion. The following example uses HEp2 cells (human epithelial cell lines from a carcinoma of the larynx), which are commonly used for *in vitro* microbial/epithelial cell adherence studies (Wyatt *et al.*, 1990).

Preparation of HEp2 cell monolayers

HEp2 cells (Flow Laboratories Ltd, Irvine, Scotland) are harvested from culture flasks after trypsin treatment. Trypsin solution, one part 0.25% w/v, is added to two parts culture medium, and after a 30-s contact time is replaced with fresh culture medium. The cell suspension so obtained is centrifuged (1000 g for 10 min) and resuspended in 2.5 ml Eagle's minimum essential medium with Earle's salts (MEM; Gibco Europe Ltd, Paisley, Scotland). This medium contains 20 mM glutamine and 2.5 mg/ml sodium bicarbonate, and is supplemented with 0.5% fetal calf serum (Flow Laboratories), 15 µg/ml penicillin and 250 µg/ml streptomycin. Of this cell suspension 0.1 ml (*ca.* 10^5 cells) is added to 4 ml MEM in the 19 mm^2 compartments of vented plastic multiwell trays (Flow Laboratories) containing sterile glass coverslips. The trays are gently rotated to disperse the cells evenly over the coverslips, placed in humidified plastic containers and incubated in air with 5% CO_2 at 37°C. The medium is replaced every 24 h and confluent cell monolayers are formed on the coverslips after 72 h. Monolayers are then washed once with PBS and covered with 1 ml of antibiotic-free MEM 6 h before the adherence assay is to be carried out.

Adherence assay

A modification of the static overlay test, as described by Tavendale *et al.* (1983), is used as the assay model. Briefly, 1 ml of washed bacterial suspension (*ca.* 10^9 cells/ml) in antibiotic-free MEM is used to replace the medium on the monolayers. The HEp2 cells are incubated statically with the bacteria at 37°C for intervals up to 2 h. At the end of the contact period, bacterial suspensions are removed by aspiration and the coverslips are washed by three replacements of fresh PBS. During each wash the multiwell plates are rotated for 15 s on an electronic mixer (Janke and Kunkel, Germany). Monolayers with attached bacteria are then fixed for 30 min in 3% glutaraldehyde and stained by a modification of the Gram method (Evans, 1992; Evans-Hurrell *et al.*, 1993). Mounted and stained coverslips are examined under oil immersion in the microscope ×1000). The bacteria attached to a minimum of 50 cells are counted and the average number attached per cell determined.

Analysis and interpretation of results

Several methods can be used to determine the extent of microbial adherence to epithelial cells *in vitro*, including direct microscopy, radiolabel uptake, fluorescent antibody staining, enzyme-linked immunosorbent assay, electronic particle counting and a nitrate reductase-based colorimetric assay (Gorman, 1991). Direct observation by microscopy, coupled with differential staining, affords the most information because it provides direct evidence of cell—microbe interaction and also allows determination of distribution frequencies after antimicrobial challenge (Gorman, 1989).

Direct manual microscopy is time-consuming and tiring. Image analysis techniques greatly reduce these difficulties and largely eliminate subjective errors (Evans-Hurrell *et al.*, 1993). The data presented in Figs 8.2 and 8.3 respectively represent the antimicrobial control of attachment of *Candida* to buccal epithelial cells and the effect of antibiotics on staphyloccoccal adhesion to HEp2 cells. They were obtained by radioisotope assay and confirmed by microscopy (Fig. 8.2) and by light microscopy combined with image analysis (Fig. 8.3).

Control of Adhesion to Model Biomaterial Surfaces

Microbial adherence to indwelling and implantable medical devices, with subsequent colonization and infection, is a major clinical problem (Bisno & Waldvogel, 1989). A possible approach to reducing this risk is to develop biocompatible medical devices with intrinsic antiadherent surfaces. An understanding of the surface modifications necessary to control microbial adhesion is at first best obtained by using surface-characterized model polymers which can be modified in a reproducible and measurable manner. In the studies outlined below hydrophobic polystyrene is the base polymer and clinical isolates of *Staphylococcus epidermidis* are the challenge organisms.

Methods of surface modification

Glow discharge treatment

Polystyrene (MW 100 000; BDH), spin-cast from a 1% solution in toluene onto silanized glass microscope slides, is oxygen glow-discharge treated in a Biorad PT 7100 RF barrel etcher (Polaron Division, Watford, UK) operating at 0.03 mbar with an oxygen flow rate of 60 μl/min. Treatment times are controlled to provide surfaces of varying hydrophobicity with water contact angles between 0° and 90° (Khan *et al.*, 1988).

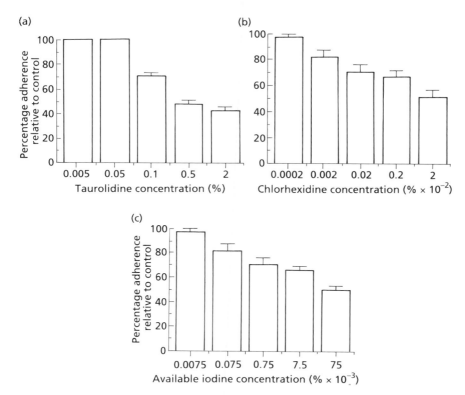

FIG. 8.2. Effect of (a) taurolidine, (b) chlorhexidine and (c) povidone–iodine on the adherence of an oral isolate of *Candida albicans* to human buccal epithelial cells.

Pluronic surface treatment

Pluronic surfactants (BASF Corporation, Parsippany, New Jersey, USA) are block copolymers of generalized structure A-B-A, where the constituent monomers A (poly(ethylene oxide); PEO) and B (poly(propylene oxide); PPO) are arranged in 'blocks' to give, for example, a copolymer (BASF code L31) of two PEO units–16 PPO units–two PEO units (shorthand nomenclature, 2/16/2). The monomeric composition of pluronics can be varied to give copolymers with very different PEO–PPO–PEO monomer ratios, thereby modifying substantially the hydrophilic/hydrophobic balance of the final surfactant molecule.

Polystyrene, dip-cast from a 1% solution in toluene on to silanized glass microscope slides, is soaked in 2% pluronic surfactant solution in distilled water at 25°C for 20 h. After treatment the pluronic-coated surfaces are

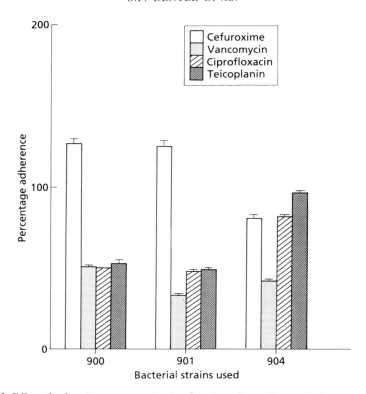

FIG. 8.3. Effect of cefuroxime, vancomycin, ciprofloxacin and teicoplanin at half the MIC on the adherence of *Staphylococcus epidermidis* strains 900, 901 and 904 grown in nutrient broth to HEp2 cell monolayers. Percentage adherence is calculated from the ratio of bacterial attachment in the presence of antibiotic to that in its absence (control). Adapted from Evans (1992).

rinsed thoroughly in distilled water before use in the adhesion assay (Bridgett *et al.*, 1991).

Adhesion studies

Treated and untreated polystyrene surfaces are challenged with washed suspensions of *S. epidermidis* in PBS at a total cell count of 5×10^8 cells/ml and incubation periods of up to 4 h. After the required incubation period, unattached cells are washed off in a stream of PBS, the slides air-dried and attached organisms heat-fixed and stained with acridine orange (250 mg/ml) in citrate phosphate buffer (0.01 mol/l; pH 6) prefiltered through a 0.2 μm membrane filter. Bacterial adhesion levels are assessed by epifluorescence microscopy and the mean surface coverage is determined by image analysis (4010 Image Analyser, Analytical Measuring Systems, Cambridge; Fig. 8.4).

Analysis and interpretation of results

Changes in polystyrene surface hydrophobicity, brought about by glow-discharge treatment, reveal the presence of apparent 'windows of maximum adhesion' where the substrate hydrophobicity that encourages maximum adhesion is closely related to the order of bacterial cell-surface hydrophobicity (Fig. 8.5). This implies that there is a substratum hydrophobicity range which is optimal for adhesion and which should be avoided in attempts to control staphylococcal attachment.

More dramatic effects of staphylococcal adhesion are seen after pluronic surface treatment, where surfactant adsorption strongly inhibits microbial attachment (Fig. 8.6). This is assumed to arise through the orientation of the adsorbed pluronic molecule with hydrophilic PEO groups protruding outwards to create a sterically stabilized surface layer (Tadros & Vincent, 1980; Lee et al., 1989). A similar steric-stabilizing effect has been reported for poly-(ethylene glycol) copolymers (Humphries et al., 1987).

Control of Adhesion to Modified Medical Devices

Microbial colonization of devices, such as peritoneal, urinary and intravascular catheters and cerebrospinal fluid (CSF) shunts, can lead to tissue damage, clinical infection and catheter occlusion. Hydrogel coatings applied to such devices have been reported to limit microbial adhesion (Kristinsson, 1989; Bridgett et al., 1992), although such coatings do not appear to control in vitro struvite and hydroxyapatite encrustation on Foley latex urinary catheters (Gorman et al., 1991).

CSF shunts

Method of surface modification

Hydrogel-coated and untreated silicone rubber CSF ventriculoperitoneal shunts were obtained from Mediplus Ltd, Bucks. The coating employed (Hydromer[TM], Hydromer Inc., New Jersey, USA) is a hydrogel material prepared by the interaction of polyvinylpyrrolidone (PVP) with one of several possible isocyanate prepolymers. This interaction renders the PVP insoluble in water but able to swell, the degree of swelling is controlled by the PVP: isocyanate ratio. When wetted, the Hydromer coating provides a hydrophilic

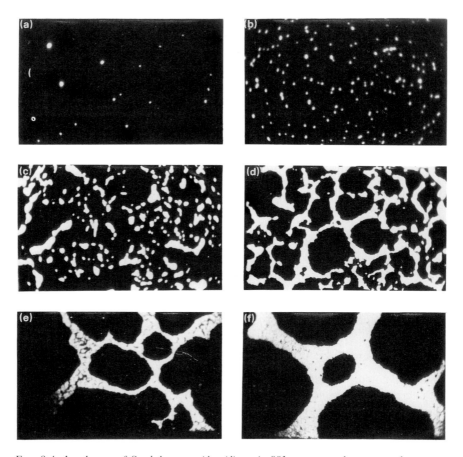

FIG. 8.4. Attachment of *Staphylococcus epidermidis* strain 901 to oxygen plasma-treated
polystyrene, water contact angle 0°, as a function of contact time. (a) $t = 15$ min; (b) $t = 30$ min;
(c) $t = 60$ min; (d) $t = 90$ min; (e) $t = 180$ min; (f) $t = 270$ min.

surface with an advancing water contact angle of approximately zero. The
Hydromer is applied by dip-coating.

Adhesion studies

Catheter segments (8 mm), opened longitudinally to expose the internal surface
and mounted on glass microscope slides (Bridgett *et al.*, 1992), are placed in
glass Petri dishes containing 1×10^9 *Staphylococcus* cells/ml in PBS at 37°C
for 4 h. After incubation, the shunt surfaces are washed in a stream of PBS

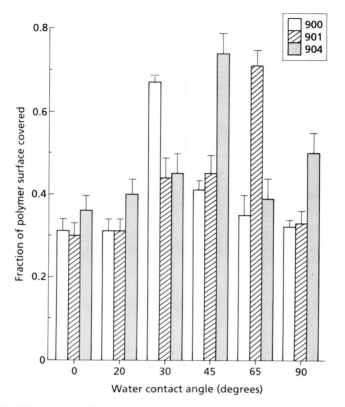

FIG. 8.5. Adhesion after 4h contact of three strains of *Staphylococcus epidermidis* 900, 901 and 904 grown in nutrient broth, to oxygen plasma-treated polystyrene with a range of water contact angles. The water contact angles for the three bacterial strains used are 900 (15°), 901 (38°), 904 (26°).

and attached bacteria are fixed, stained with acridine orange and examined by epifluorescent microscopy and image analysis as described above.

Analysis and interpretation of results

Hydromer coating reduced the adherence of all staphylococcal strains tested by between 84 and 92% of control (untreated) shunts (Fig. 8.7). It is assumed that the hydrophilic nature of the Hydromer coating renders the shunt surface energetically unfavourable to attachment and hinders the establishment of short-range hydrophobic interactions between the cell and the biomaterial surface (see Chapter 2).

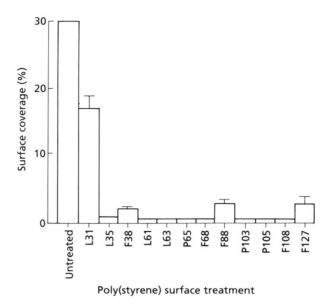

FIG. 8.6. Effect of pluronic treatment on the adhesion of *Staphylococcus epidermidis* (strain 901) grown in nutrient broth to polystyrene. Pluronics used have the following PEO/PPO/PEO block copolymer structure: L31, 2/16/2; L35, 11/16/11; F38, 43/16/43; L61, 3/30/3; L63, 10/30/10; P65, 19/30/19; F68, 76/30/76; F88, 104/39/104; P103, 17/56/17; P105, 37/56/37; F108, 129/56/129; F127, 99/67/99 (see text).

Intrauterine contraceptive devices (IUCDs)

Women using IUCDs have a 1.5−2.6-fold increased risk of developing pelvic inflammatory disease (World Health Organization, 1987) and approximately 10% of patients need to have the device removed shortly after insertion as a result of an excessive purulent discharge.

During insertion, the device carries bacteria from the heavily colonized vagina and cervix into the otherwise sterile uterus and the bacteria adhere to the device within a mucus biofilm. While the device is present within the uterus, polymer monofilament marker tails hang down through the cervix into the vagina, so providing a further access route for bacteria to the uterus. Studies with guinea-pigs have shown that planktonic bacteria introduced into the uterus survive only up to 5 days, but when attached as a biofilm they are still present 6 months after insertion (Gard *et al.*, 1992). Clearly, these sessile bacteria may become detached and initiate infection.

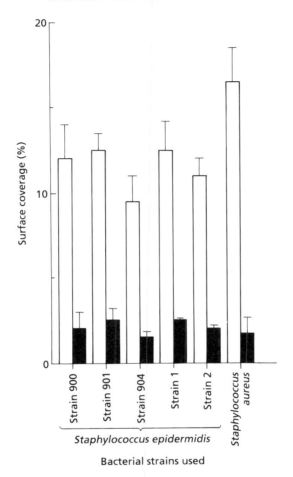

FIG. 8.7. Adhesion of a range of clinical *Staphylococcus* isolates to (□) untreated and (■) Hydromer-coated silicone rubber CSF shunts.

Bacterial adhesion to poly-(2-hydroxyethyl methacrylate) (PHEMA)-coated monofilaments

Preparation of coated threads. It has been reported that PHEMA hydrogel is non-adhesive to mammalian fibroblasts (Lydon *et al.*, 1985). PHEMA-coated IUCD threads of nylon, polyvinylidine chloride (PVDC) and polyester are prepared by dip-coating with a 5% (w/v) solution of PHEMA in 95% (v/v) ethanol. The threads are air-dried at room temperature for 1 h and stored in distilled water for at least 24 h before use.

Adherence assay. Lengths of thread, both PHEMA-coated and uncoated, are incubated with an *Escherichia coli* suspension (5×10^8 cells/ml) in 0.9% (w/v) saline at 37°C. The threads are then washed three times in 0.9% (w/v) saline to remove loosely and non-adherent bacteria. Adenosine triphosphate (ATP) is extracted from bacteria that remain firmly attached to the thread surface by immersion of the threads in 0.05% (w/v) trichloroacetic acid containing 2 mmol/l EDTA. The ATP content of these extracts is then measured by the bioluminescence assay described by Ludwicka *et al.* (1985). The light produced in the assay is proportional to the amount of ATP, which can in turn be transformed via standard curves into bacterial numbers.

Analysis and interpretation of results. Escherichia coli was found to adhere in the greatest numbers per unit area to PVDC (Fig. 8.8), whereas significantly fewer bacteria adhered to the nylon and polyester threads (two-tailed Mann−Whitney U-test; $p < 0.05$). In all cases, coating with PHEMA significantly reduced bacterial attachment to the monofilament ($p < 0.05$). The effect is similar to that reported for the Hydromer coating.

Effect of chlorhexidine incorporation into PHEMA-coated monofilaments

A number of investigators have sought to control microbial colonization by the incorporation of antimicrobial agents in materials (Isquith *et al.*, 1972; Bayston & Milner, 1981; Ikeda *et al.*, 1986; Kingston *et al.*, 1986). These developments have been extended to the application of antimicrobial coatings to medical devices and the PHEMA hydrogel offers a suitable means by which this may be achieved.

In vitro efficacy assessment. Nylon monofilaments are coated, as described above, in a 5% (w/v) solution of PHEMA in 95% (v/v) ethanol containing 1.8% (w/v) chlorhexidine diacetate. A 0.2-ml volume of an *E. coli* suspension is added to 1.8 ml of purified mucus (Lethem *et al.*, 1984) and mixed to give a final bacterial concentration of 10^8 cells/ml. An 8-cm length of impregnated thread is inserted into the inoculated mucus contained within the lumen of a sterile silicone rubber tube of internal diameter 1 mm. At various time intervals, 1-cm segments of tube are removed with a scalpel, immersed in 10 ml peptone water containing 0.1% lecithin and 1% Tween 80, and vortex-mixed vigorously. The peptone water is suitably diluted and 0.2-ml volumes of appropriate dilutions are plated on nutrient agar and incubated. Control

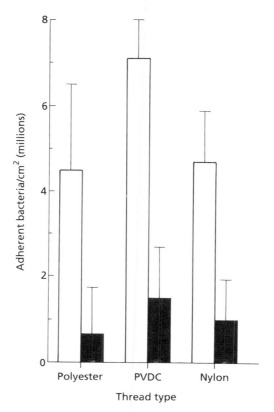

FIG. 8.8. Adhesion of *Escherichia coli* after 24 h contact with (□) uncoated and (■) PHEMA-coated threads (*n* = 6).

experiments are performed with PBS instead of mucus and PHEMA-coated threads without chlorhexidine.

In vivo efficacy assessment. Chlorhexidine-loaded monofilaments are prepared as described above, but the concentration of antimicrobial agent is increased to 20% (w/v) chlorhexidine digluconate. Virgin female Dunkin–Hartley guinea-pigs are used for the experiments, with surgical and insertion procedures as described by Gard *et al.* (1992).

At intervals, the animals are killed and their uteri examined for the presence of microorganisms. The uterine horn which held the monofilament is dissected out under aseptic conditions and washed repeatedly with 2 ml of sterile quarter-strength Ringer's solution. This is diluted and plated on nutrient agar for calculation of the bacterial concentration.

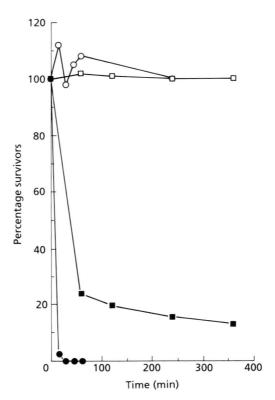

FIG. 8.9. The bactericidal effect of chlorhexidine-loaded threads in the presence and absence of mucus. (●) Loaded thread (PBS); (○) control (PBS); (■) loaded thread (mucus); (□) control (mucus).

Analysis and interpretation of results. That the chlorhexidine released from the hydrogel coating exerts significant bactericidal activity and reduces bacterial numbers in PBS by 6 logs within 30 min is clearly shown in Fig. 8.9. In the presence of mucus, however, the bactericidal activity is significantly reduced because of interactions between the antimicrobial agent and the mucus — an important consideration when attempting to control adherent biofilm bacteria by this method. This limitation was not, however, manifest when the chlorhexidine was used at a higher concentration in the guinea-pig model *in vivo* (Table 8.1). In this case, whereas the transcervical insertion of PHEMA-coated and uncoated threads led to a significantly greater uterine contamination than in the control group (two-tailed Mann−Whitney U-test, $p < 0.01$), the incorporation of chlorhexidine in the PHEMA coat ensured minimal uterine contamination.

TABLE 8.1. *Summary of uterine wash results*

Group	No. of organisms (cfu/ml) retrieved from uterine washes after 21 days			
	0–100	100–1000	1000–10 000	>10 000
A	8	–	–	–
B	5	3	–	–
C	–	–	3	5
D	8	–	–	–

Data represent the number of animals in each group with cfu/ml within the limits shown. A, Untreated controls; B, nylon monofilaments; C, PHEMA-coated nylon monofilaments; D, chlorhexidine-impregnated, PHEMA-coated, nylon monofilaments.

Conclusions

There are a number of possible strategies by which microbial adhesion and subsequent infection *in vivo* can be influenced. These include informed selection of antibiotic treatment, either therapeutic or prophylactic, biomaterial surface modification and incorporation of antimicrobial agents in medical devices. Although the effects of these approaches can be explored *in vitro*, they may at best provide only qualitative evidence of potential success. Surface-related behaviour and antimicrobial activity are both strongly affected by the host environment, and careful attention must be paid to modelling this in experimental design before extrapolation to the clinical situation.

References

BAYSTON, R. & MILNER, R.D.G. (1981) Antimicrobial activity of silicone rubber used in hydrocephalus shunts after impregnation with antimicrobial substances. *Journal of Clinical Pathology*, **34**, 1057–1062.

BISNO, A.L. & WALDVOGEL, F.A. (1989) *Infections Associated with Indwelling Medical Devices*, American Society for Microbiology, Washington.

BRIDGETT, M.J., DENYER, S.P. & DAVIES, M.C. (1991) The control of staphylococcal adhesion to poly(styrene) surfaces by polymer surface modifications with pluronic surfactants. *Journal of Pharmacy and Pharmacology*, **43**, 45p.

BRIDGETT, M.J., DAVIES, M.C., DENYER, S.P. & ELDRIDGE, P.R. (1992) *In vitro* assessment of bacterial adhesion to Hydromer-coated cerebrospinal fluid shunts. *Biomaterials*, **13**, 411–416.

COSTERTON, J.W. (1984) The etiology and persistence of cryptic bacterial infections: a hypothesis. *Reviews of Infectious Diseases*, **6** (Suppl. 3), 5608.

EVANS, J.E. (1992) *Influence of Antibiotics on Surface Characteristics and Adherence of Staphylococcus epidermidis*. PhD Thesis, Nottingham University.

EVANS-HURRELL, J.A., ADLER, J., DENYER, S.P., ROGERS, T.G. & WILLIAMS, P. (1993) A method for the enumeration of bacterial adhesion to epithellal cells using image analysis. *FEMS Microbiology Letters*, **107**, 77–82.

FRESHNEY, R.F.I. (1986) *Animal Cell Culture; A Practical Approach.* IRL Press, Oxford.

FUNFSTUCK, R., STEIN, G., FUCHS, M., BERGNER, M., WESSEL, G., KEIL, E. & SUSS, J. (1987) The influence of selected urinary constituents on the adhesion process of *Escherichia coli* to human uroepithelial cells. *Clinical Nephrology*, **28**, 244–249.

GARD, P.R., MALHI, J.S. & HANLON, G.W. (1992) Uterine contamination in the guinea-pig following transcervical uterine monofilament insertion. *Gynecological and Obstetric Investigations*, **360**, 1–7.

GORMAN, S.P. (1989) Factors affecting the distribution of micro-organisms adhered to epithelial cells. *Journal of Pharmacy and Pharmacology*, **41**, 119.

GORMAN, S.P. (1991) Microbial adherence and biofilm production. In Denyer, S.P. & Hugo, W.B. (eds.) *Mechanisms of Action of Chemical Biocides: Their Study and Exploitation*, Society for Applied Bacteriology Technical Series, No. 27, pp. 271–295. Blackwell Scientific Publications, Oxford.

GORMAN, S.P., WOOLFSON, A.D. & MACCAFFERTY, D.F. (1991) *Elemental Analysis of Latex and Polymer-Coated Urinary Catheter Encrustation by a Novel Electron Probe and Digimap Technique.* Proceedings of the 10th Pharmaceutical Technology Conference, Bologna, Italy pp. 649–651. Solid Dosage Research Unit, Liverpool.

HUMPHRIES, M., NEMCEK, J., CANTWELL, J.B. & GERRARD, J.J. (1987) The use of graft copolymers to inhibit the adhesion of bacteria to solid surfaces. *FEMS Microbiology Ecology*, **45**, 297–304.

IKEDA, T., YAMAGUCHI, H. & TAZUKE, S. (1986) Self-sterilizing materials. 2. Evaluation of surface antibacterial activity. *Journal of Bioactive and Compatible Polymers*, **1**, 301–308.

ISQUITH, A.J., ABBOTT, E.A. & WALTERS, P.A. (1972) Surface-bonded antimicrobial activity of an organosilicon quaternary ammonium chloride. *Applied Microbiology*, **24**, 859–863.

KHAN, M.A., DENYER, S.P., DAVIES, M.C. & DAVIS, S.S. (1988) Adhesion of *Staphylococcus epidermidis* to modified polystyrene surfaces. *Journal of Pharmacy and Pharmacology*, **40**, 17p.

KINGSTON, D., SEAL, D.V. & HILL, I.D. (1986) Self-disinfecting plastics for intravenous catheters and prosthetic inserts. *Journal of Hygiene*, **96**, 185–198.

KRISTINSSON, K.G. (1989) Adherence of staphylococci to intravascular catheters. *Journal of Medical Microbiology*, **28**, 249–257.

LEE, J., MARTIC, P.A. & TAN, J.S. (1989) Protein adsorption on pluronic copolymer-coated polystyrene particles. *Journal of Colloid and Interface Science*, **131**, 252–266.

LETHEM, M.I., BELL, A.E. & MARRIOTT, C. (1984) The rheological properties of cystic fibrosis sputum. In Lawson, D. (ed.) *Cystic Fibrosis: Horizons*, pp. 332–336. John Wiley, London.

LUDWICKA, A., SWITALSKI, L.M., LUNDIN, A., PULVERER, G. & WADSTROM, T. (1985) Bioluminescent assay for measurement of bacterial attachment to polystyrene. *Journal of Microbial Methods*, **4**, 169–177.

LYDON, M.J., MINETT, T.W. & TIGHE, B.J. (1985) Cellular interactions with synthetic polymer surfaces in culture. *Biomaterials*, **6**, 396–402.

SVANBORG-EDEN, C. (1978) Attachment of *Escherichia coli* to human urinary tract epithelial cells. *Scandinavian Journal of Infectious Diseases*, **15** (Suppl.), 1–69.

TADROS, T.F. & VINCENT, B. (1980) Influence of temperature and electrolytes on the adsorption of polyethylene oxide, polypropylene oxide block co-polymer on polystyrene latex and on the stability of the polymer-coated particles. *Journal of Physical Chemistry*, **84**, 1575–1580.

TAVENDALE, A., JARDINE, C.K.H., OLD, D.C. & DUGUID, J.P. (1983) Haemagglutinins and adhesion of *Salmonella typhimurium* to HEp2 and HeLa cells. *Journal of Medical Microbiology*, **16**, 371–380.

VALENTIN-WEIGAND, P., CHAATWAL, G.S., & BLOBEL, H. (1987) A simple method for quantitative determination of bacterial adherence to human and animal epithelial cells. *Microbiology and Immunology*, **31**, 1017–1023.

WORLD HEALTH ORGANIZATION (1987) *Mechanism of Action, Safety and Efficacy of Intrauterine Devices*. WHO Technical Report Series 753. WHO, Geneva.

WYATT, J.E., POSTON, S.M. & NOBLE, W.C. (1990) Adherence of *Staphylococcus aureus* to cell monolayers. *Journal of Applied Bacteriology*, **69**, 834–844.

9: Methods for the Study of Dental Plaque Formation and Control

M. Addy, M.A. Slayne and W.G. Wade

Department of Periodontology, Dental School, University of Wales College of Medicine, Heath Park, Cardiff CF4 4XN, UK

Dental plaque may be defined as the non-calcified bacterial deposit that accumulates on teeth, is free from macroscopic food debris and cannot be removed by vigorous rinsing with water. It is the association of plaque with dental caries (Axelsson & Lindhe, 1977), chronic gingivitis (Ash *et al.*, 1964; Loe *et al.*, 1965), and chronic periodontitis (van Palenstein Helderman, 1981) that has led to extensive research into plaque formation and its control.

Regular toothcleaning, by a variety of means, is performed by most, if not all peoples of the world (Frandsen, 1986) and oral hygiene methods can be broadly divided into mechanical and chemical, although the two are frequently combined. The greatest potential impact of effective plaque control would be in the prevention of gingivitis rather than caries (Frandsen, 1986). The purpose of this chapter is to describe the methods used to study bacterial attachment and plaque formation, to record the presence of gingivitis and to evaluate plaque control agents.

Models of Bacterial Attachment and Plaque Accumulation on Teeth

When choosing a model system for studies *in vitro*, it is essential to select the most appropriate system to represent the *in vivo* environment. It has been firmly established that the attachment of bacteria to oral surfaces depends on a variety of forces and physical interactions between bacteria and substrata (Gibbons, 1980, 1984). Compatible surface free energies (sfe) (van Pelt *et al.*, 1985), hydrophobicities (Rosenberg *et al.*, 1983) and specific attachment sites are important factors in bacterial adherence.

Microbial Biofilms:
Formation and Control

Hydroxyapatite adherence assay

Hydroxyapatite (HA), the major matrix component of tooth enamel, is commonly used for *in vitro* studies of bacterial attachment to teeth. Clark *et al.* (1978) first described the HA adherence model, which has since been used and modified by numerous workers (Appelbaum *et al.*, 1979; Nesbitt *et al.*, 1982; Eifert *et al.*, 1984; Wyatt *et al.*, 1987) for studies of the important primary plaque-former *Streptococcus sanguis* (Carlsson, 1970; Nyvad & Kilian, 1990).

S. *sanguis* has been used in our studies for the evaluation of antiadherence compounds. The method described here is based on the optimized *S. sanguis/* saliva-coated hydroxyapatite assay (SHA) of Eifert *et al.* (1984) and the method of Wyatt *et al.* (1987).

1 Place spheroidal HA in 20 mg amounts into 1.5 ml Eppendorf tubes and add 1 ml buffer (1 mmol/l NaH_2PO_4, 1 mmol/l $Na_2HPO_4 . 2H_2O$, 5 mmol/l KCl, 1 mmol/l $CaCl_2 . 2H_2O$, pH 6.75) to each tube.

2 Shake the tubes and equilibrate for 20 h at 4°C, then remove the buffer from the HA samples.

3 Pool saliva collected from six individuals, and vorter-mix for approximately 10 s.

4 Centrifuge the pooled saliva for 15 min at 1500 g and add 1 ml of the supernatant fluid to each HA sample.

5 Incubate the tubes on a rotating wheel at room temperature for 60 min, remove the saliva and wash the SHA samples three times in buffer.

6 Treat SHA samples for 1 min with test solutions on a rotating wheel, discard the test solution and wash the samples three times in buffer. This treatment step may be repeated if a sequential effect of consecutive treatments is required.

7 Bacteria are grown in 20 ml brain heart infusion broth (BHI) with 40 µl (2 µCi/ml) methyl-(^{3}H)-thymidine (Amersham) at 37°C for 20 h.

8 Centrifuge the culture for 30 min at 1500 g. Wash the bacterial pellet twice in buffer and resuspend in 20 ml buffer.

9 Sonicate the bacterial suspension in a water bath for 2 min, shaking at 1 min intervals, to disperse the bacterial chains. Adjust suspension to approximately 10^9 cfu/ml (absorbance = 1.2 at 550 nm).

10 Add 1 ml of radiolabelled bacterial suspension to each tube of treated SHA. Incubate the tubes on a rotating wheel for 90 min at room temperature.

11 Remove the bacterial suspensions and wash the SHA samples three times in buffer.

12 Add 0.5 ml buffer and 0.5 ml 1.25 mol/l HCl to each tube and mix the tubes for 15 min on a rotating wheel to dissolve the HA spheres.

13 Add 0.3 ml of each solution, in duplicate, to 2.7 ml OptiPhase 'HiSafe' 3 scintillation cocktail (Pharmacia-LKB) in scintillation vials. Shake the vials and scintillation-count for β-emission to 10 000 counts within 50 min. Background counts are subtracted and the results of replicates are expressed as mean values. Counts can be converted to percentage adherence values by means of a control treatment to represent 100% bacterial adherence.

The HA adherence model is a good method for quantifying bacterial adherence, although it cannot be used in qualitative studies of bacterial adherence or plaque accumulation. 'Artificial mouths' have been used to mimic the *in vivo* environment for studying dental plaque (Pigman *et al.*, 1952; Russell & Coulter, 1975). However, such systems cannot be repeatedly sampled without disturbance of the system, and they lack a means of maintaining a reproducible, defined environment.

Continuous culture

The chemostat (Marsh *et al.*, 1983) provides a reproducible means for observing the growth of mixed cultures of organisms in the oral environment. The different phases of bacterial cell growth and the complex interspecies population interactions in the oral cavity can be simulated in the chemostat. Multiple inoculations into continuous culture can be made with varied controlled nutrient supplies. McKee *et al.* (1985) developed a chemostat method for the reproducible simultaneous cultivation of nine species of plaque bacteria for up to 12 weeks. Bacteria were grown in batch culture to late exponential phase before inoculation into a chemostat containing a glucose-limited medium modified from the *Bacteroides melaninogenicus* medium of Shah *et al.* (1976). The immersion of surfaces into such a chemostat model provides a means for studying bacterial adherence. Keevil *et al.* (1987) studied the attachment of oral bacteria and plaque development on dental acrylic slabs. Dental plaque was grown by volunteers for 72 h and was used to inoculate a glucose-limited medium at a dilution rate of 0.05/h at pH 7 in a chemostat and acrylic slabs were immersed in this steady-state continuous culture which supported growth for 21 days. Microbiological analysis indicated that the development of a complex microbial community, typical of dental plaque, was largely initiated by the adherence of fusiforms and streptococci.

Perspex adherence model

Perspex can be used as a model for *in vitro* studies of bacterial adherence to acrylic dentures. Perspex has a low sfe and so bacteria of low sfe will adhere most strongly (van Pelt *et al.*, 1985). *Streptococcus oralis* is a species with a low sfe associated with primary plaque formation (Nyvad & Kilian, 1990), and has been used in our studies in the perspex adherence model to assess inhibition of bacterial adherence.

1   Immerse perspex slabs ($10 \times 10 \times 3$ mm) in 3 ml of treatment solution in 5 ml bottles for 5 min and rinse five times in flowing water. Repeat this process if sequential treatments are required.

2   Lay the slabs in 5 ml Petri dishes, cover with 4 ml bacterial suspension (10^8 cfu/ml) and leave for 1 h. Rinse the slabs five times in flowing water.

3   Immerse the slabs in 0.5% crystal violet solution for 15 s. Rinse the slabs five times in flowing water and blot dry. Crystal violet stains the bound bacteria purple.

4   Quantify the staining and reduction of staining by reading absorbance values at 530 nm.

Laboratory Methods
for the Bacteriological Examination of Plaque

Specimen collection

This is an extensive topic and a detailed discussion of different methods is beyond the scope of this review, but can be found in Tanner & Goodson (1986).

In general, plaque material should be collected with an instrument such as a curette or a spoon excavator and immediately placed in reduced transport medium. Swabs are unacceptable for oral specimens because of the effects of drying and exposure of the anaerobic portion of the flora to oxygen. For similar reasons, specimens should be handled and incubated within an anaerobic cabinet (Aranki *et al.*, 1969).

The use of a transport medium is essential to protect the bacteria in the sample from atmospheric oxygen. Reduced transport fluid, a balanced salt solution containing dithiothreitol for its reducing properties, has been shown to be superior to other media for the transport of dental plaque (Syed & Loesche, 1972). However, a simple medium (Bowden & Hardie, 1971) has been found to give satisfactory results in our studies (Wade *et al.*, 1991, 1992). This medium consists of tryptone 1 g, yeast extract 0.5 g, glucose 0.1 g, cysteine HCl 0.1 g, horse serum 2 ml and distilled water to 100 ml. The medium (pH 7.4) is sterilized by filtration through a 0.22-μm pore size disposable filter membrane and stored in a universal container with a loosened cap in an anaerobic cabinet in an atmosphere of 80% N_2, 10% CO_2 and 10% H_2.

Culture

Cultural methods remain the mainstay of plaque flora examinations. The media used are, in general, similar to those used in medical microbiology but modified for oral use. A non-selective solid medium for the cultivation of oral bacteria should be a blood agar with a good quality base such as Blood Agar base No. 2 (LabM, Bury, UK) supplemented with horse or sheep blood. The anaerobes found in plaque are often fastidious and slow-growing and require media such as Fastidious Anaerobe agar (LabM), which has been shown to be superior to other commonly used media for the cultivation of oral anaerobes (Heginbothom *et al.*, 1990). Some useful selective media with their target organisms are given in Table 9.1.

Dark-field microscopy

Dark-field microscopy has been widely used to examine plaque by means of the distribution of bacterial morphotypes, and differences have been demonstrated between the flora at healthy and diseased sites (Listgarten & Hellden, 1978; Wilson *et al.*, 1986). However, this technique is limited in that many bacterial species that differ markedly physiologically and biochemically exhibit similar cellular morphologies.

Clinical Methods
for the Measurement of Plaque Formation

Plaque identification

Plaque is, for the most part, not visible on casual examination of the teeth, although on close examination it can be seen as a creamy white deposit. Detection may be aided by the use of a dental probe or, more usually, by staining with a disclosing solution such as erythrosine.

TABLE 9.1. *Some useful selective culture media*

Medium	Target organisms	Reference
Tryptone, yeast extract, cystine agar (TYC)	Oral streptococci	de Stoppelaar *et al.* (1967)
TYC + sucrose and bacitracin (TYCSB)	*Streptococcus mutans* group	van Palenstein Helderman *et al.* (1983); Wade *et al.* (1986)
Kanamycin blood agar (KBA)	*Prevotella, Porphyromonas*	Zambon *et al.* (1981)
Crystal violet, erythromycin agar (CVE)	*Fusobacterium* spp.	Walker *et al.* (1979)
Tryptic soy, bacitracin, vancomycin agar (TSBV)	*Actinobacillus actinomycetemcomitans*	Slots *et al.* (1982)

Plaque scoring and indices

Plaque scoring is based on dimension (e.g. height or area) or quantity (e.g. dry or wet weight). With the possible exception of plaque weight measurement, most indices are subjective in nature. Numerous plaque indices have been described and only those commonly employed and in regular use by the authors will be detailed here.

1   *Debris index (Greene & Vermillion, 1960).* This is derived from area scores of individual buccal and lingual surfaces of the teeth. After plaque has been made visible by rinsing with a food dye, usually erythrosine, the following scores are assigned: 0 = no disclosed plaque; 1 = disclosed plaque up to one-third of the tooth surface; 2 = disclosed plaque from one-third to two-thirds of the surface; 3 = disclosed plaque on more than two-thirds of the surface. The index is determined by adding the scores and dividing either by the number of teeth or surfaces giving a range of 0−6 or 0−3, respectively.

2   *Turesky et al. (1970) modification of the Quigley & Hine (1962) plaque index.* This scores buccal and lingual disclosed plaque on a 0−5 scale where 0 = no plaque, 1 = isolated flecks of plaque near the gingival margin, 2 = a 1-mm band of plaque at the gingival margin, 3 = up to one-third of the surface covered with plaque, 4 and 5 = 2 and 3 respectively on the Debris Index (see above). The index is derived as described for the Debris Index.

3   *Plaque area (Addy et al. (1983) modification of the Shaw & Murray (1977) stain index).* After disclosing the plaque, the observed outline is transcribed on to fourfold magnified diagrammatic representations of the buccal and lingual surfaces of the teeth. All teeth may be used, but more usually only the buccal surfaces of upper and lower incisors, canines and premolars are employed. The plaque areas are then measured by planimetry with a graphics tablet attached to a microcomputer. Total or mean plaque area may be calculated. Variations of the method are the recording of disclosed plaque areas directly from colour photographs, in which case the area is expressed as a percentage of the respective tooth area.

4   *Plaque index (Silness & Loe, 1964).* This is semiquantitative and concerned primarily with plaque adjacent to the gingival margin. Disclosing solutions may be used, but more usually the detection is by eye and the use of a probe. The probe is drawn gently along the gingival margin in contact with the tooth. The scores are derived on the basis of 0 = no plaque; 1 = plaque not visible but detected on the probe; 2 = plaque visible; 3 = abundant plaque deposits.

Clinical Measurement of Gingivitis

Gingivitis identification

If plaque is allowed to accumulate unhindered gingivitis will develop after several days (Loe *et al.*, 1965). Chronic gingivitis, uncomplicated by modifying factors, has the features of redness, a tendency to bleed in response to minor trauma or pressure and the presence of slight swelling. These signs are exploited to varying degrees in methods used to measure gingivitis. In addition, the transudation of tissue fluid from the gingival connective tissue — usually termed gingival crevicular fluid (GCF) — can be measured. Pain is rarely, if ever, a presenting symptom of chronic gingivitis.

Gingivitis scoring and indices

As with plaque, a large number of indices have been described to record gingivitis. These fall into two groups, namely, non-invasive and invasive, with the possibility of combining the two. Non-invasive indices rely on colour and morphological change in the gingival tissues or the recording of GCF over time. Invasive indices record the tendency of the gingivae to bleed on gentle probing, either alone or together with morphological assessments. As with plaque, methods for measuring gingival inflammation which are in common usage and frequently employed by the authors will be described.

Non-invasive

Modified gingival index (Lobene et al., 1986). Visually assessed grades are given where 0 = absence of inflammation, 1 = mild inflammation (slight change in colour with little change in shape of part but not all the marginal or papillary unit), 2 = mild inflammation (criteria as 2 but involving the entire marginal or papillary unit), 3 = moderate inflammation (glazing, redness, oedema, and/or hypertrophy of the marginal or papillary unit), 4 = severe inflammation (marked redness, oedema and/or hypertrophy of the marginal or papillary unit, spontaneous bleeding, congestion or ulceration).

Gingival crevicular fluid flow measurements

The rate of flow of gingival crevicular fluid correlates with the degree of gingival inflammation and can be measured by the filter paper method of Egelberg (1964). The gingival sites to be sampled are isolated from saliva with cotton wool and the gingivae dried with a gentle stream of air. Filter paper strips (1.5 × 10 mm) are placed at or just into each crevice and left *in situ* for 3 min. The strips are allowed to bench dry and then placed into 0.2% ninhydrin solution. The length of the blue-black stain along the paper is measured with a ruler, preferably magnified. This method has been modified (Garnick *et al.*, 1979), such that the fluid taken up by strips after only a few seconds is determined electronically by recording the change in capacitance of the strip. The instrument which is commercially available for the purpose is the Periotron (Harco Medical Electronic Devices, Irvine, California, USA).

Invasive

1 *Gingival index (GI; Loe & Silness, 1963).* The criteria for scoring are: 0 = absence of inflammation; 1 = mild inflammation (slight change in colour and little change in texture); 2 = moderate inflammation (moderate glazing, redness, oedema and hypertrophy; bleeding on pressure); 3 = severe inflammation (marked redness and hypertrophy; tendency to spontaneous bleeding; ulceration).

2 *Sulcus bleeding index (SBI; Muhlemann & Son, 1971).* The criteria for this index are: 0 = healthy appearance of the papilla (P) and marginal gingiva (M), not bleeding on sulcus probing; 1 = apparently healthy P and M showing no change in colour and no swelling, but bleeding from the sulcus on probing; 2 = bleeding on probing and change in colour due to inflammation, no swelling or macroscopic oedema; 3 = bleeding on probing and change in colour and slight oedema; 4 = bleeding on probing and change in colour and/or obvious swelling; 5 = bleeding on probing and spontaneous bleeding and change in colour. Marked swelling with or without ulceration.

3 *Bleeding on probing (BOP).* This is a simple binary index in which, irrespective of colour change, the presence or absence of bleeding is noted after gentle probing (*ca.* 25 g pressure) into the gingival crevice. Modifications may be used based on the time to bleed (Cowell *et al.*, 1975) where 0 = no bleeding up to 30 s after probing, 1 = bleeding on probing within 30 s, 2 = bleeding immediately on probing, 3 = spontaneous bleeding without probing. NB: to standarize the probing, fixed pressure probes may be employed, for example to deliver a 25 g force through a 0.5 mm probe tip.

Scoring sites

Many of the indices used to score plaque and, in particular, gingivitis allow considerable flexibility of choice in the teeth or sites around each tooth from which scores are taken. Thus, at the extremes, studies of plaque and gingivitis have used a few teeth or up to six sites around each tooth. However, plaque accumulation, and therefore gingivitis, differs on different sides of the mouth, between the two arches, by tooth position and by tooth surface (Addy *et al.*, 1987). Where possible, scores for plaque and gingivitis should be recorded from teeth and surfaces which are representative of all these variables. As a suggested guide, this could be achieved by taking single measurements from the buccal and lingual surfaces of all fully erupted upper and lower permanent teeth.

Methods to Evaluate Plaque Control Agents

Plaque formation is a well ordered, continuous and dynamic process which, given time, will lead to gingivitis (Loe *et al.*, 1965). Methods for interrupting this sequence of events may be proposed, namely, inhibition of bacterial attachment by the use of antiadhesive agents, prevention of bacterial proliferation on the tooth surface with antimicrobials, and regular removal of plaque from the tooth surface by mechanical or chemical means.

Plaque control and oral hygiene practices have been the subject of frequent international symposia and reviews by several authors (Hull, 1980; Addy, 1986; Kornman, 1986). Guidelines for the testing of oral hygiene products have been established by various bodies, including the International Dental Federation (FDI), the American Dental Association (ADA) and the American Food and Drug Administration (FDA). Fundamental to the issue of evaluation must be agreement about the definition of the widely used term '*antiplaque*'. At least by implication, the guidelines referred to above support the suggestion that antiplaque should mean an effect on plaque that results in a reduction in gingivitis and/or caries (Addy *et al.*, 1983).

Clearly, there are numerous agents and infinite variations of possible formulations which make their evaluation impossible in long-term home-use trials. Thus, research and development of oral hygiene products should employ step-by-step study methods. The eventual aim is to compile a body of data to support the efficacy of a final formulation to be used by the general public. The study methods are performed *in vitro* and *in vivo*. Animal testing has also been employed, but will not be discussed here. Indeed, except for toxicological and safety testing, the need and value of animal data on oral hygiene products may be questioned on scientific and moral grounds. Most of the methods to be described, particularly *in vitro*, relate to the assessment of chemical rather than mechanical plaque control agents. However, many of the clinical methods can be adapted to study mechanical cleaning agents and regimens.

Methods in vitro

The numerous variables and the dynamic nature of the oral cavity are difficult if not impossible to mimic in the laboratory. Laboratory methods alone are not reliable predictors of plaque effects *in vivo*. Nevertheless, they can provide useful information, including: (i) the activity and availability of a specific ingredient within a formulation. Even established products have been shown, by laboratory and later clinical tests, to have little or no availability of the active ingredient; (ii) negative data, i.e. an agent without activity *in vitro*, for example antimicrobial action, is unlikely to exhibit such activity *in vivo*. Some enzymes are perhaps an exception; (iii) comparative action of the same ingredient in different formulations; (iv) possible mode or modes of action of an agent.

Tests of inhibition

These tests are analogous to those used in medical microbiology to determine appropriate antimicrobials for the treatment of infection. Their applicability to the control of normal oral flora, where the local pharmacokinetics are poorly understood and elimination of the normal flora is in any case undesirable, is unclear. The methods used include agar diffusion and broth and agar dilution techniques. There are theoretical objections to most, if not all, of these methods.

Agar diffusion methods are commonly used in clinical diagnostic microbiology to assess the susceptibility of bacterial isolates to antimicrobial drugs, usually to determine the appropriate antimicrobial for the treatment of infection. The test organism is spread over the surface of the plate and either an impregnated paper disc is placed on to the plate, or a well is cut in the plate into which a solution of the antimicrobial is placed. The diffusion of the antimicrobial into the agar gives rise to a zone where the growth of the organism is inhibited; the diameter of the zone is related to the susceptibility of the organism. Control organisms are always tested in parallel and comparison of the zones obtained with test and control strains gives information about the susceptibility of the test organism. Zone diameter is not a function of the intrinsic antimicrobial activity of a compound, but depends mainly on the rate of diffusion through the agar. Thus these tests compare the susceptibility of bacteria: they should not be used to compare antimicrobials for oral or any other use.

Agar and broth dilution methods can be criticized because they employ nutrients, including proteins and polypeptides. The ionic nature of most antiseptics makes it likely that they are partially neutralized in the presence of proteins with which they form complexes. Therefore, these tests may underestimate the antimicrobial activity of a given compound. Furthermore, the degree of this underestimation is difficult, if not impossible, to determine.

Agar dilution method for testing oral hygiene products

This method is applicable to both toothpastes (Moran *et al.*, 1988) and mouthwashes (Wade & Addy, 1989).

1　Slurries of toothpastes are prepared by mixing equal volumes of the pastes and distilled water and mixing on a rotary mixer for 1 h.

2　The slurries are then centrifuged for 30 min at 2000 g and the supernatant fluid is collected.

3　The supernatants or mouthwashes are doubly diluted in distilled water and 2 ml of each dilution is added to 18 ml of molten Wilkins–Chalgren agar (Difco, East Molesey, UK) to give final dilutions typically between 1:20 and 1:20 480.

4　Test organisms — normally a panel of oral bacteria — are suspended in phosphate buffer and inoculated on to each plate with a multipoint inoculator (LabM).

5　The plates are then incubated anaerobically at 37°C for 72 h.

6　The maximum inhibitory dilution is the highest dilution that permits no growth, growth of a single colony, or a fine visible haze.

Tests of bactericidal activity

It can be argued that tests of the ability of a substance to inhibit the growth of bacteria might be more relevant to the control of plaque, than tests for bactericidal properties. However, a reduction in the numbers of plaque or salivary bacteria as an initial response to treatment followed by bacteriostasis might be a useful result. Indeed, this scenario has been suggested as the mode of action of chlorhexidine in plaque control (Jenkins *et al.*, 1988).

Minimum bactericidal concentrations (MBC)

This can be estimated by performing an MIC (minimum inhibitory concentration) determination by the broth dilution method and then culturing the broths. The lowest concentration in which there is no bacterial growth is the MBC. These tests attract the same criticisms as tests of inhibition, in terms of the unknown influence of culture media. In addition, because of salivary flow, the time taken to kill the organisms is obviously important.

Kill curves

Kill curves are constructed by exposing the test organism to the antimicrobial and then removing portions of the suspension at various time periods and determining the viable count. The results are expressed either as the percentage of organisms killed in a given time period, or the time taken to kill a given percentage of organisms. These tests are normally conducted in distilled water and thus do not suffer from protein interactions. This has the advantage that the tests can be modified by, for example, the introduction of saliva into the system, and the effect determined.

Methods in vivo

Retention studies

As a general rule, the longer the action of an agent persists in the mouth, the more effective it will be against plaque. Most individuals can only retain preparations in the mouth for periods measured in minutes. Additionally, the dynamic nature of the oral cavity results in the rapid dilution, removal and even inactivation of agents. Compounds which can adsorb to oral surfaces and maintain activity have so far proved the most effective antiplaque agents (Hull, 1980; Addy, 1986; Kornman, 1986; Jenkins *et al.*, 1988). This so-called substantivity (Kornman, 1986) of a compound can, in part, be determined from retention studies. Two methods have been used alone or concurrently.

Buccal retention test. Volunteers receive a known dose of an agent, usually as a rinse, but toothbrushing with a paste can be employed. Swallowing is avoided during a timed delivery period, typically of 1 min. All expectorates are collected, usually including a post-treatment water rinse, and the residual agent determined by appropriate analytical techniques (Jensen & Christensen, 1971). Some studies have employed radiolabelled compounds for this method (Bonesvoll *et al.*, 1974). The major limitation of this method is the failure to differentiate between adsorbed and absorbed or swallowed compound.

Saliva/plaque levels with time. After delivery and expectoration, levels of the agent in saliva or plaque are measured at time intervals, usually to a maximum of 24 h (Bonesvoll *et al.*, 1974; Gilbert & Williams, 1987; Nabi *et al.*, 1989). The limitation of this method is that retention is only one aspect of substantivity and is not synonymous with activity. The analytical methods do not differentiate between active and inactive agent.

Antimicrobial methods

Measurement of the effects of antimicrobial agents on the oral flora provides evidence of substantivity. The method first used for chlorhexidine (Schiott *et al.*, 1970) measures the magnitude and duration of salivary bacterial count reductions following a single challenge with the antimicrobial agent.

Volunteers are requested to rinse with a measured volume (i.e. a given dose) of an agent or product for a supervised period, typically 60 s. In the case of toothpastes, delivery can be by toothbrushing or a slurry rinse (e.g. 3 g in 10 ml water) (Jenkins *et al.*, 1990). Samples of unstimulated saliva are taken before treatment and at time intervals after treatment e.g. 30, 60, 180, 300 and 420 min. For most antimicrobials, times in excess of 420 min are rarely required because salivary counts return to baseline levels within this time. Nevertheless, accepting the logistic problems, times can and have been extended (Schiott *et al.*, 1970). Saliva samples are processed immediately by vortex mixing and are then serially diluted in 1 ml volumes to 1 in 10000 and 1 in 100000 in phosphate-buffered saline (PBS). Aliquots of the dilutions are spread over blood agar plates with a spiral plater and plates are incubated anaerobically at 37°C for 48 h. Counts are finally determined from the grid system for the spiral plater. In the absence of a spiral plater, 0.1 ml volumes may be spread on plates and counts of the whole plate made after suitable incubation (Roberts & Addy, 1981).

Washout periods of a few days are usually sufficient to prevent carryover effects. When possible, a negative control cell, such as water, and a positive control cell, namely chlorhexidine, should be included. The method is labour-intensive but causes little inconvenience to volunteers. Multiple product comparisons are therefore possible.

Experimental plaque methods

Short-term plaque regrowth studies are, perhaps, the most commonly used clinical experiments for the screening of oral hygiene products. The aim is to determine the chemical action of the agent separately from mechanical tooth-cleaning. Originally the method was intended for mouthwashes, but it was modified for toothpastes by delivering the preparation on to the teeth in trays or as water slurry rinses (Gjermo & Rolla, 1970; Addy *et al.*, 1983; Saxton, 1986) The duration of the test can vary from 16 h to several days (Harrap, 1974; Addy *et al.*, 1983; Saxton *et al.*, 1987). The validity of very short-term studies has been questioned, primarily because of the possibility of producing false-positive or alpha errors (Addy *et al.*, 1983).

A study period of 4–5 days — Monday to Friday, for example — has commonly been used. On day 1 all subjects are rendered plaque- and calculus-free by scaling and polishing of the teeth. Normal toothcleaning is suspended and subjects apply the formulation for a timed period, as stated usually by rinsing, and for the prescribed number of times per day. When possible, to ensure compliance, the regimens should be supervised, although it must be accepted this may not be possible. On day 5 plaque is scored by one or more of the indices described. Normal oral hygiene is then resumed. In crossover studies washout periods of 48–72 h are usually sufficient.

Experimental gingivitis methods

This method is a modification (Loe & Schiott, 1970) of the 'experimental gingivitis in man' protocol (Loe *et al.*, 1965) ued to study plaque formation and its aetiological role in gingivitis. Study periods range from 12 to 28 days, and record the development of plaque and gingivitis in healthy volunteers. At commencement, gingivitis levels should be close to zero and plaque scores are reduced to zero by prophylaxis. In crossover designs, washout periods of at least 14–20 days should be provided to prevent gingivitis carryover. The formulations under test are used according to the respective regimen — supervised if possible — and in the absence of normal toothcleaning methods. At the end of each experimental period, plaque and gingivitis are recorded by one or more of the indices described. Volunteers should then receive a prophylaxis and return to normal toothcleaning habits. The number of formulations that can be tested by this method at any one time are limited: in parallel group designs the number of volunteers is the limiting factor; in crossover studies time and volunteer compliance have to be considered.

Toothbrushing home-use methods

Most agents for the control of plaque and, therefore, gingivitis are, for the foreseeable future, likely to be used by the general public as adjuncts to, rather than replacements for, mechanical toothcleaning procedures. Therefore, it is essential that efficacy can be demonstrated when used with toothbrushing. Such clinical trials may extend over periods of a few days to 6 months or longer (Clerehugh *et al.*, 1989; Stephen *et al.*, 1990), depending on the stage of product research and development. Nevertheless, proven efficacy in a 6-month home-use study is eventually necessary, if 'antiplaque' activity is to be claimed for a formulation or product (Council on Dental Therapeutics, 1986). Most study methods in this category use parallel groups of subjects but, depending on the aim, can be of two slightly different designs. First are studies which evaluate the therapeutic value of an agent or oral hygiene regimen (Johansen *et al.*, 1975). In this case, subjects are entered into the trial only if they have a minimum level of gingivitis, e.g. mean GI equal to or greater than 1. The subjects may receive baseline prophylaxis (this is not essential) and then use the allocated formulation and/or regimen for the prescribed period. During and at the end of the trial subjects are scored for plaque and gingivitis according to the appropriate indices. Second are studies that evaluate the preventive role of agents or regimens (Stephen *et al.*, 1990). Subjects in these cases enter with a level of established disease and receive professional treatment and oral hygiene advice during a prestudy period. Those achieving an improved baseline gingival health are entered into the study and scored for plaque and gingivitis at appropriate times during the study and/or at the end of the study.

Study design and analysis

Study designs can be most varied, but within the context of those described here certain general points can be made. Most clinical trials are based on the null hypothesis that no significant difference exists between the active formulations and the control or placebo formulation. The important factors that determine whether a statistically significant difference, if present, can be shown are the size of the expected difference, the measurement methods employed, the number of subjects recruited to the study, and — related to the latter — whether a parallel or a crossover design is used. For these reasons it is advisable to seek professional statistical advice whenever possible at the planning stage of any study.

As with all studies, plaque control trials should conform to sound scientific principles. Thus, they should be at least single- or operator-blind and, when feasible, double-blind. Subjects should be randomly allocated to treatments or orders of treatments. The study should be suitably controlled, although the subject of controls is still open to debate. There are several possibilities for the choice of controls, namely, a placebo, a formulation minus the active ingredient, a benchmark commercial product and a positive control. Clearly, more than one — if not all — of these controls can be used in studies. Placebos for potential plaque inhibitors are of value particularly at the early screening stage. 'Minus active' controls are necessary to demonstrate the efficacy of the active ingredient, particularly when contained in a complex formulation such as a toothpaste. Benchmark controls are particularly valuable at some stage — preferably the final stage — of product evaluation. Oral hygiene products, particularly toothpastes, are not new entities and it would appear logical, therefore, to demonstrate that a new formulation is at least as effective for the control of plaque and gingivitis as products already available to the public. The positive control most commonly used in the evaluation of plaque inhibitors is chlorhexidine, which is considered the most effective antiplaque agent and has yet to be superseded. Chlorhexidine (0.2%) is typically used as a 10 ml rinse, or the equivalent dose, twice per day as one of the cells of the antimicrobial, plaque regrowth or experimental gingivitis types of investigation.

The analysis of plaque and gingivitis trials has been reviewed (Chilton & Fleiss, 1986). Importantly, data collected according to most plaque and gingivitis indices is non-parametric, with a non-Gaussian distribution, and requires the appropriate statistical tests (see Chapter 19). On occasions normally distributed data or suitably transformed data may be analysed by parametric methods.

References

ADDY, M. (1986) Chlorhexidine compared with other locally delivered antimicrobials. A short review. *Journal of Clinical Periodontology*, 13, 957−964.

ADDY, M., WILLIS, L. & MORAN, J. (1983) The effect of toothpaste and chlorhexidine rinses on plaque accumulation during a 4-day period. *Journal of Clinical Periodontology*, 10, 89−98.

ADDY, M., GRIFFITHS, G., DUMMER, P.M.H., HUNTER, M.L., KINGDON, A. & SHAW, W.C. (1987) The distribution of plaque and gingivitis and the influence of toothbrushing had in a group of South Wales 11−12 year old children. *Journal of Clinical Periodontology*, 14, 564−572.

APPELBAUM, B., GOLUB, E., HOLT, S.C. & ROSAN, B. (1979) *In vitro* studies of dental plaque formation: adsorption of oral streptococci to hydroxyapatite. *Infection & Immunity*, 25, 717−728.

ARANKI, A., SYED, S.A., KENNEY, E.B. & FRETER, R. (1969) Isolation of anaerobic bacteria from human gingiva and mouse cecum by means of simplified glove box procedure. *Applied Microbiology*, 17, 568−576.

ASH, M., GITLIN, B.N. & SMITH, N.A. (1964) Correlation between plaque and gingivitis. *Journal of Periodontology*, 35, 425−429.

AXELSSON, P. & LINDHE, J. (1977) The effect of plaque control programme on gingivitis and dental caries in school children. *Journal of Dental Research*, 56, 142−153.

BONESVOLL, P., LOKKEN, P., ROLLA, G. & PAUS, P.N. (1974) Retention of chlorhexidine in the human oral cavity after mouthrinses. *Archives of Oral Biology*, 19, 209−212.

BOWDEN, G.H. & HARDIE, J.M. (1971) Anaerobic organisms from the human mouth. In Shapton, D.A. & Board, R.G. (eds.) *Isolation of Anaerobes*, pp. 177−205. Academic Press, London.

CARLSSON, J., GRAHNEN, H., JONSSON, G. & WIKNER, S. (1970) Establishment of *Streptococcus sanguis* in the mouth of infants. *Archives of Oral Biology*, 15, 1143−1148.

CHILTON, N.W. & FLEISS, J.L. (1986) Design and analysis of plaque and gingivitis clinical trials. *Journal of Clinical Periodontology*, 13, 400−406.

CLARK, W.B., BAMMAN, L.L. & GIBBONS, R.J. (1978) Comparative estimates of bacterial affinities and adsorption sites on hydroxyapatite surfaces. *Infection and Immunity*, 19, 846−853.

CLEREHUGH, V., WORTHINGTON, H. & CLARKSON, J. (1989) The effectiveness of two test dentifrices on dental plaque formation: a one week clinical study. *American Journal of Dentistry*, 2, 221−224.

COUNCIL ON DENTAL THERAPEUTICS (1986) Guidelines for acceptance of chemotherapeutic products for the control of supragingival dental plaque and gingivitis. *Journal of the American Dental Association*, 112, 529−532.

COWELL, C.R., SAXTON, C.A., SHEIHAM, A. & WAGG, B.J. (1975) Testing therapeutic measures for controlling chronic gingivitis in man: a suggested protocol. *Journal of Clinical Periodontology*, 2, 231−240.

DE STOPPELAAR, J.D., VAN HOUTE, J. & DE MOOR, C.E. (1967) The presence of dextran-forming bacteria resembling *Streptococcus bovis* and *Streptococcus sanguis* in human dental plaque. *Archives of Oral Biology*, 12, 1199−1201.

EGELBERG, J. (1964) Gingival exudate measurements for evaluation of inflammatory changes of the gingivae. *Odontologisk Revt*, 15, 381−398.

EIFERT, R., ROSAN, B. & GOLUB, E. (1984) Optimization of an hydroxyapatite adhesion assay for *Streptococcus sanguis*. *Infection and Immunity*, 44, 287−291.

FRANDSEN, A. (1986) Mechanical oral hygiene practices. State of the science review. In Loe, H. & Kleinman, D.V. (eds.) *Dental Plaque Control Measures and Oral Hygiene Practices: Proceedings From a State of the Science Workshop*, pp. 93−116. IRL Press, Oxford.

GARNICK, J.J., PEARSON, R. & HARRELL, D. (1979) The evaluation of the periotron. *Journal of Periodontology*, **50**, 424–426.

GIBBONS, R.J. (1980) Adhesion of bacteria to surfaces of the mouth. In Berkeley, R.C.W., Lynch, J.M., Melling, J., Rutter, P.R. & Vincent, B. (eds.) *Microbial Adhesion to Surfaces*, pp. 351–388. Ellis Horwood, Chichester.

GIBBONS, R.J. (1984) Adherent interactions which may affect microbial ecology in the mouth. *Journal of Dental Research*, **63**, 378–385.

GILBERT, R.J. & WILLIAMS, P.E.O. (1987) The oral retention and antiplaque efficacy of triclosan in human volunteers. *British Journal of Clinical Pharmacology*, **23**, 579–583.

GJERMO, P. & ROLLA, G. (1970) Plaque inhibition by antibacterial dentifrices. *Scandinavian Journal of Dental Research*, **78**, 464–470.

GREENE, J.C. & VERMILLION, J.R. (1960) Oral hygiene index: a method of classifying oral hygiene status. *Journal of the American Dental Association*, **61**, 172–179.

HARRAP, G.J. (1974) Assessment of the effect of dentifrices on the growth of dental plaque. *Journal of Clinical Periodontology*, **1**, 166–174.

HEGINBOTHOM, M., FITZGERALD, T.C. & WADE, W.G. (1990) Comparison of solid media for cultivation of anaerobes. *Journal of Clinical Pathology*, **43**, 253–256.

HULL, P.S. (1980) Chemical inhibition of plaque. *Journal of Clinical Periodontology*, **7**, 431–442.

JENKINS, S., ADDY, M. & WADE, W. (1988) The mechanism of action of chlorhexidine: a study of plaque growth on enamel inserts *in vivo*. *Journal of Clinical Periodontology*, **15**, 415–424.

JENKINS, S., ADDY, M. & NEWCOMBE, R. (1990) Comparative effects of toothpaste brushing or toothpaste rinsing on salivary bacterial counts. *Journal of Periodontal Research*, **25**, 316–319.

JENSEN, J.E. & CHRISTENSEN, F. (1971) A study of the elimination of chlorhexidine from the oral cavity using a new spectro-photometric method. *Journal of Periodontal Reseach*, **6**, 306–311.

JOHANSEN, J.R., GJERMO, P. & ERIKSEN, H.M. (1975) Effect of two years' use of chlorhexidine-containing dentifrices on plaque, gingivitis, and caries. *Scandinavian Journal of Dental Research*, **83**, 288–292.

KEEVIL, C.W., BRADSHAW, D.J., DOWSETT, A.B. & FEARY, T.W. (1987) Microbial film formation: dental plaque deposition on acrylic tiles using continuous culture techniques. *Journal of Applied Bacteriology*, **62**, 129–138.

KORNMAN, K.S. (1986) Antimicrobial agents. In Loe, H. & Kleinman, D.V. (eds.) *Dental Plaque Control Measures and Oral Hygiene Practices*, pp. 121–142. IRL Press, Oxford.

LISTGARTEN, M.A. & HELLDEN, L. (1978) Relative distribution of bacteria at clinically healthy and periodontally diseased sites in humans. *Journal of Clinical Periodontology*, **5**, 115–132.

LOBENE, R.R., WEATHERFORD, T., ROSS, N.M., LAMM, R.A. & MENAKER, L. (1986) A modified gingival index for use in clinical trials. *Clinical Preventive Dentistry*, **8**, 3–6.

LOE, H. & SCHIOTT, C.R. (1970) The effect of mouthrinses and topical application of chlorhexidine on the development of plaque and gingivitis in man. *Journal of Periodontal Research*, **5**, 79–83.

LOE, H. & SILNESS, J. (1963) Periodontal disease in pregnancy. 1. Prevalence and severity. *Acta Odontologica Scandinavica*, **21**, 532–551.

LOE, H., THEILADE, E. & JENSEN, S.B. (1965) Experimental gingivitis in man. *Journal of Periodontology*, **36**, 177–187.

McKEE, A.S., McDERMID, A.S., ELLWOOD, D.C. & MARSH, P.D. (1985) The establishment of reproducible, complex communities of oral bacteria in the chemostat using defined inocula. *Journal of Applied Bacteriology*, **59**, 263–275.

MARSH, P.D., HUNTER, J.R., BOWDEN, G.H., HAMILTON, I.R., McKEE, A.S., HARDIE, J.M. & ELLWOOD, D.C. (1983) The influence of growth rate and nutrient limitation on the microbial composition and biochemical properties of a mixed culture of oral bacteria grown in a chemostat. *Journal of General Microbiology*, **129**, 755–770.

MORAN, J., ADDY, M. & WADE, W.G. (1988) Determination of minimum inhibitory concentrations of commercial toothpastes using an agar dilution technique. *Journal of Dentristry*, **16**, 27−31.

MUHLEMANN, H.R. & SON, S. (1971) Gingival sulcus bleeding − a leading symptom in initial gingivitis. *Helvetica Odontologica Acta*, **15**, 107−113.

NABI, N., MUKERJEE, C., SCHMID, R. & GAFFAR, A. (1989) *In vitro* and *in vivo* studies on triclosan/PVM/MA copolymer/NaF combination as an antiplaque agent. *American Journal of Dentistry*, **2**, 197−206.

NESBITT, W.E., DOYLE, R.J. & TAYLOR, K.G. (1982) Hydrophobic interactions and the adherence of *Streptococcus sanguis* to hydroxyapatite. *Infection and Immunity*, **38**, 637−644.

NYVAD, B. & KILIAN, M. (1990) Comparison of the initial streptococcal microflora on dental enamel in caries-active and in caries-inactive individuals. *Caries Research*, **24**, 267−272.

PIGMAN, W., ELLIOTT, H. & LAFFRE, R. (1952) An artificial mouth for caries research. *Journal of Dental Research*, **31**, 627−633.

QUIGLEY, G. & HEIN, J. (1962) Comparative cleansing efficiency of manual and power brushing. *Journal of the American Dental Association*, **65**, 26−29.

ROBERTS, W.R. & ADDY, M. (1981) Comparative *in vivo* and *in vitro* antibacterial properties of antiseptic mouthrinses containing chlorhexidine, alexidine, cetyl pyridinium chloride, and hexetidine. *Journal of Clinical Periodontology*, **8**, 220−230.

ROSENBERG, M., ROSENBERG, E., JUDES, H. & WEISS, E. (1983) Bacterial adherence to hydrocarbons and to surfaces in the oral cavity. *FEMS Microbiology Letters*, **20**, 1−5.

RUSSELL, C. & COULTER, W.A. (1975) Continuous monitoring of pH and Eh in bacterial plaque growth on a tooth in an artificial mouth. *Applied Microbiology*, **29**, 141−144.

SAXTON, C.A. (1986) The effects of a dentifrice containing zinc citrate and 2,4,4,-trichloro-2-hydrodiphenyl ether. *Journal of Periodontology*, **57**, 555−561.

SAXTON, C.A., LANE, R.M. & VAN DER OUDERAA, F. (1987) The effects of a dentifrice containing zinc salt and a non-cationic antimicrobial agent on plaque and gingivitis. *Journal of Clinical Periodontology*, **14**, 144−148.

SCHIOTT, C.R., LOE, H., JENSEN, S.B., KILIAN, M., DAVIES, R.M. & GLAVIND, K. (1970) The effect of chlorhexidine mouthrinses on the human oral flora. *Journal of Periodontal Research*, **5**, 84−89.

SHAH, H.N., WILLIAMS, R.A.D., BOWDEN, G.H. & HARDIE, J.M. (1976) Comparison of the biochemical properties of *Bacteroides melaninogenicus* from human dental plaque and other sites. *Journal of Applied Bacteriology*, **41**, 473−492.

SHAW, L. & MURRAY, J.J. (1977) A new index for measuring extrinsic stains in clinical trials. *Community Dentistry and Oral Epidemiology*, **5**, 116−120.

SILNESS, J. & LOE, H. (1964) Periodontal disease in pregnancy II. Correlation between oral hygiene and periodontal condition. *Acta Odontologica Scandinavica*, **22**, 121−135.

SLOTS, J. (1982) Selective medium for the isolation of *Actinobacillus actinomycetemcomitans*. *Journal of Clinical Microbiology*, **15**, 606−609.

STEPHEN, K.W., SAXTON, C.A., JONES, C.L., RITCHIE, J.A. & MORRISON, T. (1990) Control of gingivitis and calculus by a dentifrice containing a zinc salt and triclosan. *Journal of Periodontology*, **61**, 674−679.

SYED, S.A. & LOESCHE, W.J. (1972) Survival of human dental plaque flora in various transport media. *Applied Microbiology*, **24**, 638−644.

TANNER, A.C.R. & GOODSON, J.M. (1986) Sampling of microorganisms associated with periodontal disease. *Oral Microbiology and Immunology*, **1**, 15−20.

TURESKY, S., GILMORE, N.D. & GLICKMAN, I. (1970) Reduced plaque formation by the chloromethyl analogue of Victamine C. *Journal of Periodontology*, **41**, 41−43.

VAN PALENSTEIN HELDERMAN, W.H. (1981) Microbial aetiology of periodontal disease. *Journal of Clinical Periodontology*, **8**, 261−280.

VAN PALENSTEIN HELDERMAN, W.H., IJSSELDIJK, M. & HUIS IN'T VELD, J.H.J. (1983) A selective medium for the two major subgroups of the bacterium *Streptococcus mutans* isolated from human dental plaque and saliva. *Archives of Oral Biology*, **28**, 599–603.

VAN PELT, A.W.J., WEERKAMP, A.H., UYEN, M.H.W.J.C., BUSSCHER, H.J., DE JONG, H.P. & ARENDS, J. (1985) Adhesion of *Streptococcus sanguis* CH3 to polymers with different surface free energies. *Applied and Environmental Microbiology*, **49**, 1270–1275.

WADE, W.G. & ADDY, M. (1989) *In vitro* activity of a chlorhexidine-containing mouthwash against subgingival bacteria. *Journal of Periodontology*, **60**, 521–524.

WADE, W.G., ALDRED, M.J. & WALKER, D.M. (1986) An improved medium for isolation of *Streptococcus mutans*. *Journal of Medical Microbiology*, **22**, 319–323.

WADE, W.G., GRAY, A.R., ABSI, E.G. & BARKER, G.R. (1991) Predominant cultivable flora in pericoronitis. *Oral Microbiology and Immunology*, **6**, 310–312.

WADE, W.G., MORAN, J., MORGAN, J.R. & ADDY, M. (1992) The effects of antimicrobial acrylic strips on the subgingival microflora in chronic periodontitis. *Journal of Clinical Periodontology*, **19**, 127–134.

WALKER, C.B., RATLIFF, D., MULLER, D., MANDELL, R. & SOCRANSKY, S.S. (1979) Medium for selective isolation of *Fusobacterium nucleatum* from human periodontal pockets. *Journal of Clinical Microbiology*, **10**, 844–849.

WILSON, R.F., WOODS, A. & ASHLEY, F.P. (1986) Dark-field microscopy of dental plaque. A clinical and laboratory evaluation. *British Dental Journal*, **159**, 114–120.

WYATT, J.E., HESKETH, L.M. & HANDLEY, P.S. (1987) Lack of correlation between fibrils, hydrophobicity and adhesion for strains of *Streptococcus sanguis* biotypes 1 and 2. *Microbios*, **50**, 7–15.

ZAMBON, J.J., REYNOLDS, H.S. & SLOTS, J. (1981) Black-pigmented *Bacteroides* species in the human oral cavity. *Infection and Immunity*, **32**, 198–203.

10: Sensitivity of Bacteria in Biofilms to Antibacterial Agents

W.W. NICHOLS

Department of Infection Research, ZENECA Pharmaceuticals, Mereside, Alderley Park, Macclesfield SK10 4TG, UK

The sensitivity to antibacterial agents of bacteria in biofilms is an important topic in medicine, dentistry, water distribution and a range of industries (Costerton *et al.*, 1987; Nichols, 1989, 1991). It seems timely to describe in one place an illustrative group of methods that have been used for assessing such sensitivities, and this chapter is primarily concerned with bacteria and antibacterial agents of medical importance. Most of the methods have been culled from the literature. In Table 10.1 there is a brief summary of the appropriateness of the different methods for various purposes.

Colony on Filter: A Simple Method for Exposure of Bacterial Colonies to Antibacterial Agents

Forming the colonies

Sterile nitrocellulose membrane filters (0.45 µm pore size; 25 mm diameter), five to a plate, are placed on to fresh agar plates. When wetted, the surface of each filter is inoculated with 1 µl of a suspension of cells grown overnight at 37°C in the same liquid medium; care should be taken to apply the inoculum as a single droplet. The plates with their inoculated filters are then incubated for 16 h in air at 37°C in the case of *Pseudomonas aeruginosa* (Nichols *et al.*, 1989). A similar method has been described by Millward & Wilson (1989) for assessing the effect of chlorhexidine on aggregates of *Streptococcus sanguis*.

Measurement of the effect of the antibacterial agent

To expose organisms to antibacterial agents, a filter with its colony is removed from the culture plate with sterile forceps and transferred to a fresh culture

Microbial Biofilms:
Formation and Control

TABLE 10.1. *Methods for producing a biofilm or (micro)colony for testing the susceptibility of constituent cells to antibacterial agents and their use in teaching and research*

| | Teaching | | | |
Biofilm model	Hands-on	Demonstration	Research	Reference
Colony-on-filter	+	−	−	Nichols *et al.* (1989)
Modified Robbins device	−	+	+	Nickel *et al.* (1985)
Discs in batch culture	−	+*	+	Gristina *et al.* (1989)
Tiles in iron-restricted chemostat culture	−	−	+	Anwar *et al.* (1989)
Controlled growth rate perfused biofilms	−	+	+	Evans *et al.* (1990)
Glass slide microcolonies	+	+	+	Widmer *et al.* (1990)
Animal model of biofilm infection	−	−	+	Widmer *et al.* (1991)

* Laboratory-made polymethylmethacrylate discs. Other surgical materials are not available in disc form.

plate (e.g. IsoSensitest agar) containing the requisite concentration of the agent. A filter is also placed on a control plate without the agent. These plates are then incubated as required. Viable counts are also made immediately as a zero-time control on three of the filters.

The viable counts of the colonies on the filters before or after antibiotic exposure are made as follows (Nichols *et al.*, 1989). Each filter with its adherent colony is placed in 2 ml of buffer (pH 7.4) at 37°C containing 0.1 mol/l NaCl, 5 mmol/l $MgCl_2$ and 25 mmol/l MOPS (3-[morpholino]-propanesulphonic acid), and the bacteria are dispersed by vigorous vortex mixing. This buffer was designed for use with *Ps. aeruginosa*. A suitable general buffer would be phosphate-buffered saline (PBS: g/l NaCl, 8.00; K_2HPO_4, 1.21; KH_2PO_4, 0.34; pH 7.3; see below). An antibiotic-neutralizing agent should be included in the buffer. For this purpose sodium polyanethol sulphonate (0.05% w/v; Krogstad *et al.*, 1981) has been used in the case of tobramycin and broad-spectrum β-lactamase (2.5×10^4 units/ml; Centre for Applied Microbiology and Research, Porton Down, Salisbury, Wilts, UK) in the case of the β-lactam cefsulodin (Nichols *et al.*, 1989). Where an enzyme is used to inactivate antibiotic, the colony is left in the initial resuspension buffer (e.g. for 30 min at 37°C) before dilution, to allow time for enzyme action. In all cases, serial dilutions of the dispersed suspensions are then made in the same buffer at 37°C and volumes of 0.1 ml are spread on to predried culture plates (e.g. IsoSensitest agar), which are then incubated at 37°C for 16 h.

Comments on the colony-based method

This procedure is very simple but is limited in its capacity to model biofilms that occur outside the laboratory. A major difference between these bacterial colonies and biofilms in patients (such as those on vascular, urinary or peritoneal catheters; or on orthopaedic prostheses) or in aqueous environmental locations, is that the antibacterial agent-containing medium to which the colonies are exposed is static. Because of this lack of mixing, the antibacterial agent in the immediate vicinity of the colonies will rapidly be depleted, and this can be expected to cause an apparent reduction in antibacterial suscepti-bility. However, even in the case of biofilms bathed in a well mixed liquid medium, there is still a diffusive mass-transfer region close to the film surface (Palenik *et al.*, 1989). Despite the above reservation, the method used for *Ps. aeruginosa* and tobramycin (Nichols *et al.*, 1989) agreed quantitatively, in terms of the concentration that was severely bactericidal to the biofilm bacteria, with data obtained in a different laboratory with the modified Robbins device (Nickel *et al.*, 1985).

The Modified Robbins Device

Establishing a biofilm with the modified Robbins device

The modified Robbins device is the classic apparatus for growing easily re-covered biofilms and is described in detail in Chapter 1.

A study of tobramycin and *Ps. aeruginosa* with this apparatus is described by Nickel *et al.* (1985). Sterile medium from a reservoir held at 37°C in a thermostatically controlled water bath is pumped through the device by a peristaltic pump at 60 ml/h. Bacteria are allowed to attach to the surfaces of discs of sample material by inoculating a reservoir of medium with 2% (v/v) of an exponential-phase batch culture, grown at the same temperature. After a colonization period (e.g. 8 h) the tube pump inlet is switched to fresh medium. At times thereafter, the tube to the pump can be aseptically changed to medium that contains the antibacterial agent. By removing the sampling plugs and replacing these with fresh sterile plugs, discs of the material being studied can be removed at timed intervals as determined by the experimenter.

A similar method, but involving the formation of a biofilm community without an inoculum of *in vitro* grown bacteria, has been described by Exner *et al.* (1987). A silicone tube is perfused with tapwater for 50 days to form the biofilm, and then portions of the tube are immersed in disinfectant solutions for various times as required.

Measurement of antibiotic effect with the modified Robbins device

The medium reservoir is replaced with one containing the growth medium plus the antibacterial agent (e.g. tobramycin 100 or 1000 µg/ml). At intervals studs are removed and the sample discs of material removed aseptically. The discs are washed with sterile growth medium and then surface material is aseptically scraped into a measured volume of sterile diluent (e.g. phosphate-buffered saline (PBS) at 37°C, containing an antibacterial–neutralizing agent; see above). This suspension, together with the scraped disc, is sonicated at low power in a bath-type sonic cleaner and mixed by vortexing. Dilutions are then made in PBS at 37°C, and 0.1-ml volumes are spread on nutrient agar recovery plates. The decrease in numbers of viable cells/cm^2 of biofilm can thus be monitored as a function of time.

Comments on the modified Robbins device

As in any continuous-culture method in which growth medium lines are changed, and where medium reservoirs are kept in water baths, rigorous aseptic technique is necessary to avoid bacterial contamination. The growth rates of the cells in the biofilm are not controlled in the modified Robbins device. However, because infecting biofilms *in vivo* probably also contain bacteria of differing individual growth rates, this should be seen as an advantage of the model rather than a disadvantage. The growth medium can be made low in iron, if desired, in order more closely to simulate the low availability of iron in the host (see below).

Batch-Culture Studies of Coagulase-Negative Staphylococcal Biofilms on Biomedical Materials

Biomaterials

This type of method has been widely used (e.g. Prosser *et al.*, 1987) and the method described here is taken from Gristina *et al.* (1989). Biofilms are grown on 1.5-mm thick discs of biomaterial that are cut with a lathe from 7-mm diameter rods of surgical grade stainless steel or ultrahigh molecular weight polyethylene. Gristina *et al.* (1989) obtained their materials from Howmedica Inc., Rutherford, NJ, USA, but the firm does not supply these rods generally. After cutting, the discs are lightly polished and passivated with HNO$_3$. Poly-methylmethacrylate bone-cement components (Howmedica, London, UK) are mixed according to the manufacturer's instructions and packed into a 7 mm internal diameter polypropylene cylinder. When the material has polymerized, discs 1.5 mm thick are cut on the lathe and soaked overnight in distilled water.

Discs of all three materials are shaken in urea (8 mol/l, 500 ml) at room temperature for 20 h before use. They are then placed in $CaCl_2$ (5 mol/l, 500 ml) at 4°C for 1 h, followed by 5% nitric acid (500 ml) at room temperature for 1 h and then two changes of distilled water (2×500 ml) at room temperature for 1 h each. The dried discs are then sterilized by exposure to ethylene oxide for 24 h.

Biofilm formation in batch culture

In order to replicate exactly the growth conditions used by Gristina *et al.* (1989), bacteria (*Staphylococcus epidermidis, Staphylococcus hominis, Staphylococcus hyicus*) are grown into early-exponential to midexponential phase in cation-supplemented Mueller−Hinton broth (containing Mg^{2+} at $20-35$ mg/l and Ca^{2+} at $50-100$ mg/l), by direct inoculation (e.g. into 10 ml volumes) from brain heart infusion agar slopes, followed by incubation at 37°C for 6 h. The bacteria are then diluted to *ca.* 10^6 cfu/ml (0.5 McFarland turbidity standard) and the suspensions distributed into sterile six-well cell-culture plates (e.g. 5 ml/well). Discs are placed into these suspensions and incubated at 37°C for 24 h on a platform rocker. The discs are then removed with sterile forceps, non-adherent bacteria are removed by vigorous mixing in two changes of 5 ml of PBS at 37°C, and the discs are placed in 1-ml volumes of the same growth medium containing antibiotic. Exposure to antibiotic takes place in 26-well plates, which are similarly incubated at 37°C, on a platform rocker. Controls without antibiotic are set up, and two discs are examined to determine the viable counts of biofilms before exposure to antibiotic.

Viable counting of the batch-cultured biofilms

After antibiotic exposure, discs are again removed with sterile forceps and non-adherent cells removed by shaking in two 5-ml changes of sterile PBS at 37°C. The disks are then placed in 10 ml PBS (containing an appropriate antibiotic-neutralizing agent) at room temperature and sonicated for 5.5 min in a low-output cleaning sonicator (e.g. an Ultramet III sonic cleaner; Buehler Ltd, Evanston, Ill., USA), followed by vortex mixing for 30 s. Each suspension is then diluted tenfold in PBS at room temperature to 10^{-6}, and 0.1-ml volumes are spread on to Columbia agar base containing 5% horse blood. Colonies are counted after incubation at 37°C for 24 h.

Comments on the batch-cultured biofilm method

This is a rapid, convenient method that can be applied to large numbers of samples of biomaterials, and/or large numbers of different bacteria, and antibacterial agents. The only constraint is the limited availability of discs of surgical-grade stainless steel and ultrahigh molecular weight polyethylene. Polymethylmethacrylate components are, however, readily available.

A very similar method for use with biofilms on discs of catheter material has been described by Prosser *et al.* (1987). Their method differed slightly in that, after antibiotic exposure, discs were not washed but the surface film was scraped with a sterile scalpel into 5 ml of PBS. The disc was added to the scrapings in the PBS, and then the whole sample was mixed by vortexing for 1 min. The sample was then gently sonicated for 5 min in an Ultramet III sonic cleaner. Viable counts were determined by spiral plating.

Chemostat-Grown Biofilms under Low-Iron Conditions

Forming the biofilms in a low-iron medium

Iron is tightly bound by iron-binding proteins in human tissues, which has led to the hypothesis that this contributes to 'nutritional immunity' (Weinberg, 1975) whereby systemic bacteria multiply less rapidly than they would do under iron-replete conditions (see Hershko *et al.*, 1988, for a more recent review). Anwar *et al.* (1989) and Anwar & Costerton (1990) have reiterated these observations by pointing out that the vanishingly low concentration of free Fe^{3+} in host tissues can be expected to affect the physiology of bacteria that colonize the surfaces of internal medical devices. These authors have therefore used a method that involves continuous chemostat cultivation of a biofilm, in the presence of dispersed cells, in a medium from which iron has been removed by passage through a column of CHELEX 100 ion exchange resin (Bio-Rad Laboratories Ltd, Bio-Rad House, Maylands Avenue, Hemel Hempstead, HP2 7TD, UK). The chemostat used is the 50-ml, small-scale, all-glass chemostat described by Gilbert & Stuart (1977).

All glassware is washed in 5% Decon 75 (BDH Chemicals, Poole, Dorset, UK) followed by rinsing once in distilled water, once in 1% (v/v) HCl and six more times in distilled water. This is done in order to minimize iron leaching from the glass during the experiment. For the same reason, plastic vessels and pipettes are used where possible. The growth medium is depleted of iron by the method of Kadurugamuwa *et al.* (1987). The medium is made up at double strength in 1-l batches, each of which is passed three times through a pretreated (see below) CHELEX-100 column (30.5×2.4 cm in glass) at a flow-rate of 2 ml/min/cm^2. The medium is then passed twice through a regenerated column of the same resin, the first 100 ml of the first passage being discarded. Columns are pretreated and regenerated in the same way: two bed-volumes of 1 mol/l HCl are passed through the column, then five volumes of double-distilled water, then two bed-volumes of 1 mol/l NaOH, five more bed-volumes of double-distilled water, and finally 0.66 mol/l sodium phosphate buffer at pH 7.4 is passed through the column until the eluate reaches the same pH. The iron-depleted medium is sterilized by filtration. Mg^{2+} is returned as a filter-sterilized solution of $MgSO_4 . 7H_2O$ (10 mg/ml) to a final concentration of 0.1 mg/ml. The treated medium is used immediately after iron-depletion.

Rectangular tiles that provide the surface for attachment of the biofilm ($0.5 \times 2 \times 0.1 - 0.2$ cm) are cut from methylmethacrylate dental resin (Product RR, De Trey Ltd, Weybridge, UK) cast in a glass Petri dish overnight. The tiles are suspended in the chemostat liquid by nylon thread. Steel thread was used originally for this method in the study of a model of dental plaque formation (Keevil *et al.*, 1987). The suspended tile and the chemostat apparatus are sterilized together by autoclaving. Freshly iron-depleted tryptic soy broth at 37°C is fed into the water-jacketed chemostat vessel, which is also held at 37°C. Bacteria are inoculated as 5 ml of an exponential-phase culture grown in the same medium at the same temperature. Fresh medium is then pumped in at the required dilution rate. With mucoid *Ps. aeruginosa*, the biofilm takes 5 days to reach an almost-constant cell density of 2×10^9 cfu/cm^2 at a dilution rate of 0.05/h, whereas the dispersed cell population remains at about 4×10^9 cfu/ml from day 1 onwards.

Measurement of the effects of antibiotics on chemostat-grown biofilms

For exposure to antibiotic, tiles are removed from the chemostat, washed three times with 10 ml of PBS to remove non-adherent bacteria, and then placed in 10 ml of iron-depleted tryptic soy broth containing antibiotic, at 37°C. After the requisite exposure time, the tiles are washed three times with 10 ml PBS at 37°C, placed in 1 ml PBS and vortexed for 3 min. Serial dilutions are made in PBS at 37°C and 0.1-ml volumes spread on nutrient agar plates, which are incubated at 37°C for 16 h before counting.

Comments on the chemostat system

As with the modified Robbins device, the method is labour- and time-intensive and demanding of experimental skills, care being necessary to avoid bacterial contamination.

Growth Rate-Controlled Continuously Perfused Biofilms

Forming the continuously perfused biofilm

Gilbert *et al.* (1989) have used *Escherichia coli* ATCC 8739 to establish continuously perfused bacterial films adherent to cellulose acetate membranes.

The originators of the method, Helmstetter & Cummings (1963), used *Escherichia coli* B/r (ATCC 12407). Bacteria are first grown for 16 h at 37°C in aerated chemically defined liquid medium with glycerol as the limiting nutrient (glycerol, 10 mmol/l; $(NH_4)_2SO_4$, 6 mmol/l; $MgSO_4$, 0.5 mmol/l; KCl, 13.5 mmol/l; KH_2PO_4, 28 mmol/l; K_2HPO_4, 72 mmol/l: pH 7.4). Organisms are subcultured into fresh medium and grown to exponential phase (*ca.* 10^8 cfu/ml) under identical conditions. A 60-ml volume of this culture is vacuum-filtered on to a cellulose acetate membrane (Oxoid: 0.22 μm pore size, 47 mm diameter, prewashed with 50 ml of sterile medium). The filter is then inverted and placed in the continuous-perfusion apparatus maintained at 37°C by a water jacket. Fresh medium at 37°C percolates through the filter from above and the biofilm adherent to the underside is thereby continuously perfused. The liquid medium builds up a hydrostatic head above the filter until a steady state is reached, when the perfusion rate becomes equal to the rate of addition of fresh medium. Viable counts are determined in the perfusate by serial dilution in sterile physiological saline at 37°C and plating 0.1-ml volumes in triplicate on predried Oxoid CM3 agar. The plates are incubated at 37°C for 16 h and the colonies counted. The number of bacteria released from the film per millilitre of perfusate becomes constant after about 2 h. At this steady state, the specific growth-rate constant becomes equal to the rate of dilution of the small volume of bacteria plus intercellular water adhering to the membrane (see Chapter 3).

Measurement of the effects of antibiotic on growth rate-controlled cells

The method has been used in studies with tobramycin (Evans *et al.*, 1990). Bacteria from the biofilm on the membrane filter are resuspended in physiological saline at 37°C to a density of *ca.* 5×10^8 cfu/ml. A 0.1-ml volume of this suspension is mixed with 9.9 ml of a solution of tobramycin (50 ng/ml) in water at 37°C. The exposure to tobramycin is continued for 1 h at 37°C with gentle shaking, and 0.1-ml samples are diluted in 9.9 ml of physiological saline for viable counting, as described above. It is prudent to have an inactivating (neutralizing) agent in the diluting solution in order to prevent further antibacterial effects (Krogstad *et al.*, 1981; see above). Results are expressed as percentage reductions in viable counts, relative to controls not exposed to tobramycin.

Comments on the perfused-biofilm method

This procedure essentially measures the susceptibility of attached bacteria, rather than biofilm bacteria as conventionally understood. Continuous perfusion of fresh medium through the film will result in the nutrient concentration being similar for each cell. In a biofilm on an inert surface, on the other hand, where organic solutes, ions and gases are obtained from a bathing aqueous phase, nutrient gradients and heterogeneous spatial distributions of cell physiology can be expected (Revsbech, 1989; Wimpenny *et al.*, 1989). Moreover, because of the perfusion in this method, one can expect an absence of the small molecule-mediated cell–cell signalling that occurs in non-perfused bacterial aggregates (Shapiro & Hsu, 1989), again leading to a different physiology.

In this model system, the bacteria resuspended from the film were as sensitive as — or more sensitive than — dispersed cells grown in chemostats at identical growth rates. This leads to the inference that the lower susceptibilities to antibacterial agents observed in most biofilm studies (Nichols, 1991) are not due to the attachment *per se* of the constituent bacteria. Rather, the lowered susceptibilities are likely to be due in part to the physiology of the constituent bacteria, which results from the secondary gradients of organic compounds, inorganic ions, oxygen and excreted products, that predictably occur within biofilms (van Loosdrecht *et al.*, 1990, and see above). A further interpretative problem exists in the antibiotic-exposure conditions used in the perfused-biofilm method. That is, that the bacteria are dispersed before exposure, rather than being tested *in situ* in the attached film (Evans *et al.*, 1990).

Finally, the system of Gilbert *et al.* (1989) is not a realistic model of biofilms in nature or in infections, but it is a convenient experimental device for answering questions about bacterial physiology associated with attachment to, or growth at, a surface.

Glass Slides and Animal Models

These two experimental procedures are considered together because they were developed as complementary methods in one laboratory (Zimmerli *et al.*, 1982; Widmer *et al.*, 1990).

Exposure of microcolonies on glass slides to antibiotics

Sterile glass microscope slides held in a rack are immersed in a bacterial suspension, at an approximate density of 5×10^3 cfu/ml, in PBS containing glucose (0.25%, w/v). The inoculum density is checked by plating 0.1 ml on predried solid medium (e.g. Mueller–Hinton agar) and incubation at 37°C for 16 h. Both *S. epidermidis* (Widmer *et al.*, 1990) and *E. coli* (Widmer *et al.*, 1991) have been studied by this procedure. The slides are exposed to the bacteria for 24 h at 35°C with agitation at 90 rpm. During this time, adherent microcolonies form on the surfaces of the glass slides. The slides are removed from the suspension and washed three times in sterile 0.85% (w/v) NaCl. One slide is immediately examined for viable colonies, as described below. Test slides are then immersed in growth medium (trypticase soy broth supplemented with Ca^{2+} at 50 mg/l) containing various concentrations (e.g. double the minimum bactericidal concentration, MBC) of antibiotics, and including an antibiotic-free control. After a further 24 h of agitation at 35°C in these new conditions, the slides are removed, washed three times with 0.85% (w/v) NaCl at 35°C, and viable adherent colonies are counted as follows. The washed slides are gently pressed with sterile forceps on to the surface of antibiotic-free Mueller–Hinton agar, taking care to ensure that bubbles are not formed between the glass and agar surfaces. Both sides of the slides can be treated in this way, but because of the difficulty of avoiding smudging the second side while 'printing' the first, we have used one side only (Thain & Nichols, unpublished work). The plates are then incubated at 37°C for 16 h and the colonies counted. The result is rejected if there are fewer than 100 or more than 1000 cfu/slide before exposure to antibiotic.

Comments on the glass slide method

Although the numbers of colonies before and after antibiotic exposure are determined, the sizes of these (micro)colonies, in terms of cfu/colony, have not been measured. For a quantitative appreciation of the difference between adherent colonies and dispersed bacteria, it would be useful to determine the sizes of colonies at the time of antibiotic exposure, and the sizes of the surviving colonies after antibiotic exposure. However, such extra measurements would complicate the method.

Tissue cages in whole animals

S. epidermidis (Widmer *et al.*, 1990) and *E. coli* (Widmer *et al.*, 1991) have been used in these model foreign-body infections. Ethylene oxide-sterilized Teflon tubes (32 mm long, 12 mm external diameter, 10 mm internal diameter) perforated by 130 regularly spaced holes (1 mm diameter) and sealed at each end by a Teflon cap, are implanted aseptically, four per animal, into the flanks of albino guineapigs weighing 600–1100 g. Experiments are started after complete healing of wounds, 2–4 weeks following surgery. Before experiments are begun, the sterility of the interstitial fluid that accumulates in the tissue cages is confirmed by aspirating 50 μl and plating on Mueller–Hinton agar. At day 0, the infection is established by local inoculation of bacteria, e.g. *S. epidermidis* 10^4 cfu; *E. coli* 7×10^4 cfu. After 24 h, during which time the infection becomes established, antibiotic therapy is started at a site remote from the infection, for example by intraperitoneal injection. The regimen used by Widmer *et al.* (1990, 1991) was to treat at 12-h intervals for 4 days. Tissue-cage fluid is aspirated to determine viable counts in the fluid during and after therapy, and for up to 16 days of follow-up.

At times determined by the experimenter, the chambers are removed aseptically from anaesthetized animals and cultured on Mueller–Hinton agar by rolling the tube over the agar surface at least four times and incubating the plates at 37°C for 16 h (Maki *et al.*, 1977). Quantitative viable counts can also be obtained by sonication and vortexing in PBS at 37°C, as described above under batch-culture studies.

Lucet *et al.* (1990) have described a tissue-chamber model of methicillin-resistant *Staphylococcus aureus* infection in the rat. This allows the study of chronic biofilm infections that last for more than 2 months. Advantages over the guineapig model are claimed for studying curative therapy, as opposed to the direct antimicrobial action studies that are the subject of this review.

Comments on the tissue-chamber biofilm infection model

A biofilm infection model such as that described here would seem to be essential to demonstrate the *in vivo* efficacy of antibiotics believed to be active against bacteria in biofilms. The studies so far reported show that compounds kill or inhibit the growth of bacteria in the tissue-cage model with the same potency ranking as for bacteria in microcolonies *in vitro* (Widmer *et al.*, 1990).

Methods Predicted to be of Importance in Future Studies

Although the methods discussed below are likely to prove important in the future, they are not described in full because of their specialist nature and because studies with antibacterial agents have not yet been published.

Laser confocal scanning microscopy with image analysis

Not many papers have been published that describe the application of this method to biofilm studies (Lawrence *et al.*, 1991, Zanyk *et al.*, 1991; and see Chapters 1 and 5). The method requires special facilities depending on the needs of the experimenter. It involves monitoring the fluorescence of a dye through an optical microscope so that bacteria within a biofilm can be observed, either by virtue of them taking up the dye and altering its fluorescence, or by virtue of their excluding the dye and therefore being non-fluorescent. The biofilms are grown in a flow cell on the microscope stage and sectioned optically by the scanning system. An experimental model of biofilm development from attachment, through microcolony formation, to microcolony coalescence, can thus be analysed in real time. With choice of appropriate fluorophores, it should be possible to monitor aspects of bacterial physiology *in situ*, including viability. In principle, it should also be possible to use this method to monitor individual biofilm-cell physiology on exposure of the film to antibacterial agents in the bathing medium.

Constant-depth biofilm fermenter

In this apparatus a biofilm is produced and maintained in a steady state at a constant thickness (e.g. 0.3 mm) by continuous shaving of the top of the film (Wimpenny *et al.*, 1989; and see Chapter 3). Nutrients enter the film at the same face from which excess bacteria are removed. Biofilms in this model have yet to be exposed to antibacterial agents. However, samples of the biofilm are recoverable for viable counting, so the method is amenable to such studies.

The 'bioelectric effect'

Khoury *et al.* (1992) have reported that passing a low-density electric current across a biofilm potentiates antibacterial agents bathing the film. This occurs at current densities that do not themselves have any effect on the viability of the bacterial cells in the film. The mechanism of this 'bioelectric effect' is unknown; indeed, other groups have yet to confirm the basic observation. Although methods are described by Khoury *et al.* (1992), newer methods are to be expected in the future for exploring and refining such an important discovery.

Acknowledgement

It is a pleasure to acknowledge the expert technical assistance and advice of J.L. Thain.

References

ANWAR, H. & COSTERTON, J.W. (1990) Enhanced activity of combination of tobramycin and piperacillin for eradication of sessile biofilm cells of *Pseudomonas aeruginosa*. *Antimicrobial Agents and Chemotherapy*, **34**, 1666–1671.

ANWAR, H., DASGUPTA, M., LAU, K. & COSTERTON, J.W. (1989) Tobramycin resistance of mucoid *Pseudomonas aeruginosa* biofilm grown under iron limitation. *Journal of Antimicrobial Chemotherapy*, **24**, 647–655.

COSTERTON, J.W., CHENG, K.-J., GEESEY, G.G., LADD, T.I., NICKEL, J.C. DASGUPTA, M. & MARRIE, T.J. (1987) Bacterial biofilms in nature and disease. *Annual Review of Microbiology*, **41**, 435–464.

EVANS, D.J., BROWN, M.R.W., ALLISON, D.G. & GILBERT, P. (1990) Susceptibility of bacterial biofilms to tobramycin: role of specific growth rate and phase in the division cycle. *Journal of Antimicrobial Chemotherapy*, **25**, 585–591.

EXNER, M., TUSCHEWITZKI, G.-J. & SCHARNAGEL, J. (1987) Influence of biofilms by chemical disinfectants and mechanical cleaning. *Zentralblatt für Bakteriologie und Hygiene B*, **183**, 549–563.

GILBERT, P. & STUART, A. (1977) Small-scale chemostat for the growth of mesophilic and thermophilic microorganisms. *Laboratory Practice*, **26**, 627–628.

GILBERT, P., ALLISON, D.G., EVANS, D.J., HANDLEY, P.S. & BROWN, M.R.W. (1989) Growth rate control of adherent bacterial populations. *Applied and Environmental Microbiology*, **55**, 1308–1311.

GRISTINA, A.G., JENNINGS, R.A., NAYLOR, P.T., MYRVICK, Q.N. & WEBB, L.X. (1989) Comparative *in vitro* antibiotic resistance of surface-colonizing coagulase-negative staphylococci. *Antimicrobial Agents and Chemotherapy*, **33**, 813–816.

HELMSTETTER, C.E. & CUMMINGS, D.J. (1963) An improved method for the selection of bacterial cells at division. *Biochimica et Biophysica Acta*, **82**, 608–610.

HERSHKO, C., PETO, T.E.A. & WEATHERALL, D.J. (1988) Iron and infection. *British Medical Journal*, **296**, 660–664.

KADURUGAMUWA, J.L., ANWAR, H., BROWN, M.R.W., SHAND, G.H. & WARD, K.H. (1987) Media for study of growth kinetics and envelope properties of iron-deprived bacteria. *Journal of Clinical Microbiology*, **25**, 849–855.

KEEVIL, C.W., BRADSHAW, D.J., DOWSETT, A.B. & FEARY, T.W. (1987) Microbial film formation: dental plaque deposition on acrylic tiles using continuous culture techniques. *Journal of Applied Bacteriology*, **62**, 129–138.

KHOURY, A.E., LAM, K., ELLIS, B. & COSTERTON, J.W. (1992) Prevention and control of bacterial infections associated with medical devices. *ASAIO Journal*, **38**, M174–M178.

KROGSTAD, D.J., MURRAY, P.R., GRANICH, G.G., NILES, A.C., LADENSON, J.H. & DAVIS, J.E. (1981) Sodium polyanethol sulfonate inactivation of aminoglycosides. *Antimicrobial Agents and Chemotherapy*, **20**, 272–274.

LAWRENCE, J.R., KORBER, D.R., HOYLE, B.D., COSTERTON, J.W. & CALDWELL, D.E. (1991) Optical sectioning of microbial biofilms. *Journal of Bacteriology*, **173**, 6558–6567.

LUCET, J.-C., HERRMANN, M., ROHNER, P., AUCKENTHALER, R., WALDVOGEL, J.A. & LEW, D.P. (1990) Treatment of experimental foreign body infection caused by methicillin-resistant *Staphylococcus aureus*. *Antimicrobial Agents and Chemotherapy*, **34**, 2312–2317.

MAKI, D.G., WEISE, C.E. & SARAFIN, H.W. (1977) A semiquantitative culture method for identifying intravenous catheter-related infection. *New England Journal of Medicine*, **296**, 1305–1309.

MILLWARD, T.A. & WILSON, M. (1989) The effect of chlorhexidine on *Streptococcus sanguis* biofilms. *Microbios*, **58**, 155–164.

NICHOLS, W.W. (1989) Susceptibility of biofilms to toxic compounds. In Characklis, W.G. & Wilderer, P.A. (eds.) *Structure and Function of Biofilms*, pp. 321–331. John Wiley and Sons Ltd, Chichester.

NICHOLS, W.W. (1991) Biofilms, antibiotics and penetration. *Reviews in Medical Microbiology*, **2**, 177–181.

NICHOLS, W.W., EVANS, M.J., SLACK, M.P.E. & WALMSLEY, H.L. (1989) The penetration of antibiotics into aggregates of mucoid and non-mucoid *Pseudomonas aeruginosa*. *Journal of General Microbiology*, **135**, 1291–1303.

NICKEL, J.C., RUSESKA, K., WRIGHT, J.B. & COSTERTON, J.W. (1985) Tobramycin resistance of *Pseudomonas aeruginosa* cells growing as a biofilm on urinary catheter material. *Antimicrobial Agents and Chemotherapy*, **27**, 619–624.

PALENIK, B., BLOCK, J.-C., BURNS, R.G., CHARACKLIS, W.G., CHRISTENSEN, B.E., GHIORSE, W.C., GRISTINA, A.G., MOREL, F.M.M.; NICHOLS, W.W., TUOVINEN, O.H., TUSCHEWITZKI, G.-J. & VIDELA, H.A. (1989) Biofilms: properties and processes. In Characklis, W.G. & Wilderer, P.A. (eds.) *Structure and Function of Biofilms*, pp. 351–366. John Wiley and Sons, Chichester.

PROSSER, B. LA T., TAYLOR, D., DIX, B.A. & CLEELAND, R. (1987) Method of evaluating effects of antibiotics on bacterial biofilm. *Antimicrobial Agents and Chemotherapy*, **31**, 1502–1506.

REVSBECH, N.P. (1989) Microsensors: spatial gradients in biofilms. In Characklis, W.G. & Wilderer, P.A. (eds.) *Structure and Function of Biofilms*, pp. 129–144. John Wiley and Sons, Chichester.

SHAPIRO, J.A. & HSU, C. (1989) *Escherichia coli* K-12 cell–cell interactions seen by time-lapse video. *Journal of Bacteriology*, **171**, 5963–5974.

VAN LOOSDRECHT, M.C.M., LYKLEMA, J., NORDE, W. & ZEHNDER, A.J.B. (1990) Influence of interfaces on microbial activity. *Microbiological Reviews*, **54**, 75–87.

WEINBERG, E.D. (1975) Nutritional immunity. Host's attempt to withhold iron from microbial invaders. *Journal of the American Medical Association*, **231**, 39–41.

WIDMER, A.F., FREI, R., RAJACIC, Z. & ZIMMERLI, W. (1990) Correlation between *in vivo* and *in vitro* efficacy of antimicrobial agents against foreign body infections. *Journal of Infectious Diseases*, **162**, 96–102.

WIDMER, A.F., WIESTNER, A., FREI, R. & ZIMMERLI, W. (1991) Killing of nongrowing and adherent *Escherichia coli* determines drug efficacy in device-related infections. *Antimicrobial Agents and Chemotherapy*, **35**, 741–746.

WIMPENNY, J.W.T., PETERS, A. & SCOURFIELD, M. (1989) Modeling spatial gradients. In Characklis, W.G. & Wilderer, P.A. (eds.) *Structure and Function of Biofilms*, pp. 111–127. John Wiley and Sons, Chichester.

ZANYK, B.N., KORBER, D.R., LAWRENCE, J.R. & CALDWELL, D.E. (1991) Four-dimensional visualization of biofilm development by *Pseudomonas fragi. Binary*, **3**, 24–29.

ZIMMERLI, W., WALDVOGEL, F.A., VAUDAUX, P. & NYDEGGER, U.E. (1982) Pathogenesis of foreign body infection: description and characteristics of an animal model. *Journal of Infectious Diseases*, **146**, 487–497.

11: *Legionella* Biofilms and their Control

C.W. KEEVIL, A.B. DOWSETT AND J. ROGERS
*Pathology Division, PHLS Centre for Applied Microbiology
and Research, Porton Down, Salisbury SP4 0JG, UK*

Legionella pneumophila is the aetiological agent of Legionnaire's disease, which can be fatal, and also of the non-fatal Pontiac fever. The organism can be isolated from a wide range of aquatic environments and is considered to be ubiquitous. Within such environments of low nutrient content, the legionellas grow in association with the other organisms that colonize the system, including flavobacteria (Wadowsky & Yee, 1983), cyanobacteria (Tison *et al.*, 1985) and amoebae (Rowbotham, 1980). Outbreaks have been associated with potable water systems, including cooling towers, humidifiers, domestic showers and taps. Infection occurs when the organism is inhaled in small-particle — less than 5 μm in diameter — aerosols which are able to penetrate beyond the lung bronchioles (Fitzgeorge *et al.*, 1983).

The growth of microorganisms at the interface between water and the drinking-water system surface has long been recognized (Whipple *et al.*, 1927). Bacteria are able to gain sufficient nutrients by forming a consortium and developing into a biofilm (Ellwood *et al.*, 1982) that consists of a range of microorganisms, including bacteria, fungi, protozoa and, possibly, viruses, along with their extracellular products. The materials on which the biofilm forms may become incorporated into the biofilm, including metal corrosion products (Keevil *et al.*, 1990; Walker *et al.*, 1991). The biofilm may serve not only to allow the growth of bacteria in water systems (Geldritch *et al.*, 1972) but may also provide a haven that protects the microorganisms from excesses of oxygen concentration, redox potential and biocide treatment. Waterborne pathogens of public health concern, such as aeromonads, coliforms, campylobacters and cryptosporidia, may shelter within the biofilm haven. Their subsequent release in sloughed-off clumps or when disinfection practices are relaxed, may lead to infection in man.

The survival and growth of legionellas has often been studied under

Microbial Biofilms:
Formation and Control

conditions inappropriate for modelling growth within water systems. The objective of this chapter is to detail a reproducible method for growing legionellas under conditions that realistically simulate the environment. The main objections to previous studies include the use of pure cultures, since legionellas are incapable of growing alone in water, the use of growth media rather than water, the supplementation of water with additives, particularly rich nutrients that encourage excessive growth, and the use of recirculating systems which are rarely appropriate. Continuous culture is arguably the best tool for investigating microbial ecology in the laboratory and for modelling open-flow nutrient systems. With these problems in mind, a chemostat model was developed with a natural inoculum and water as the nutrient.

Development of the Chemostat Biofilm Model

Inoculum

The inoculum for the mixed culture work was derived from a calorifier that harboured an indigenous population of pathogenic *L. pneumophila* serogroup 1 Pontiac, which had been responsible for an outbreak of Legionnaire's disease. The microflora from the calorifier was concentrated by filtration and resuspended into sterile water before inoculation into the chemostat. The inoculum, which was not subcultured before inoculation in order to avoid artificial selection of the population, consisted of a diverse range of bacteria, fungi and protozoa, including principally *Alcaligenes, Acinetobacter, Achromobacter, Aeromonas, Flavobacterium, Methylobacter, Vibrio, Pseudomonas* spp. and *Actinomycetes*. The protozoa included amoebae and ciliates.

Chemostat model

The chemostat model consisted of two or three fermenter vessels linked in series to reproduce the conditions found in a water system. The first vessel was analogous to a reservoir or a storage or holding tank within a building; the second and third vessels represented a distribution system or a cold or hot water plumbing system within a building. The chemostat design was modified from that originally used for modelling caries or periodontal disease and dental plaque formation in anaerobic environments (Keevil *et al.*, 1987; Keevil, 1989). Careful selection of the materials for the chemostat construction prevented chemical modification of the water. For example, fermenters constructed with stainless steel leach iron, manganese, chromium etc. into the cultures. The vessel top plates were therefore constructed of titanium, the bases were glass and the tubing was silicone rubber. The top plate had eight large ports (22 mm diameter) and four small ports (10 mm diameter) which allowed the insertion of silicone bungs (which could be replaced aseptically with bungs containing the test materials suspended from titanium wires) and electrodes (Fig. 11.1).

The chemostat was supplied with potable water for the growth of the microorganism consortium. The water used was tapwater, river-derived from a lowland catchment and of moderate hardness, and which had previously been found able to support the growth of legionellas in mixed microbial populations (West *et al.*, 1989). The water was transported in 20 l high-density polyethylene containers, stored at 4°C and filter-sterilized with 0.2 μm nylon membranes to prevent alteration of the water chemistry (Colbourne *et al.*, 1988).

The first chemostat vessel had a retention volume of 500 ml and the flow rate of the sterile water into the vessel resulted in a dilution rate of 0.05/h, equivalent to a mean generation time of 13.9 h. When the retention volume was exceeded, the effluent was pumped via an overhead weir system into the second vessel. Additional sterile water was pumped into the second vessel along with the effluent from the first vessel, to maintain a total dilution rate of 0.2/h. The effluent from the second vessel was pumped via an overhead weir into a third vessel, where the efficacy of biocides was evaluated.

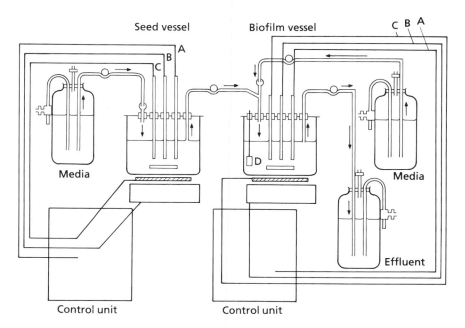

FIG. 11.1. The two-stage version of the continuous-culture biofilm model. A, pH monitoring; B, oxygen/stirrer control; C, temperature control; D, plumbing material.

The environmental parameters of the chemostats were controlled with Anglicon microprocessor control units (Brighton Systems, Newhaven) linked to an IBM-compatible computer. The temperature of the vessels was maintained between 20 and 60°C with proportional integral derivative (PID) controllers and the temperature was measured with a quartz-sheathed temperature probe inserted into the aqueous phase of the vessel; the temperature was corrected with an external heater pad beneath the fermenters. The dissolved oxygen tension (DOT) was maintained at 20% of air saturation via feedback control of the stirrer speed. The fluid velocity in the vessels remained between 1 and 2 m/s and this maintained the Reynolds number well below the transition zone for turbulent flow. This was necessary to keep the chemostat model within the parameters of rates of flow that exist in pipes. E_h and pH were monitored but not controlled throughout the experiments, in order to avoid the need to add titrants that would change water chemistry.

Biofilm was generated on the surfaces of 1 cm^2 coupons of various plumbing construction materials used to contain water. A 1-mm hole was drilled into one end of the coupons so that they could be suspended at the end of titanium wires which were inserted through the silicone bungs of the fermenter top plate. The coupons were cleaned with acetone and the coupon/bung assembly autoclaved before aseptic insertion into the culture. Glass control tiles were inserted simultaneously with the test materials to compare the effects of the individual growth environments investigated. The coupons were periodically removed from the vessel in order to determine biofilm development. Some coupons were examined directly by microscopy, and others were washed gently in 10 ml of sterile water and the biofilm removed from the materials by scraping with a dental probe. Subsequent staining of the coupons with crystal violet or alcian blue confirmed the complete recovery of microorganisms and exopolymers, respectively. The recovered biofilm was resuspended into 1 ml of sterile water and vortexed to disperse the microorganisms. These samples were serially diluted before plating on to various agar media. The planktonic phase was also monitored when biofilms were to be removed.

Microbial quantification and identification

The microbiological assessment of biofilm and planktonic growth included the use of selective and non-selective media. Total microbial populations were enumerated on low-nutrient R2A medium (Reasoner & Geldritch, 1985) to avoid substrate shock to oligotrophic species. Buffered charcoal yeast extract (BCYE) agar was used to culture fastidious bacteria, including legionellas (Pasculle *et al.*, 1980). This medium was supplemented with glycine, vancomycin, polymixin and cycloheximide to produce GVPC agar, a selective medium for legionellas (Dennis *et al.*, 1984). Plates were incubated at 30°C for 7 days and then counted.

Biofilm development was assessed by scanning electron microscopy and differential interference contrast microscopy, to evaluate protozoal populations. Legionellas and other bacteria isolated from the biofilm and planktonic phases were identified with the Biolog Gram-negative identification system and database (Mauchline & Keevil, 1991).

Uses for the Model

Assessment of materials used in water systems

The successful generation of biofilm on a wide range of plumbing materials has made it possible to investigate the effect of the substratum. The current British Standard 6920 (Anon, 1988) uses a batch method and does not include the growth of bacterial pathogens, including legionellas. In the trials of the materials used within water systems the negative control surface was glass and the positive control was latex, which has failed the BS6920 test. Biofilms were generated in River Thames-derived water at 30°C over 28 days, and monitored after 1, 4, 7, 14, 21 and 28 days. Biofilm developed on the materials at different rates and was composed of differing populations of microorganisms. By this method it has been shown that the materials can affect the growth of the legionellas by providing nutrients to the microbial consortium, as is the case for latex, or by inhibiting microbial growth, as for copper.

Copper supported sparse microbial growth and the legionellas comprised a very low proportion of the population (Table 11.1). The low colonization of

TABLE 11.1. *Comparison of the ability of plumbing materials to support biofilm formation and colonization by Legionella pneumophila serogroup 1 Pontiac*

Materials	Maximum colonization		Colonization ratio	
	Flora*	*L. pneumophila**	Flora	*L. pneumophila*
Copper (aged)	70	0.7	1	1
Glass	150	1.5	2.6	2.9
Polybutylene	180	2.0	2.1	2.1
Polyethylene	960	23.0	13.7	33.0
uPVC	1070	11.0	15.3	15.7
cPVC	1700	78.5	24.3	112.1
Ethylene propylene	27 000	500	386	714
Latex	89 000	550	1271	785

* Colonization units are 10^3 cfu/cm². The colonization ratio is the cfu of the total microbial flora or legionellas recovered from each of the materials referred to the copper data. Data from Rogers *et al.* (1991).

copper surfaces was presumed to be due to the inhibitory effect of copper ions, either by selectively inhibiting the legionellas or by inhibiting the organisms that support their growth.

The most prolific biofilms were those that developed on the surface of elastomeric materials; these biofilms consisted of extracellular polymer in which the microorganisms were embedded (Fig. 11.2). The biofilm covered the entire elastomer surfaces after only 24 h and contained more than 8.9×10^6 cfu/cm^2. The elastomeric materials did not appear to support the growth of protozoa, though they were clearly present on the glass control surface. This may indicate that these particular materials favoured the growth of legionellas, in the apparent absence of amoebae and other protozoa, by supporting the presence of a consortium of bacteria which provided essential nutrients for the pathogen.

The colonization of plastic surfaces was intermediate between the copper and the elastomeric surfaces (Table 11.1), with a diverse range of colonization characteristics within the group. The cPVC biofilm supported the most abundant growth of the plastics, and the increase in the numbers of legionellas in the biofilm was disproportionately higher than the increase in total numbers of microbes. The amoebae present on the surface of the material may have contributed to the increase in the numbers of legionellas by acting as hosts for bacterial growth. Their presence on this plastic surface is probably due to the presence of bacterial species favoured as food sources by the protozoa. Polyethylene and polypropylene exhibited the same kind of increase in numbers of legionellas and total flora, but to a lesser extent than the cPVC. The uPVC and polypropylene surfaces formed biofilms that had proportional increases in both the total flora and the legionellas.

The laboratory biofilm model, in its contained Class III cabinet facility, allows rapid assessment under 'worst-case' conditions, since the numbers of legionellas that grow in the planktonic and sessile biofilm phases are probably high enough, under the appropriate environmental conditions, to cause outbreaks if they occur in a cooling tower. The model studies described here indicate that some of the plastic pipe plumbing materials, which are beginning to rival copper, may be inappropriate and are a cause of concern for the public health in some physicochemical environments. For example, polyethylene is used routinely for cold water supplies, whereas polybutylene has found favour for hot water systems in the USA and parts of Europe. Although these materials are acceptable according to BS6920, neither appears particularly suitable when assessing colonization by pathogenic legionellas. Several of the plastics supported biofilms, more than 4% of the population of which was *L. pneumophila*, at 30°C.

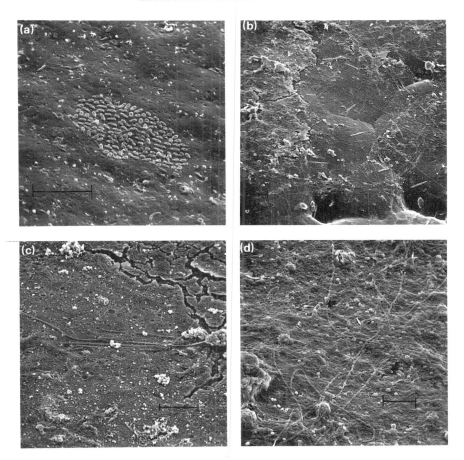

FIG. 11.2. Scanning electron micrographs of biofilms developed in the chemostat. (a) A microcolony growing on a newly immersed cPVC tile. (b) A 21-day-old biofilm supported by polpropylene illustrating the basal layer of the biofilm with thicker layers occurring in some areas. (c) After only 24 h an apparent confluent layer of biofilm had developed on the elastomer ethylenepropylene: the cells are visible within the extracellular polymer. (d) A mature biofilm on latex, showing that the microorganisms recolonize the top of the extracellular products of previous growth (bar = 10 μm).

Studying the ecology of Legionella pneumophila

The above results demonstrate that there is no direct relationship between the total biofilm count and the number of pathogens incorporated into the biofilm. This depends on the bacteria and protozoa, which may either aid the survival and growth of the legionellas, or inhibit them. The effect of environmental parameters that affect the ecology of *L. pneumophila* in biofilms has been investigated by altering the temperature of the model and the water chemistry. This allows the ecology of the growth of legionellas to be examined in greater detail than has previously been possible. The reproducible nature of the chemostat model means that the results are directly comparable. Thus, the environmental conditions of the second or subsequent vessels can be altered to adverse conditions, for example by investigating extremes of temperature or biocide concentration, without modifying the complex microbial inoculum entering from the first 'seed' vessel. This ensures reproducibility and, equally importantly, sustains a continual challenge of microorganisms through a treated environment analogous to that occurring in the field. Conversely, batch-culture modelling systems cannot be resuscitated after treatment regimens and are inappropriate to model open flow systems. Moreover, an accumulation of metabolic products and changes in pH in the closed system may occur, leading to either inhibition or modification of the microbial population.

The effect of water chemistry on legionella ecology was investigated by supplying the model with differing source waters, including a soft, upland catchment water, a river-derived water of medium hardness and a borehole-derived hard water. Marked differences in the proportion of different species within the microbial populations were observed, depending on the water chemistry (Rogers *et al.*, 1992). In particular, the numbers of *L. pneumophila* were fivefold lower in the medium-hard river water than in the hard borehole water. When the hard water was further investigated at different temperatures, the biofouling of polybutylene pipe was maximal at 50°C. However, the colonization by *L. pneumophila* was maximal at 40°C. Nevertheless, the pathogen was still detected in the biofilm at 50°C, contrary to the widespread belief that it does not survive at this temperature.

The comparative resistance to heat of biofilms containing legionellas such as *L. bozemanii* in hot water systems operating at less than 50°C was noted during field studies in Scottish institutional buildings (Keevil *et al.*, 1989a). The continuous-culture biofilm model confirmed the correlation between increasing water temperatures to over 50°C and biofilm reduction observed in the institutional buildings (Walker *et al.*, 1991). The easiest procedure might, therefore, be to 'pasteurize' systems intermittently. For example, the water of large institutional buildings is hardly used at night and could be pumped from top to bottom at 60°C, as recommended by the UK DHSS Code of Practice (1988). The finding of *L. bozemanii* in an institutional building at the bottom of a calorifer, running at 60°C at the outlet, shows the need to open all outlets in the system, if only for a few minutes each, to ensure that heat penetrates all deadlegs. Nevertheless, the institutional building survey and the laboratory biofilm model indicate that some aquatic bacteria can survive for many weeks at 60°C. Thus, although the Code of Practice may be sufficient to control pathogenic species, it may be necessary to heat water systems to perhaps 70°C for, say, 1 h to control generalized biofouling. These heating procedures may have merit in being less labour-intensive than cleaning alone. The rate of biofilm buildup arising between treatments could be assessed by incorporating a continuous oxygen concentration monitor into the system and watching for waters that become anoxic. If other measures are taken to reduce the nutrient input to the hot water systems — such as tank cleaning, reduced storage capacity and water treatment — the frequency of heat treatment could be still further reduced.

Evaluation of biocides

Biofilm formation in drinking water is responsible for such diverse processes as corrosion (Walker *et al.*, 1991), the maintenance of legionellas, other pathogens and indicator organisms (Keevil *et al.*, 1989b; see also Chapter 12), and nitrite formation by oxidation of ammonia (Mackerness and Keevil, unpublished data). It is important to apply preventative measures to minimize these problems. Routine cleaning is clearly important, but will not prevent viable bacteria recolonizing the systems. Regular physical cleaning, such as forcing a tight-fitting plug through a pipe ('pigging'), would help control the extent of biofilm formation, particulary if combined with the application of dispersants (organic acids or alkalis), but this is not practical in the tortuous plumbing of institutional buildings. The only viable alternative currently available is chemical disinfection.

To help overcome problems with pathogenic legionellas, the UK DHSS Code of Practice (1988) recommends that hospital cooling towers are cleaned twice a year. This requires draining out the water and stripping down the tower and its contents, including louvres, baffles, drift eliminators and the filler pack. The cleaned systems are then flushed and refilled with clean water, and usually disinfected with chlorine to give a residual concentration of 15 ppm free chlorine. For cold and hot water potable supplies, including hot water calorifiers and services and whirlpool spa baths, residual disinfection is commonly practised. The most used disinfectant is chlorine, but chlorine dioxide and monochloramine have also been used for drinking water, whereas ozone or bromination have been used to treat swimming baths and whirlpools. Chlorination is a relatively cheap and practical method, but it should not be forgotten that chlorine is very reactive and will combine readily with organic compounds within the water, making it necessary to add more than theoretical calculation would suggest. Furthermore, the biocidal activity of chlorine is very sensitive to the pH of water and decreases rapidly for pH values over 7. Thus a threefold increase in the measured free chlorine is required at pH 8 to provide a similar biocidal activity to that at pH 7. In practice, the maintenance of an effective free residual chlorine concentration in a water system cannot be relied upon to prevent biofilm formation, and legionellas have been recovered from water containing up to 7 ppm free chlorine (Tobin *et al.*, 1986). Such caveats have encouraged the development of proprietary biocides, such as isothiazolones, to control biofouling and to minimize the presence of pathogens such as legionellas in recirculating cooling towers and air-conditioning humidifiers (McCoy *et al.*, 1986). Nevertheless, many biocide studies disregard the important of the presence of biofilm in systems and are therefore flawed.

The most commonly used model for the inactivation of microorganisms by disinfectants is based on the 'Chick−Watson law', which states that $\ln (N/N_o) = -kC^n t$, where N/N_o is the ratio of surviving organisms at time t, C is the disinfection concentration, and k and n are empirical constants; n is known as the concentration exponent). This implies that disinfection concentration and contact time, the $(C \times T)$ factor, are the two key variables that determine disinfection efficacy and it has had a profound influence on disinfection regulations. However, applications of the 'law' have assumed complete and uniform mixing of microorganisms and disinfectant, which ignores that diffusion may be rate-limiting and that disinfectant concentration may decrease with time. Indeed, most data have been obtained from laboratory studies with monodispersed organisms. Importantly, Ridgway & Olson (1982) have shown that the majority of viable bacteria in chlorinated potable water are attached to surfaces such as particles, where they presumably survive in biofilms that do not behave as assumed in Chick−Watson disinfection kinetics. In more recent work, the 'biofilm factor' has been regarded as important and the efficacy of alternative disinfectants such as chlorine dioxide and monochloramine have been investigated. Although hypochlorous acid was a superior disinfectant for heterotrophic and coliform bacteria in the planktonic water phase, especially when compared with monochloramine, the latter was far more potent against sessile biofilm bacteria growing on activated carbon granules or steel (LeChavellier *et al.*, 1988) or bitumen-painted cast iron (Keevil *et al.*, 1990; see also Chapter 12). This may indicate that greater attention should be paid to new strategies for the choice and evaluation of disinfectants, their method of delivery and contact time in water systems.

The continuous-culture biofilm model offers a reproducible means for determining the efficiency of biocides in specific physicochemical environments on both preformed biofilm or clean surfaces before colonization by pioneer bacteria from planktonic microbial communities. The biocides may be evaluated by constant addition to the vessel or by pulsing predetermined concentrations. This methodology is particularly important for testing disinfectants against waterborne pathogens such as legionellas since the most appropriate concentrations and timing of additions can be evaluated without the need for extensive pilot-scale cooling tower rigs and the associated danger of releasing infectious agents to the environment.

TABLE 11.2. *Resistance to biocides of planktonic and sessile biofilm populations containing Legionella pneumophila on mild steel*

Biocide	Planktonic		Biofilm	
	Non-legionella	*L. pneumophila*	Non-legionella	*L. pneumophila*
Control	2×10^5	2×10^3	5×10^6	5×10^5
DIDAC*				
300 ppm	0	0	1×10^5	1×10^3
600 ppm	0	0	7×10^4	0
BCHD†				
10 ppm	0	0	4×10^3	1×10^2

* Diisodecyl dimethylammonium chloride.
† 2-2′-bromo-chloro-dimethylhydantoin.
Mature biofilms were exposed to DIDAC for 48 h or BCHD for 30 h; planktonic populations were exposed to each biocide for 1 h.

Mild steel coupons began to rust when immersed in the biofilm model and provided an excellent substratum for biofouling and colonization by *L. pneumophila* (Table 11.2). The legionellas continued to grow well in the planktonic phase but were killed, along with other members of the microbial population, when treated with 300 ppm of the proprietary biocide, diisodecyl dimethylammonium chloride. This high concentration was not, however, sufficient to eradicate the biofilm and kill the legionellas within it. Even a concentration of 600 ppm only reduced the biofouling by two orders of magnitude, but legionellas were eradicated at this concentration. Similarly, 2-2′-bromo-chloro-dimethylhydantoin is now a widely used oxidizing disinfection and, at its recommended concentration of 10 ppm, it was able to kill the planktonic populations of the microbial microflora and the legionellas. Nevertheless, the bacteria within the biofilm survived a 30-h exposure to the recommended concentration. These data re-emphasize the refractory nature of biofilms and the care needed in selecting the appropriate eradication regimen.

The ability of new inhibitory surfaces to prevent biofouling and colonization by pathogens, such as legionellas, can be evaluated by comparing them with control materials immersed in the model system with defined microbial communities. Moreover, novel paints with antifouling or biocidal properties can be compared with unpainted substrata, by immersion in the model. For example, a paint containing titanium dioxide particles coated with silver was able to suppress biofouling and colonization by *L. pneumophila* serogroup 1 Pontiac for several weeks until the silver had leached out of the system (Rogers and Keevil, unpublished data).

Investigating biofilm topology

The large numbers of biofilm samples that have been examined after development in a variety of defined physicochemical environments have led to a greater understanding of how biofilm develops on materials and the structure that is formed. Microscopical evaluation of biofilm development has shown that the microorganisms initially attach preferentially to the surfaces if they offer a nutrient supply, and growth then occurs to form small microcolonies. After about 14 days the microcolonies produce a film over the surfaces. The biofilm, when fully mature, consists of a low basement of microorganisms from which tall stacks of microorganisms rise. The basal background is approximately 5 μm in depth and the stacks reach up to 100 μm in height (Fig. 11.3). When the unstained biofilm is viewed under low-power light microscopy, it is evident that the microorganisms exist in microcolonies which are visible because of their various colours of white, grey, yellow, pink and red. Some of the stacks of microorganisms are closely associated layers of two or more species growing in bands, possibly in some form of symbiosis.

The heterogeneous biofilm mosaic supports the growth of a range of protozoa that can be observed grazing on the microcolonies. The protozoa have a disruptive effect on the biofilm and may enhance the loss or 'sloughing' of particles of biofilm into the aqueous phase. They may selectively graze on the biofilm population and in turn affect fastidious pathogen numbers by selecting a microbial population that inhibits or encourages pathogen growth. In addition, the protozoa may directly affect pathogen numbers by acting as a host for the organism.

Recent work has shown that a monoclonal antibody (MAb) raised against the lipopolysaccharide of *L. pneumophila* can be used to 'tag' and assist in

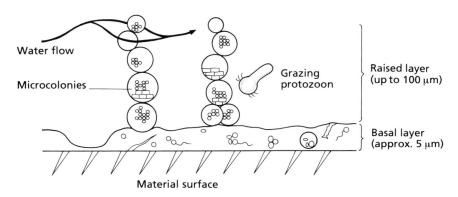

FIG. 11.3. Diagrammatic representation of a mature biofilm formed in the chemostat.

locating the pathogen in biofilms (Rogers & Keevil, 1992). The biofilms were recovered from the continuous-culture model and incubated with the MAb for 8 h at 4°C to avoid non-specific binding. Subsequent studies have cut this incubation time considerably. After gentle washing, the captured MAb was tagged with an antimouse polyclonal antibody linked to fluorescein isothio-cyanate (FITC) or 5 nm gold particles. These markers were subsequently revealed by examining the biofilms by episcopic fluorescence or differential interference contrast (DIC) microscopy, respectively. A novel configuration of a Nikon microscope equipped with 150× magnification, non-contact, long working-distance objective permits visualization of the probes. The episcopic design of the ultraviolet lighting assembly has allowed the observation of biofilms on opaque substrata such as metals and plastic pipe materials used to construct plumbing systems (Keevil & Walker, 1992). The FITC-labelled legionellas fluoresce apple-green in the fluorescence mode, whereas the immunogold-labelled legionellas appear silver-gold in DIC mode (Rogers & Keevil, 1992). The former procedure allows rapid scanning of materials for individual fluorescing legionellas against a black background, and DIC permits excellent visualization of the biofilm detail in which the immunogold-labelled legionellas are revealed.

These advances in modelling biofilm formation and rapid visualization of pathogenic *L. pneumophila* within their biofilm haven are beginning to provide new insights into the ecology of legionellas and their interaction with amoebae and other protozoa. This knowledge should allow the development of new construction materials and better biocide design strategies for the future.

Acknowledgement

The technical assistance of Mrs E. Elphick and the computational skills of Mr R. Knowles are gratefully acknowledged.

References

ANON. (1988) *Suitability of Non-Metal Products for Use in Contact with Water Intended for Human Consumption with Regard to their Effect on the Quality of the Water*. British Standards Institute BS6920, Section 2.4 Growth of Aquatic Microorganisms. British Standards Institution, London.

COLBOURNE, J.S., TREW, R.W. & DENNIS, P.J. (1988) Treatment of water for aquatic bacterial growth studies. *Journal of Applied Bacteriology*, **65**, 1–7.

DENNIS, P.J., BARTLETT, C.L.R. & WRIGHT, A.E. (1984) Comparison of isolation methods for *Legionellae* spp. In Thornsbury, C., Balows, A., Feeley, J.C. & Jakubowski, W. (eds.) *Legionella: Proceedings of the 2nd International Symposium*, pp. 294–296. American Society for Microbiology, Washington.

DHSS (1988) *Code of Practice: The Control of Legionella in Health Care Premises*. HMSO, London.

ELLWOOD, D.C., KEEVIL, C.W., MARSH, P.D. & WARDELL, J.N. (1982) Surface-associated growth. *Philosophical Transactions of the Royal Society of London B*, **297**, 513–532.

FITZGEORGE, R.B., BASKERVILLE, A., BROSTER, M.G., HAMBLETON, P. & DENNIS, P.J. (1983) Areosol infection of animals with strains of *Legionella pneumophila* of different virulence: comparison with intraperitoneal and intranasal route of infection. *Journal of Hygiene*, **90**, 81–90.

GELDRICH, E.E., NASH, H.D., REASONER, D.J. & TAYLOR, R.H. (1972) The necessity of controlling bacterial populations in potable water: community water supply. *Journal of the American Water Works Association*, **64**, 92–96.

KEEVIL, C.W. (1989) Chemostat models of human and aquatic corrosive biofilms. In T. Hattori (ed.) *Recent Advances in Microbial Ecology*, pp. 151–156. Japan Scientific Societies Press, Tokyo.

KEEVIL, C.W. & WALKER, J.T. (1992) Normarski DIC microscopy and image analysis of biofilms. *Binary*, **4**, 92–95.

KEEVIL, C.W., BRADSHAW, D.J., DOWSETT, A.B. & FEARY, T.W. (1987) Microbial film formation: dental plaque deposition on acrylic tiles using continuous culture techniques. *Journal of Applied Bacteriology*, **62**, 129–138.

KEEVIL, C.W., WALKER, J.T., McEVOY, J. & COLBOURNE, J.S. (1989a) Detection of biofilms associated with pitting corrosion of copper pipework in Scottish hospitals. In Gaylarde, C.C. & Morton, L.H.G. (eds.) *Biocorrosion*, pp. 99–117. Biodeterioration Society, Kew.

KEEVIL, C.W., WEST, A.A., WALKER, J.T., LEE, J.V., DENNIS, P.J. & COLBOURNE, J.S. (1989b) Biofilms: detection, implications and solutions. In Wheeler, D., Richardson, M.L. & Bridges, J. (eds.) *Watershed 89: The Future of Water Quality in Europe*, Vol. II, pp. 367–374. Pergamon Press, Oxford.

KEEVIL, C.W., MACKERNESS, C.W. & COLBOURNE, J.S. (1990) Biocide treatment of biofilms. *International Biodeterioration*, **26**, 169–179.

LECHAVELLIER, M.W., CAWTHORN, C.D. & LEE, R.G. (1988) Inactivation of biofilm bacteria. *Applied and Environmental Microbiology*, **54**, 2492–2499.

McCOY, W.F., WIREMAN, J.W. & LASHEN, E.S. (1986) Efficacy of methylchloro/methylisothiazolone biocide against *Legionella pneumophila* in cooling tower water. *Journal of Industrial Microbiology*, **1**, 49–56.

MAUCHLINE, W.S. & KEEVIL, C.W. (1991) Development of the BIOLOG substrate utilization system for identification of *Legionella* spp. *Applied and Environmental Microbiology*, **57**, 3345–3349.

PASCULLE, A.W., FEELEY, J.C., GISBSON, R.J., CORDES, L.G., MYEROWITZ, P.L., PATTON, C.M., GORMAN, G.W., CARMACK, C.L., EZZELL, J.W. & DOWLING, J.N. (1980) Pittsburgh pneumonia agent: direct isolation from human lung tissue. *Journal of Infectious Diseases*, **141**, 727–732.

REASONER, D.J. & GELDRICH, E.E. (1985) A new medium for the enumeration and subculture of bacteria from potable water. *Applied and Environmental Microbiology*, **49**, 1–7.

RIDGWAY, H.F. & OLSON, B.H. (1982) Chlorine resistance patterns of bacteria from two drinking water distribution systems. *Applied and Environmental Microbiology*, **44**, 972–987.

ROGERS, J. & KEEVIL, C.W. (1992) Immunogold and fluorescein immuno-labelling of *Legionella pneumophila* within an aquatic biofilm visualised using episcopic differential interference contrast microscopy. *Applied and Environmental Microbiology*, **58**, 2326–2330.

ROGERS, J., DOWSETT, A.B., LEE, J.V. & KEEVIL, C.W. (1991) Chemostat studies of biofilm development on plumbing materials and the incorporation of *Legionella pneumophila*. In Rossmoore, H.W. (ed.) *Biodeterioration and Biodegradation*, Vol. 8, pp. 458–460. Elsevier Applied Science, New York:

ROGERS, J., DENNIS, P.J., LEE, J.V. & KEEVIL, C.W. (1992) The effect of water chemistry and temperature on the survival and growth of *Legionella pneumophila* in potable water systems. In Barbaree, J.M. (ed.) *Fourth International Symposium on Legionella*. American Society for Microbiology, Washington.

ROWBOTHAM, T.J. (1980) Preliminary report on the pathogenicity of *Legionella pneumophila* for freshwater and soil amoebae. *Journal of Clinical Pathogenesis*, 33, 1179–1183.

TISON, D.L., POPE, D.H., CHERRY, W.B. & FLIERMANS, C.B. (1985) Growth of *Legionella pneumophila* in association with blue-green algae (Cyanobacteria). *Applied and Environmental Microbiology*, 39, 456–459.

TOBIN, R.S., EWEN, P., WALSH, K. & DUTKA, B. (1986) A survey of *Legionella pneumophila* in water in twelve Canadian cities. *Water Research*, 20, 495–500.

WADOWSKY, R.M. & YEE, R.B. (1983) Satellite growth of *Legionella pneumophila* with an environmental isolate of *Flavobacterium breve*. *Applied and Environmental Microbiology*, 46, 1447–1449.

WALKER, J.T., DOWSETT, A.B., DENNIS, P.J.L. & KEEVIL, C.W. (1991) Continuous culture studies of biofilm associated with copper corrosion. *International Biodeterioration*, 27, 121–134.

WEST, A.A., ARAUJO, R., DENNIS, P.J., LEE, J.V. & KEEVIL, C.W. (1989) Chemostat models of *Legionella pneumophila*. In Flannigan, B. (ed.) *Airborne Deteriogens and Pathogens*, pp. 107–116. Biodeteriation Society, Kew.

WHIPPLE, G.C., FAIR, G.M. & WHIPPLE, M.C. (1927) *The Microscopy of Drinking Water*, 4th edn. Wiley, New York.

12: Formation and Control of Coliform Biofilms in Drinking Water Distribution Systems

C.W. Mackerness[1], J.S. Colbourne[2], P.L.J. Dennis[2], T. Rachwal[2] and C.W. Keevil[1]

[1]*Pathology Division, PHLS Centre for Applied Microbiology and Research, Porton Down, Salisbury SP4 0JG; and* [2]*Thames Water Utilities PLC, Water and Environmental Sciences, Newgent House, Vastern Road, Reading RG1 2GB, UK*

Under the Water Quality and Public Health legislation of the UK, water undertakings are obliged to provide wholesome water and to monitor the wholesomeness and sufficiency of supplies. The presence of coliform bacteria in general, and *Escherichia coli* in particular, in drinking water is therefore a public health concern, because it is assumed that these organisms are the result of faecal contamination of human or animal origin. Water quality regulations (Anon., 1982) state that, ideally, *E. coli* should not be detectable in a 100-ml sample of drinking water, that no sample should contain more than three coliforms, that coliform organisms should not be detectable in any two consecutive samples, and that 'for any given distribution system, coliform organisms should not occur in any more than 5% of routine samples'.

Coliform bacteria are defined as organisms that produce acid and gas from lactose or mannitol at 37°C or 44°C. The presence of *E. coli* can be established by indole production at 44°C, or by the use of other appropriate identification methods.

It is at present assumed that the presence of coliform bacteria in drinking water may be due to (i) a loss of residual disinfectant (e.g. chlorine); (ii) back-siphonage, cross-connections, line breaks, and/or repair in the distribution main; (iii) survival and recovery of injured organisms; and (iv) failure of the treatment plant.

The presence of coliforms in drinking water, when there are no known

Microbial Biofilms:
Formation and Control

breaches in treatment barriers, and the presence of these organisms in the absence of any evidence of faecal contamination, continues to be a major problem for the water industry, and has emerged as a critical regulatory issue.

A more serious consideration is the growth and/or survival in potable water supplies of bacteria that belong to the coliform group, and that may mask the presence of other indicator organisms as a result of *true* treatment plant failure. The chronic presence of coliform bacteria in drinking water supplies is characterized by the absence of coliforms from water that leaves the treatment plant; the routine presence of coliforms in distribution supply samples at various points; the persistence of coliforms in the system in spite of the maintenance of a disinfectant residual that is assumed to be effective; and the persistence of the problem over a long time.

The successful control of such situations has been reported infrequently, and little research has been done on potential causes of the problem.

After leaving the treatment plant, drinking water is not completely free from microorganisms and the distribution supply has a diverse bacterial flora. These flora are only considered in terms of their influence on aesthetic or operational properties, such as taste and odour, and it is assumed that, *per se*, they do not constitute a significant health hazard.

Although residual disinfection methods for drinking water vary throughout the world, in the UK residual chlorine between 0.1 and 0.5 mg/l has been regarded as sufficient to preserve the bacterial quality of water during distribution. Nevertheless, maintenance of an effective free residual chlorine does not prevent biofilm formation. Biofilms on the pipework of the distribution supply occur as a result of attachment of microorganisms to surfaces, followed by growth, formation of products such as extracellular polysaccharide, and, finally, sloughing of bacteria into the system. There are a number of important ecological considerations in relation to biofilm formation:

1   the accumulation of macromolecules at solid/liquid interfaces to create a favourable environment in a situation otherwise deficient in nutrient;

2   the involvement of extracellular bacterial polymers in the attachment of bacteria to surfaces and entrapment of nutrients;

3   the protection of biofilm bacteria from residual chemical disinfectants.

The question arises as to whether coliforms can become part of such a biofilm community. If so, the sloughing of such organisms from the biofilm into the flowing water could account for their presence, in the absence of evidence of contamination or plant failure. There is also some evidence that regrowth, or aftergrowth, of total coliforms and aeromonads takes place in fully treated water where there has been no evidence of accidental faecal contamination (Payment *et al.*, 1988).

Although *E. coli* has attracted the greatest interest, in terms of the assessment of microbiological quality of drinking water, other bacteria have recently been considered as waterborne pathogens. For example, although they are not coliforms, aeromonads are ubiquitous in freshwater environments and some *Aeromonas* spp. are increasingly recognized as the aetiological agents of gastro-intestinal disease (reviewed by Khardori & Fainstein, 1988; Altwegg & Geiss, 1989; Janda, 1991). Although cases of infection by *Aeromonas* spp. have been described (Burke *et al.*, 1984a), evidence for their pathogenicity remains contentious, because there is no adequate animal model and Koch's postulates remain to be satisfied. Significantly, the study by Burke *et al.* (1984a) provided some evidence that *Aeromonas*-associated diarrhoea in Australia correlated with the isolation of *Aeromonas* spp. from the chlorinated tap water supply, since both were greatest in summer. It was further suggested that the isolation of this organism was related to water temperature, residual chlorine and the interaction between these two variables. In contrast, another study by the same group (Burke *et al.*, 1984b) showed that, for an unchlorinated domestic supply, both clinical and environmental isolations continued in the winter. This suggests that observations of clinical incidence are independent of temperature but are related to another variable, such as bacterial counts.

Distribution System Biofilm Formation

Van der Kooij (1991) has suggested that the multiplication of aeromonads in drinking water during distribution is by the formation of biofilm within the pipework, but there is, as yet, no direct evidence of this. Since the examination of distribution system biofilms is difficult, a laboratory model of bacterial biofilm formation is described here, with particular reference to colonization by *E. coli* and *Aeromonas hydrophila*, and some strategies for the control of the biofilm are discussed.

Continuous-Culture Model

The biofilm model consisted of two continuous-culture glass vessels in series (Fig. 12.1), each with a working volume of 500 ml. The top plate and sampling ports were made of titanium and all connectors and probes in the vessels were made of glass, so that there was no leaching of metal ions into the cultures. The dilution rate of the first vessel, which was used as the control, was 0.05/h and the effluent from this vessel supplied the second vessel, which was supplemented with additional medium to give a final dilution rate of 0.2/h. The medium for the growth of bacteria in both vessels was nylon membrane filter-sterilized tapwater from the London distribution system. Previous studies had shown that this simple membrane filtration technique provided water that was both sterile and chemically unaltered (Colbourne *et al.*, 1988). The cultures were maintained aerobically at a constant temperature of 25°C and

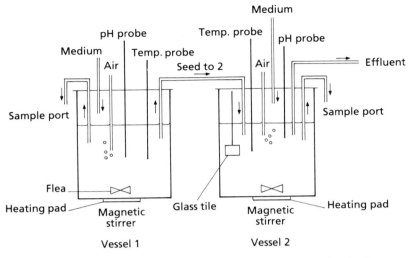

F<small>IG</small>. 12.1. The two-stage continuous-culture model for the study of biofilm development.

the pH was monitored but not controlled, because the addition of acid or alkali might affect the water chemistry.

The inoculum for the model

Tap water (10 l) was passed through a 0.2 μm nylon membrane and the organisms trapped on the membrane were resuspended in Page's amoebal saline (PAS; Page, 1967) and used to inoculate the chemostats — that is, the organisms were not subcultured. The bacteria of the inoculum were Gram-negative rods and some of these were identified by the API 2ONE (API Biomeriux, Basingstoke, UK) and BIOLOG GN (Biolog Inc., Haywood California, USA) system according to the manufacturers' instructions. The predominant identifiable organisms of the inoculum and biofilm are listed in Table 12.1. *Escherichia coli* and *A. hydrophila* were environmental isolates obtained from Thames Water Utilities plc, and were inoculated as 5-ml suspensions of a 10^9 cfu/ml culture into the model when it was at steady state.

Formation of biofilms in the model

Biofilms were grown on 1 cm^2 glass tiles, as an inert control surface, and bitumen-painted (Stanton and Stavely, Nottingham, UK) mild steel as used in line distribution mains. The paint thickness was 70 μm and all tiles were 'cured' by immersion for 7 days in sterile water before use. Tiles were

TABLE 12.1. *Bacteria identifiable in the inoculum and biofilm in the model*

Inoculum	Biofilm
Flavobacterium spp.	*Flavobacterium* spp.
Alcaligenes spp.	*Alcaligenes latus*
Acinetobacter spp.	*Acinetobacter* spp.
Achromobacter spp.	*Achromobacter xyloxidans*
Pseudomonas spp.	*Pseudomonas fluorescens*
	Pseudomonas paucimoblis
	Pseudomonas vesicularis
	Pseudomonas stutzeri
	Pseudomonas corrugata

suspended in the vessels by titanium wire and were sampled after immersion for various times in the culture. The tiles were removed aseptically and gently washed in PAS, and a sterile dental probe was used to detach the biofilm, which was resuspended in 2 ml sterile water. The sample was rigorously vortexed, serially diluted in sterile water and spread plates were inoculated in duplicate for each dilution and incubated at 30°C for 7 days. At the same time as a biofilm sample was taken, a sample port allowed for the removal of culture for the determination of viable counts of the planktonic (i.e. liquid) phase.

The media

Heterotrophic biofilm bacteria were counted on low-nutrient R2A medium (Reasoner & Geldrich, 1985) which had previously been shown to minimize substrate shock for nutrient-stressed organisms. *Escherichia coli* was counted on MacConkey agar (Oxoid), and *A. hydrophila* on ampicillin−dextrin agar (Havelaar *et al.*, 1987) which contained (g/l): tryptone 5, dextrin 10, yeast extract 2, NaCl 3, KCl 2, $MgSO_4 .7H_2O$ 0.2, $FeCl_3 .6H_2O$ 0.1, bromothymol blue 0.04, ampicillin 0.1% (w/v) 1 ml, sodium desoxycholate 1% (v/v) 10 ml, agar 15.

Control of coliform biofilms

A heterotrophic biofilm was established at steady state, and coliforms were then inoculated as described above. A stock monochloramine solution was prepared and added to the medium of the second chemostat, to give a monochloramine concentration of 0.3 mg/l in the vessel, in order to determine the effect of this residual disinfectant on heterotrophic and coliform biofilms.

Biofilm Development

The inoculum contained a great variety of bacteria and the predominant organisms were identified to genus level. The biofilms also contained a wide species diversity, and organisms from similar genera were identified in both the inoculum and the biofilm (Table 12.1). The model is, therefore, an accurate representation of the organisms in the water distribution system. The majority of the organisms in the biofilm were pseudomonads, in particular *Pseudomonas paucimobilis* and *Pseudomonas fluorescens*.

As was expected, the second chemostat at the higher dilution rate of 0.2/h had higher counts for both planktonic and biofilm bacteria, as compared with the chemostat at the lower dilution rate of 0.05/h (Table 12.2). These slightly higher counts in the second vessel may have been due to the increased rate of flow of nutrients at the higher dilution rate, or the fact that this vessel was constantly fed by the effluent from the first vessel, which contained bacterial metabolites. Nonetheless, there were no fluctuations in either the planktonic or the biofilm counts, which implies that both phases are stable.

Glass tiles were used as control surfaces, since it was assumed that glass would be an inert surface with regard to bacterial attachment: bitumen-painted mild steel is the approved material for use in line distribution mains. There was little difference between the planktonic and the biofilm counts for both materials, which suggests that no materials inhibitory or stimulatory for bacterial attachment or growth leached from the bitumen paint (Table 12.3). From the point of view of water companies, this means that bitumen-painted mild steel is a good material for lining mains, because this material appears not to affect the overall bacteriological quality of the water for distribution.

Once a stable heterotrophic biofilm had been established, after 7 days on bitumen-painted mild steel tiles, the model was challenged with *A. hydrophila*

TABLE 12.2. *The effect of dilution rate on the heterotrophic bacteria of the biofilm and planktonic phases*

Time (days)	Dilution rate			
	0.2/h		0.05/h	
	Biofilm	Planktonic	Biofilm	Planktonic
1	6.3	6.7	4.8	5.7
7	6.0	6.4	5.7	5.7
14	6.2	6.2	5.8	5.6
21	6.6	6.0	5.8	5.0

Viable counts are expressed as $\log_{10}$. Cultures were grown aerobically at 25°C and the biofilm was developed on glass tiles.

TABLE 12.3. *Comparison of the viable heterotrophic planktonic and biofilm bacteria when biofilm was developed on glass or bitumen-painted mild steel*

| Time (days) | Viable counts ($\log_{10}$ cfu/ml) | | | |
| | Glass | | Bitumen-painted mild steel | |
	Biofilm	Planktonic	Biofilm	Planktonic
1	6.3	6.7	6.9	5.7
7	6.0	6.4	5.1	5.8
14	6.2	6.2	6.1	6.0
21	6.6	6.0	6.1	5.7

Cultures were grown aerobically at 25°C and at a dilution rate of 0.2/h.

and *E. coli*. Both organisms became members of the biofilm community, and the viable counts were constant after 7 days (Table 12.4). The fluctuations in the counts for 1 and 4 days were probably due to the re-establishment of the steady state.

After *A. hydrophila* and *E. coli* had become established in the biofilm, monochloramine (0.3 mg/l) was added to the medium of the vessel; this is the concentration regularly used by water undertakings. In previous studies of the action of monochloramine on heterotrophic bacteria in the model (Keevil *et al.*, 1990) there was only a small reduction in the viable counts for both the biofilm and planktonic bacteria. Similarly, at this monochloramine concentration there was no decrease in the viable counts of *A. hydrophila* or *E. coli* — rather, the counts rose slightly (Table 12.5).

TABLE 12.4. *Colonization of a heterotrophic biofilm by Aeromonas hydrophila and Escherichia coli*

| Time (days) | Viable counts ($\log_{10}$ cfu/ml) | | |
	Heterotrophs	*A. hydrophila*	*E. coli*
1	6.1	4.8	5.6
4	5.8	3.3	5.0
7	4.9	3.7	3.4
14	5.8	3.8	3.3
21	5.6	3.8	3.1

Cultures were grown aerobically at 25°C at a dilution rate of 0.2/h, and biofilm was developed on bitumen-painted mild steel.

TABLE 12.5. *The effect of monochloramine (0.3 mg/l) on a heterotrophic biofilm containing Aeromonas hydrophila and Escherichia coli*

Time (days)	Viable counts ($\log_{10}$ cfu/ml)		
	Heterotrophs	*A. hydrophila*	*E. coli*
1	4.9	3.3	2.8
4	5.9	3.0	5.1
7	5.3	4.3	4.6
14	5.6	4.2	4.2
21	5.3	4.5	4.3

The viable counts are expressed as $\log_{10}$. Cultures were grown aerobically at 25°C at a dilution rate of 0.2/h, and biofilm was developed on bitumen-painted mild steel. Monochloramine was added to the culture medium at a concentration of 0.3 mg/l.

Strategies for Biofilm Control

Clearly, ordinary tap water contains sufficient nutrients to support the growth of the heterotrophic bacteria *A. hydrophila* and *E. coli* as biofilms in this unique model system. That the model is an accurate description of a distribution supply is demonstrated by the fact that the model maintained bacterial species similar to those found in field investigations. It can therefore be regarded as representative of the distribution system. It is important to note that, in the first instance, survival of coliforms in the distribution system biofilm requires a stable multispecies heterotrophic biofilm. It is unlikely that under these conditions a biofilm will consist of a single bacterial species; rather, it is the diversity of the biofilm population that makes the community stable.

The data show that bitumen-painted mild steel provided an excellent substratum for the establishment of a stable heterotrophic biofilm and, what is more important, this material did not result in an increase or decrease in the total bacterial flora when compared with glass, as representative of an inert material. The components of the bitumen paint could have leached material to be utilized by or to be inhibitory to biofilm bacteria, but the data show that there was little difference between the biofilm and planktonic viable counts for both glass and bitumen-painted mild steel.

The model has demonstrated that *E. coli* and *A. hydrophila* can, indeed, become part of the autochthonous heterotrophic biofilm, albeit in low numbers. The biofilm bacteria are resistant to residual chemical disinfectants, and this is consistent with evidence from field investigations in the USA, where even relatively high chlorine concentrations (12 mg/l) were insufficient to control coliform problems (Ridgway & Olson, 1982). The precise mechanisms by which bacteria are resistant to disinfection within the water distribution system are poorly understood. However, the best hypothesis would be one consistent with the attachment of bacteria to a surface — biofilm formation — and subsequent capsule formation; any residual disinfectant would then be assumed to be unable to penetrate this capsule layer. Alternatively, the disinfectant may penetrate the biofilm at sublethal doses, which would select for resistant bacteria within the biofilm. Therefore, the occurrence of coliforms in a distribution system which has no known faecal contamination might be due to the detachment of these organisms from the biofilm, from where they can then be transported around the system. This could account for non-compliance of drinking water standards with respect to *E. coli*.

Since the model has demonstrated that *A. hydrophila* and *E. coli* were established in the biofilm, it is appropriate to use the model to investigate suitable methods of biofilm control. If it is assumed that these bacteria are already present in the drinking water supply system, the only course available for their control by the supplier is to use a residual disinfectant. The ideal disinfectant would be one which persists; is non-toxic to man; has a wide mode of action, preferably bactericidal but at least bacteriostatic; does not taint the water with a taste or odour; and is cheap and easy to add to the system.

Monochloramine and free chlorine have for some time been the popular choices for residual disinfectants by many water undertakings. The concentrations of monochloramine routinely used range between 0.1 and 0.5 mg/l, which is thought to be sufficient to control coliform problems. There is, however, increasing evidence that even relatively high concentrations of these disinfectants are ineffective in controlling either the heterotrophic bacteria or coliforms. This study shows that *E. coli* and *A. hydrophila* survive at a monochloramine concentration of 0.3 mg/l for 21 days.

Acknowledgement

The financial support of the Department of the Environment by project grant PECD 7/7/310 is acknowledged.

References

ALTWEGG, M. & GEISS, H.K. (1989) *Aeromonas* as a human pathogen. *Critical Reviews in Microbiology*, **16**, 253–286.

ANON. (1982) *The Bacteriological Examination of Drinking Water Supplies*. Reports on Public Health and Medical Subjects No 71. HMSO, London.

BURKE, V., ROBINSON, J., GRACEY, M., PETERSON, D. & PARTRIDGE, K. (1984a) Isolation of *Aeromonas hydrophila* from a metropolitan supply: seasonal correlation with seasonal isolates. *Applied and Environmental Microbiology*, **48**, 361–366.

BURKE, V., ROBINSON, J., GRACEY, M., PETERSON, D., MEYER, N. & HALEY, V. (1984b) Isolation of *Aeromonas* spp. from an unchlorinated domestic supply. *Applied and Environmental Microbiology*, **48**, 367—370.

COLBOURNE, J.C., TREW, R.M. & DENNIS, P.L.J. (1988) Treatment of water for aquatic bacterial growth studies. *Journal of Applied Bacteriology*, **65**, 79—85.

HAVELAAR, A.H., DURING, M. & VERSTEEGH, J.F.M. (1987) Ampicillin dextrin agar medium for the enumeration of *Aeromonas* species in water by membrane filtration. *Journal of Applied Bacteriology*, **62**, 279—287.

JANDA, J.M. (1991) Recent advances in the study of the taxonomy, pathogenicity and infectious syndromes associated with the genus *Aeromonas*. *Clinical Microbiology Reviews*, **4**, 397—410.

KEEVIL, C.W., MACKERNESS, C.W. & COLBOURNE, J.C. (1990) Biocide treatment of biofilms. *International Biodeterioration*, **26**, 169—179.

KHARDORI, N. & FAINSTEIN, V. (1988) *Aeromonas* and *Plesiomonas* as etiological agents. *Annual Review of Microbiology*, **42**, 395—419.

PAGE, F.C. (1967) Taxonomic criteria for limax amoeba with description of three new species of *Harmannella* and three of *Vohlkampfia*. *Journal of Protozoology*, **14**, 499—521.

PAYMENT, P., GRAMADE, F. & PAQUETTE, G. (1988) Microbiological and virological analysis of water from two water filtration plants and their distribution systems. *Canadian Journal of Microbiology*, **34**, 1304—1309.

REASONER, D.J. & GELDRICH, E.E. (1985) A new medium for the enumeration and subculture of bacteria from potable water. *Applied and Environmental Microbiology*, **49**, 1—7.

RIDGEWAY, H.F. & OLSON, B.H. (1982) Chlorine resistance patterns of bacteria from two drinking water distribution systems. *Applied and Environmental Microbiology*, **44**, 972—987.

VAN DER KOOIJ, D. (1991) Nutritional requirements of aeromonads and ther multiplication in drinking water. *Experientia*, **47**, 444—446.

13: The Control of Biofilm in Recreational Waters

*Microbiology Department, Mater Infirmorum Hospital, Crumlin Road,
Belfast BT14 6AB, UK*

Recreational waters are those found in swimming pools, whirlpools, spas and
hot tubs. There are many variations of the last three, but what they all have in
common is an elevated water temperature and some form of agitation or water
aeration. Hot tubs are generally confined to the USA, and are often small and
made of wood.

Swimming is one of the most popular leisure activities in the UK, and will
be further encouraged by the recent pronouncement of the UK Secretary of
State for Education of the need for all children to be able to swim. The
number of individuals who use swimming pools is very difficult to estimate
but, according to the Sports Council's facilities database, there are 1462 pools
in England (Pool Water Treatment Advisory Group, 1990).

Since the 1970s there has been a very rapid rise in the popularity of
commercial spas, and they are now commonplace in many leisure facilities. In
addition, there has been a great increase in the number of domestic whirlpools
being installed. Figures are not available for the UK, but at the beginning of
1983 in the USA it was estimated that 560 000 whirlpools were in use (Favero,
1984).

Recreational Water Facilities

Domestic whirlpools

Domestic whirlpools are conventional baths modified by the addition of pumps
for water and air (Fig. 13.1). They are given a variety of names, ranging from
whirlpool, turbojet, spa jet, whirl spa, multispa, multijet, hydrojet, to retrojet.
During operation, water is circulated via a pump and introduced through a
number of angled jets placed in the sides and bottom of the bath. Warm air is

Microbial Biofilms:
Formation and Control

Copyright © 1993 by the Society for Applied Bacteriology
All rights of reproduction in any form reserved
0-632-03753-9

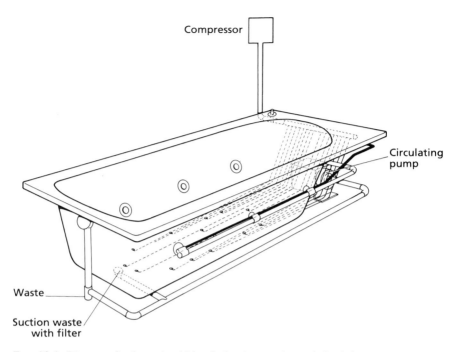

Compressor

Circulating
pump

Waste

Suction waste
with filter

FIG. 13.1. Diagram of a domestic whirlpool, showing aeration and circulation systems.

compressed and introduced through the water jets or injectors on the bottom of the bath. After use the bath is drained through the standard bath outlet and a minimum amount of residual water is left in the pipework. In addition, filters may be included to protect the pump from damage by hair and other material.

Commercial spas

These are designed to achieve the same effect as the domestic whirlpool, but they are much larger, with a seating arrangement under the water to accommodate between eight and 12 persons. Spas are about 4 ft deep and normally operate at temperatures between 30°C and a maximum of 40°C. In most cases they have their own water treatment plant (Fig. 13.2) and, because of the high rate of usage, the turnover rate is generally between eight and ten water changes per hour. Water is filtered through diatomaceous earth, cartridge, medium or slow sand filters and then heated, disinfected with 3−5 mg/l chlorine or 4−6 mg/l bromine, and adjusted to a pH of 7.4−7.6 (Swimming Pools and Allied Trade Association, 1986).

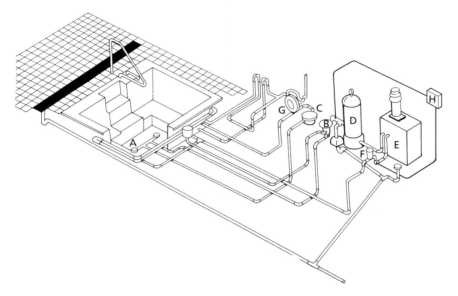

FIG. 13.2. Diagram of a commercial whirlpool to show the filtration, aeration and circulation systems. A, Drain; B, circulation pump; C, hydrotherapy pump; D, filter; E, heater; F, chlorinator; G, air compressor; H, chlorine monitor (adapted from Davis, 1985).

Swimming pools

Swimming pools vary in size from small learner pools to 50-m eight-lane competitive pools. Thus, pool usage and the volume of water present is very variable. In the treatment plant (Fig. 13.3) water is first filtered through sand filters (a few pools still have multimedia filters), heated, and disinfected to approximately 2 mg/l free chlorine. The pH is then adjusted to between 7.4 and 7.6 and the water returned to the pool. In some highly used leisure pools, the average turnover time is 3 h or less (Pool Water Treatment Advisory Group, 1990).

Features of Recreational Waters

Recreational waters have the following features in common.
1 *Temperature.* Waters are held at elevated temperatures: that of the domestic whirlpool depends on the temperature of the domestic hot water supply.
2 *Recirculation.* In all systems there is a recirculation of water. In swimming pools and spas, water is filtered during this process.

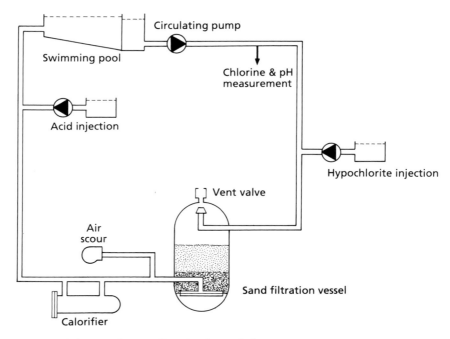

FIG. 13.3. Schematic diagram of a swimming pool plant.

3 *Disinfection.* In all systems, except the domestic whirlpool, water is disinfected. A variety of biocides is used, chlorine being the most popular.

4 *Sources of contamination.* All systems are open to the atmosphere and subject to aerial contamination by dust and debris. In addition, users contribute organisms from the skin, gastrointestinal, upper respiratory and urinary tracts and chemical contamination with ammonia and other nitrogenous compounds from sweat and urine (Warren *et al.*, 1981).

The organisms that cause problems are similar and are mainly *Pseudomonas* spp., which are common in both the environment and the human gastrointestinal tract, and *Legionella* spp., which are common environmental organisms (Jones & Bartlett, 1985). Further discussion will be confined to commercial spas and swimming pools, because these are more likely to cause outbreaks of infection. Most difficulties have been experienced with spas, because they have a smaller volume of water, tend to be run at a higher temperature and are used much more intensively. This makes the load of organic material larger and creates difficulties in maintaining the correct pH and biocide level (Jones & Bartlett, 1985). In addition, the high temperature and the often prolonged periods of immersion cause dilatation of the hair follicles of users, and allows pathogens to enter. Problems have arisen with domestic whirlpools, but these have generally been on a small scale and reflect those seen in spas. Flotation pools are relatively new and will not be specifically considered. However, they are likely to be subject to similar problems with respect to water quality.

Microorganisms Associated with Recreational Water

One difficulty in considering the many organisms associated with recreational water is to know whether they are associated with biofilm growth. In some cases, such as *Pseudomonas aeruginosa* and *Legionella pneumophila* the relationship is well established, whereas in others, such as viruses and protozoa, the relationship has not been investigated.

Non-biofilm-associated organisms

A summary of organisms not associated with biofilms in recreational water but causing illness among users is shown in Table 13.1.

Biofilm-associated organisms

The main biofilm-associated organisms are *Pseudomonas* spp. and *Legionella* spp. The infective conditions they have caused are summarized in Table 13.2.

Pseudomonas aeruginosa is the organism most frequently isolated from spas, and is associated with outbreaks of dermatitis and folliculitis. This takes the form of a rash which, depending on the presence of hair follicles, is more

TABLE 13.1. *Pathogens associated with recreational water not associated with biofilm*

Bacteria	*Mycobacterium marinum*
Virus	Adenovirus
	Herpes simplex
	Parainfluenza virus 1
	Papillomavirus
Fungi	*Trichophyton* spp.
	Epidermophyton spp.
Protozoa	*Naegleria fowleri*
	Cryptosporidium spp.
	Giardia lamblia
	Trichomonas vaginalis
	Acanthamoeba spp.
Chlamydia	*Chlamydia trachomatis*

TABLE 13.2. *Biofilm-associated pathogens in recreational waters and the conditions they cause*

Legionella pneumophila	Legionnaire's disease
	Pontiac fever
Pseudomonas aeruginosa	Skin rashes
	Otitis externa
	Mastitis
	Respiratory tract infections
	Urinary tract infections
	Keratitis

pronounced on the buttocks, groin, hips and axillae. It is most severe on areas of skin covered by the bathing costume (Jones & Bartlett, 1985; Highsmith *et al.*, 1985). In the USA the first outbreak of spa-associated disease caused by *Pseudomonas* spp. was reported by McCausland & Cox (1972) and, over the next 10 years, 74 outbreaks, mostly caused by *Ps. aeruginosa*, were recognized (Highsmith *et al.*, 1985). The number of *Pseudomonas* organisms required to cause folliculitis has not been determined. In pools with inadequate biocide levels, between 10^4 and 10^6 cfu/ml have been found; in one study, bathers suffered no ill effects at these levels (Ratnam *et al.*, 1986). However, Price & Ahearn (1988) felt that concentrations greater than 10^3 cfu/ml constituted a potential hazard for healthy individuals. They acknowledged, however, that host factors and exposure time were probably more important than the number of organisms present.

Legionnaire's disease is the pneumonia caused by *L. pneumophila* and has been well publicized because of the original Philadelphia outbreak in 1967 and, more recently, in hospitals and the British Broadcasting Corporation in London. Pontiac fever is the less common non-pneumonic form of illness caused by the same organism. In all outbreaks of Legionnaire's disease the common factor has been the generation of aerosols containing the organism. Whereas Pontiac fever has been associated with spas (Jones & Bartlett, 1985), incidents of Legionnaire's disease have not so far been observed. To penetrate to the alveoli the droplet diameter must be less than 10 μm, and droplets of up to 20 μm are deposited in the tracheobronchial tree. As the size of the organism is $0.3-0.9$ μm in diameter and 2 μm long, the minimum droplet diameter to contain one organism must be 2 μm (Baron & Willeke, 1986). These workers showed, by means of an aerodynamic particle sizer suspended at different heights above the water surface, that droplets of this size can be found in the air above spas, particularly when the jets and aeration are switched on. Although one of the spas they investigated had been incriminated in an outbreak of Pontiac fever, they did not, unfortunately, determine whether the aerosols actually contained viable bacteria. Little work has been done to determine the numbers of organisms necessary to initiate an infection. This may prove difficult to establish, because host factors are likely to be very important.

Physiological Aspects
of Biofilm-Associated Organisms

Many bacteria can form biofilms on solid surfaces and they generally adhere by means of fimbriae and an exopolysaccharide glycocalyx. The film alters the environment of the organisms within it by concentrating nutrients and providing protection from biocides in the external environment (Gorman, 1991).

Pseudomonas spp.

Though much is known about these organisms, little work has been done on their association with recreational water. They grow well in water because of their simple nutritional requirements: they will grow even in distilled water (Favero et al., 1971) and recreational waters will provide ample nutrients. Residual water in pipes or filters will encourage growth of Pseudomonas, particularly if the level of biocide is low and there are cracks and crevices — particularly in filters—to provide a niche in which the organism can become established. Pseudomonads attach easily to hydrophobic substrates, such as the polystyrene widely used in pipes and filters (Price & Ahearn, 1988). The predominant serotype is 011, although others including 03, 04, 06, 07 and 09 have also been found (Favero, 1984; Jones & Bartlett, 1985; Highsmith et al., 1985).

The attachment of organisms is affected by a combination of several factors that have been discussed elsewhere. When growing in biofilm, Ps. aeruginosa is far more resistant to antibiotics and biocides than when growing in the planktonic state (Gorman, 1991). A similar phenomenon was observed when other pathogenic Gram-negative organisms, including Escherichia coli, Salmonella typhimurium and Yersinia enterocolitica, were attached to carbon particles and exposed to chlorine (LeChevallier et al., 1984). These workers suggested that attachment alone may be the main mechanism by which Gram-negative organisms survive disinfection, because of the reduced surface area of the organism exposed to biocide; slime production adds to this resistance by further reducing biocide penetration. Although it has been suggested that organisms growing in biofilm are intrinsically more resistant to biocides, this is not the case when they are subsequently grown in the planktonic state (Highsmith et al., 1985; but see Chapter 10).

The natural history of a biofilm also affects its biocide resistance. The older the film and if grown in nutrient-limited conditions, the greater the chlorine resistance of *Klebsiella pneumoniae* (LeChevallier *et al.*, 1984). Similarly, *Ps. aeruginosa* grown in deionized water was found to be far more resistant to chlorine and other biocides than strains grown on agar, but this resistance was lost after a single passage on nutrient agar medium (Carson *et al.*, 1972). These observations point to physiological changes that probably involve alterations in the composition of the outer membrane.

Legionella spp.

Since these organisms have only recently become important, their biofilm potential in recreational waters has not been widely explored. They have, however, been extensively examined in relation to cooling towers and other heat-exchange systems. *Legionella* shares many of the characteristics of other Gram-negative organisms, including the ability to attach and grow as a biofilm. It multiplies in tap water at temperatures as high as 42°C and, in one study (Groothius *et al.*, 1985), it was more commonly isolated from spas between 35°C and 40°C than from swimming pools between 8°C and 30°C.

As with other Gram-negative organisms, the chlorine resistance of *Legionella* is related to its natural history, and organisms grown in tap water are more resistant to chlorine than those grown on rich media. This resistance disappears on agar subculture (Kuchta *et al.*, 1985). These organisms are also much more resistant to chlorine than *E. coli*, the standard indicator organism for water quality (Winn, 1988). However, they do not grow in sterile tap water unless associated with other environmental bacteria and environmental amoebae (Winn, 1988). Although the former are likely to be present in recreational water, little is known about the presence of amoebae.

Biocides Used in Recreational Waters

Since swimming pools and spas are closed systems, the addition of biocides is essential to prevent the build-up of microorganisms. On the other hand, biocide levels must not be excessive because toxic effects, such as dermatoses in bathers, can occur particularly with brominated pools (Jones & Bartlett, 1985). In addition, maintenance of the correct water balance — the term given to the combination of mineral concentration, pH, total alkalinity, calcium hardness, total dissolved solids and temperature — is also important (Edlich *et al.*, 1988). If the water balance is incorrect there may be scale formation, which can then promote the attachment of biofilm. The main biocides used in recreational waters are shown in Table 13.3.

TABLE 13.3. *Biocides used in recreational waters*

Biocide	Swimming pool	Spa
Chlorine compounds		
Sodium hypochlorite	+	+
Sodium dichloroisocyanurate	+	+
Calcium hypochlorite	+	+
Chlorine dioxide gas	+	−
Bromine compounds		
Elemental	+	+
Bromo-chloro-dimethylhydantoin	+	+
Ozone		
Ozone + chlorine	+	−
Ozone + bromine	−	+

Chlorine

Chlorine is the most commonly used disinfecting agent. In water chlorine forms hypochlorous acid which, depending on the pH, dissociates to the hypochlorite ion and hydrogen ion (Seyfried & Fraser, 1980; Edlich *et al.*, 1988). Since hypochlorous acid in its molecular form is a much more effective biocide than the slow-acting hypochlorite ion, the control of pH between 7 and 8 is most important. At pH 7 approximately 75% of free chlorine occurs as hypochlorous acid, but this falls to 20% at pH 8, whereas at pH 7.5 only about 50% is present as hypochlorous acid. The recommended pH range for swimming pools and spas is between 7.2 and 7.8 (Swimming Pools and Allied Trade Association, 1986; Pool Water Treatment Advisory Group, 1990).

The proportion of ammonium compounds present in the water is also important. These are derived from the hydrolysis of urea present in the pool and derived from sweat and urine; they combine with hypochlorous acid to give chloramines (Penny, 1991). Although the latter are bactericidal, their action is rather limited (LeChevallier *et al.*, 1988). Chloramines are responsible for the 'chlorine smell' present in swimming pools at times of high usage. Shock disinfection or superchlorination to free chlorine levels of 10 ppm is often required to control the build-up of chloramines, particularly in spas (Davis, 1985).

Although the disinfection activity of chlorine increases with temperature, more of it is required at higher temperatures because of thermal decomposition. Muraca *et al.* (1987) found that, at 43°C − just above the normal spa temperature − 120% more chlorine was required for disinfection than that required at 25°C.

Bromine

Bromine chemistry is similar to that of chlorine: in water it forms hypobromous acid and the hypobromite ion. It is not, however, widely used as a disinfectant for swimming pools in the UK and is not thought suitable for disinfecting spas, because of the possible toxic effects for users (Jones & Bartlett, 1985; Penny, 1991). In addition, the chemistry of bromine in spa conditions may be less predictable than that of chlorine, particularly the lack of stability of hypobromous acid in the concentrations of ammonia likely to be encountered (Jones & Bartlett, 1985). This has led to suggestions of poor activity against *Ps. aeruginosa*, which allows it to grow particularly in filters (De Jonckheere, 1982). In addition, liquid bromine cannot be used because of its toxicity and it is more commonly used as a solid organic complex such as bromo-chloro-dimethylhydantoin. However, this is more expensive and less effective than chlorine (Jones & Bartlett, 1985).

Ozone

Ozone is a much more powerful oxidizing agent than chlorine. It is generated by electrical discharge but does not produce any residual activity in the water. Thus it has an extremely short life and, in practice, is usually used in combination with chlorine.

Methods for Investigating Water Quality

Choice of indicator organism
for assessing the quality of recreational water

There are two schools of thought about the best indicator organism for assessing the quality of water in spas. On the one hand it is argued that the currently used total coliform, faecal coliform and total plate counts, as used for drinking water, are satisfactory; on the other, it is contended that it would be more logical to measure different kinds of organisms, because the use of recreational water does not correlate with the incidence of enteric disease. For swimming pools these organisms would include skin and nasal organisms such as staphylococci, and throat organisms such as streptococci (Favero, 1984; Tosti & Volterra, 1988; Edlich *et al.*, 1988). In spas, *Ps. aeruginosa* should also be measured.

Measurement of bacteriological water quality

For the routine sampling of chlorinated pools, sample bottles should contain

25 mg sodium thiosulphate and have a capacity of at least 250 ml. Sampling points should have a water depth of at least 1 m and the operator should wear rubber gloves during collection to prevent sample contamination. The bottle should be plunged into the pool to a depth of at least 20 cm, taking care that the sodium thiosulphate is not washed out (Zura *et al.*, 1990). Several samples are necessary, particularly for spas, and it has been suggested that they should be collected from the surface, at mid-depth and from the bottom of the pool, preferably during heaviest use (Edlich *et al.*, 1988). For experimental sampling pipettes are often used, since this permits more precise identification of the sample point (Price & Ahearn, 1988).

It is important to appreciate that samples taken in this way measure organisms in the planktonic state, and may not reflect organisms present in the biofilm. For example, *L. pneumophila* may be present in biofilm but may not be detected in the planktonic water because of adequate biocide levels.

The most commonly used method for examining waters from both swimming pools and spas is determination of the most probable number (MPN) of organisms with MacConkey broth. Samples that give a growth of *E. coli* or any coliform organism in 100 ml after 24 h incubation are considered to be unsatisfactory. After the same period of incubation, the plate count must show fewer than 100 cfu/ml and ideally fewer than 10 cfu/ml (Pool Water Treatment Advisory Group, 1990). The results are often commented on individually, e.g. 10 organisms/ml, satisfactory; 11−25, slightly raised plate count; 26−100, raised plate count; 101−1000, high plate count; 1000/ml, very high plate count.

Samples are not at present routinely examined for *Ps. aeruginosa* but, if specificallly requested, this can be done with an MPN method. No standard method is prescribed and we use an asparagine/ethanol medium. Since there are no recommended standards, the number of organisms present in the specimen are reported.

An alternative to the MPN method, membrane filtration, can be used and the membranes are incubated on pads containing sodium lauryl sulphate broth. To encourage the growth of damaged organisms, membranes are generally allowed to 'repair' by incubation at 30°C for 4 h before incubation for growth at 37°C and 44°C. Other media used include, Tergitol 7 agar supplemented with 2.5% 2,3,5-triphenyltetrazolium chloride and covered with a thin layer of blue bile agar for total and faecal coliforms; m-enterococcus agar for faecal streptococci; Baird−Parker medium supplemented with 5% Egg Yolk Tellurite Emulsion for *Staphylococcus aureus* and Pseudomonas selective medium for *Ps. aeruginosa* (De Jonckheere, 1982). LeChevallier *et al.* (1987) used m-endo LES agar (BBL Microbiology Systems, Cockeysville, MD, USA) and m-T7 agar (Difco) for coliforms and R2A agar for heterotrophic bacteria.

Examination of organisms in the biofilm is more difficult and is not usually practised in recreational waters. LeChevallier *et al.* (1987) examined films in drinking water distribution systems and found coliforms mainly confined to iron tubercles, possibly because of iron-stimulated growth. Although this may be relevant to old plant, particularly in swimming pools, in more modern installations with plastic piping tubercles do not tend to form. In these situations growth will tend to be more uniform over the surface, and in the joints and crevices between the piping.

Methods for specific organisms

Historically, the effect of biocides on organisms associated with recreational water has always been studied with the organisms growing in the planktonic state, that is, in suspension. Although such studies are important, their direct application to growth in biofilms can be misleading, primarily because in this state organisms tend to be much more resistant. Occasionally this has been incorrectly interpreted as the organisms having become resistant to the biocide.

Organisms in the planktonic state

Pseudomonas aeruginosa

Because of the presence of contaminating organisms in recreational water, selective media are usually required to detect *Pseudomonas*. A modified XLD (xylose-lysine-desoxycholate) agar was used by Havelaar *et al.* (1985) and a Pseudosel medium (BBL) modified by the addition of sodium thiosulphate 0.3 g/l by Price & Ahearn (1988). The latter investigated the number of *Ps. aeruginosa* in aerosols above spas by turning the jets on and holding a Petri dish containing appropriate agar surface-downwards about 4–6 cm above the surface, for 1 min.

In a study of *Ps. aeruginosa* isolated from 19 outbreaks of spa-associated infection, Highsmith *et al.* (1985) found that serotyping by slide agglutination, with commercially available specific antisera (Difco), was the most reliable method for distinguishing strains, as compared with pyocine typing, antibiograms obtained with antibiotic sensitivity discs, and the production of several extracellular enzymes.

Legionella spp.

The most commonly used medium for isolating *Legionella* spp. from water is Buffered Charcoal Yeast Extract Agar supplemented with α-ketoglutarate, cysteine and iron, and modified by the addition of antibiotic supplements (Winn, 1988).

The taxonomy of the legionellas is complex, and a direct or indirect fluorescent antibody technique is the main method used (Winn, 1988). Biochemical typing has proved of little value, as this group of organisms is relatively inert.

Amoebae

In view of the link between the presence of *Legionella* spp. and amoebae, surprisingly few studies have investigated pools for the presence of the latter. One study on hydrotherapy pools, in which amoebae were obtained by incubating concentrated water samples with live *E. coli* (De Jonckheere, 1982), found that halogenated pools were not highly contaminated but those disinfected with UV radiation contained very high numbers of thermophilic *Naegleria lovaniensis*. However, *Naegleria fowleri*, the main pathogenic species, was not found.

Organisms in biofilm

Microbial adherence and biofilm production has been reviewed by Gorman (1991), and is extensively discussed in this volume. The various techniques described for the production, examination and indirect estimation of biofilms *in vitro* are equally applicable to investigating biofilms in recreational water, but have not yet been extensively used.

Biofilms in filter media

Sand is the usual filter medium used for recreational water systems, but few studies have examined organisms on these filters. However, granulated activated carbon filters, as used in drinking water treatment, have been investigated (Camper *et al.*, 1986) although the particles are likely to be finer. The procedure is to remove the fine carbon particles remaining in the water after filtration on gauze filters. The gauze is then shaken vigorously in sterile water to dislodge the particles. To inactivate planktonic organisms, suspended particles are chlorinated with sodium hypochlorite 2 g/l for 30 min at 4°C. After the addition of sodium thiosulphate they are again shaken vigorously by hand, before counting the organisms present by the spread-plate or membrane-filter methods. The main drawback of this technique is that it is not quantitative, because it does not take into account the number of bacteria adhering to each particle. To overcome this, Camper *et al.* (1985, 1986) and LeChevallier *et al.* (1987) compared this technique with a homogenization method. The particles were homogenized for 2 or 3 min in a blender at $16\,000-22\,000$ rpm in a solution of buffered peptone, ethylene glycol-bis-(β-aminoethyl ether)-N,N,N',N'-tetraacetic acid (EGTA) and Zwittergent $3-12$ for 3 min at 4°C before counting the number of organisms. Homogenization improved the yield of heterotrophic bacteria by between 8.6- and 50-fold, compared with the conventional manual shaking method.

Distribution systems

Biofilms in drinking water distribution systems can be sampled by vigorous flushing with high-pressure water and either collecting the effluent water in large vessels or filtering the water coming out of the pipe through nylon netting (*ca.* 300 μm pore size). Tubercles and other material dislodged from the pipes by mechanical cleaning methods (pigging) can be trapped in the same way (LeChevallier *et al.*, 1987). Alternatively, pipes can be cut open and scraped with a sterile spatula. The dislodged material is then crushed in a mortar and pestle before homogenization (Camper *et al.*, 1986).

Methods for Investigating the Effects of Biocides on Organisms

As has already been emphasized, whether or not organisms are growing in the planktonic or the biofilm state has a considerable effect on their biocide susceptibility. In the past, failure to take this into account has led to misleading interpretations and suggestions that organisms can become resistant to the biocides tested.

Planktonic methods

To evaluate disinfectants for use in swimming pools, the quantitative suspension test (Van Klingeren *et al.*, 1980) can be used. This is based on the principle of the Dutch Standard Suspension Test and uses buffered bovine albumin solution (BBAS) as an artificial swimming pool water. BBAS contains (g/l): KH_2PO_4, 4; K_2HPO_4, 14; and bovine albumin, 3; pH 7.2. A known concentration of chlorine solution, either as sodium hypochlorite or sodium dichloroisocyanurate, is added to suspensions of the organism. After different exposure times at 25°C, the number of surviving organisms can be calculated. Although used in the Netherlands, this method has not been widely cited elsewhere.

To imitate a hospital water supply system, Muraca *et al.* (1987) constructed a model plumbing system from copper piping, brass spigots, a Plexiglas reservoir, ball valves with rubber seats, an electric hot water tank and a pump. They examined the effects of chlorine, ozone, heat and UV radiation on planktonic *L. pneumophila*. A suspension of the organism was added to the system to a concentration of 10^7 cfu/ml and the water circulated for approximately 1 h to allow equilibration. Biocides were then added and the survival of the organisms monitored with viable counts taken at intervals over a 6-h period. The effects of turbid water were evaluated by adding suspended solids concentrated from hot-water tank effluent at 4−5 mg/l. Although this system was designed to model a hospital water supply system, it could, for example, be modified by the inclusion of more plastic pipes and an aeration system to make it more typical of a spa.

Chlorine resistance

The chlorine resistance of *Ps. aeruginosa* was tested by exposing organisms to various levels of free chlorine for different times and measuring their survival (Highsmith *et al.*, 1985). Chlorine demand free water (CDFW) was prepared from membrane filter-sterilized tapwater. This was first hyperchlorinated with 5% sodium hypochlorite to give a level of $2-4$ mg/l and then dechlorinated by exposure to direct sunlight for $3-4$ days. A more practical method, particularly in the UK, would be to boil the water (Kuchta *et al.*, 1983). The absence of chlorine demand and residual chlorine was confirmed by the amperometric method. Such water was used for all experiments and for washing apparatus and the preparation of solutions. To test for chlorine susceptibility, a stock solution of sodium hypochlorite was diluted to give concentrations in the range $0.2-2$ mg/l. A suspension of the test strain that had been well washed in CDFW and containing between 10^3 and 10^4 cfu/ml was then added. The suspensions were incubated at 42°C and tested for viable organisms at various times by plating on suitable medium. Before plating, the residual chlorine was measured and neutralized with sodium thiosulphate. Alternatively, membrane filtration may be used. One of the difficulties with this procedure was to maintain an accurate level of free chlorine in the presence of organisms up to a concentration of 10^3 cfu/ml. With this number of organisms the chlorine dissipated rapidly, compared with CDFW alone.

Biofilm methods

Although techniques have not been specifically designed for the investigation of biofilm control in recreational waters, it is possible to speculate on methods that might be suitable. Organisms could be grown as biofilm on strips (5 mm², 0.5 mm thick) of different plastic materials, tiles etc., as used in recreational water systems, and exposed to biocides at different concentrations and times. The effects on the adherence of the organisms could be examined by light microscopy, scanning electron microscopy, fluorescence microscopy and bioluminescence. Survival could be determined by the shake method to dislodge the organisms, which can then be counted. This procedure can be made more accurate by the homogenization method of Camper *et al.* (1986) and LeChevallier *et al.* (1987), described above. Alternatively, the Robbins device, which has been used to determine the number of viable bacteria in a biofilm, both in the laboratory and in particular installations, could be employed (see

Chapter 1). This method has, however, been criticized because of the large number of planktonic bacteria present, which may inactivate the agent under study (Gorman, 1991).

Methods of Overcoming the Problems of Biofilms

Biofilms are likely to be present even in the best-maintained pools, and it is probably unrealistic in general practice to try to prevent their establishment. Once established, however, they are extremely difficult to eliminate completely, and they present a continuous potential risk, particularly if they contain pathogens. For example, *L. pneumophila* is known to have survived in hotel water supplies — presumably in biofilm — for at least 3 years of correct operation before a single lapse in procedure allowed the organisms into the water to cause a further outbreak of Legionnaire's disease. Good pool practice will keep the level of biofilm low, and an effective level of biocide in the water will ensure that any organisms released from the film are rapidly inactivated. This will also ensure that, when a lapse in procedure occurs, the reduced amount of biofilm present can only release small numbers of organisms into the planktonic state. The elements of good practice have been well documented for swimming pools (Pool Water Treatment Advisory Group, 1990) and for spas, but the American recommendations (US Department of Health and Human Services, 1981) differ in fine detail from those for the UK (Swimming Pools and Allied Trade Association, 1986). The elements of good practice, include the following.

1   Automatic, rather than manual, addition of biocide and the automatic monitoring of free chlorine levels in the water by the amperometric rather than the redox method.

2   Adequate maintenance.

3   Maintenance of adequate biocide levels. For example, free chlorine levels should be between 1 and 3 mg/l in swimming pools and between 3 and 5 mg/l in spas. These should be maintained at a minimum $2 : 1$ ratio with the combined residual chlorine, which should not exceed 1 mg/l. Frequent superchlorination to 10 mg/l and holding for between 1 and 4 h is recommended for spas, although not for swimming pools. The pH should be maintained between 7.2 and 7.8, but preferably between 7.4 and 7.6. The USA recommendations suggest that, for spas, all these measurements should be checked at hourly intervals, whereas in the UK the recommendation is for checks every 15 min. Swimming pools should be checked before the pool opens and at 2-hourly intervals when non-automated dosing equipment is used, and three times per day when such equipment is used. Swimming pool filters should be backwashed at least once per week and, if heavily used, spa filters should be backwashed daily.

4 Maintenance of a good water balance. This is important so that total dissolved solids in swimming pools are kept below 3000 mg/l, but preferably nearer 1500 mg/l. The recommended levels for spas are much lower, with 300 mg/l suggested as the lower level and 1500 mg/l as the upper. The total alkalinity in swimming pools, as a buffer to prevent pH fluctuation, should be maintained in the range of 120–150 mg/l $CaCO_3$ when sodium hypochlorite is used, and between 60 and 200 mg/l in spas.

5 Turnover time for water in swimming pools should not exceed 3 h, and should be less than 1.5 h in learner or shallow pools. In spas the system should be capable of a turnover time between 6 and 30 min. To prevent the build-up of dissolved solids and to reduce the level of chloramines in both systems, it is necessary to replace some or all of the pool water at intervals. In spas, because of the high rate of evaporation resulting from the high temperature and aeration, it is recommended that the pool is emptied and refilled at intervals varying between daily and a few months, depending on usage. In swimming pools, on the other hand, complete replacement is not generally practical and regular replacement of water at the rate of 30 l/bather/day is recommended.

6 Regular bacteriological sampling. For both types of pool this is important. For swimming pools it should be done at monthly intervals, or more frequently in the event of an unsatisfactory result. Frequencies have not been defined for spas, but once a fortnight, which has been suggested in the USA, may prove burdensome in the UK. Sampling should take place at peak usage times rather than when pools are empty. Good results are indicated by the absence of coliforms, *E. coli*, *Pseudomonas*, staphylococci or faecal streptococci in 100 ml of water. The plate count should be fewer than 100 cfu/ml and preferably fewer than 10 cfu/ml.

7 Restricting the number of bathers. Each pool, depending on its size and type, should have a recommended bather load. For swimming pools a formula for calculating this is given by the Pool Water Treatment Advisory Group (1990).

8 Adequate staffing levels, written procedures for all aspects of the pool management and a high level of training and supervision.

Conclusions

Biofilm formation, physiology and control share common features wherever they are found. The organisms that cause problems may, however, be different, and are affected by environmental circumstances. In recreational waters the problems caused by biofilms are more prominent in spas, because of the reduced water volume, higher running temperature and greatly intensified bather usage.

Acknowledgements

The author wishes to thank Mr C. Leckey of the Bacteriology Department, Belfast City Hospital, for water analysis details; Mr I. Lyons of MNM Plc, Chancery Bridge Industrial Estate, Thornes Lane, Wakefield, Yorkshire WF1 5QN, for permission to reproduce Fig. 13.1; Mr K. Mulhern of the Eastern Health and Social Services Board for Fig. 13.2 (adapted from Davis, 1985), and Mr G. C. Carey of Messrs Barr and Wray, 324, Drumoyne Road, Glasgow G51 4DY, for Fig. 13.3.

References

BARON, P.A. & WILLEKE, K. (1986) Respirable droplets from whirlpools: measurement of size distribution and estimation of disease potential. *Environmental Research*, **39**, 8–18.

CAMPER, A.K., LeCHEVALLIER, M.W., BROADWAY, S.C. & McFETERS, G.A. (1985) Growth and persistence of pathogens on granular activated carbon filters. *Applied and Environmental Microbiology*, **50**, 1378–1382.

CAMPER, A.K., LeCHEVALLIER, M.W., BROADAWAY, S.C. & McFETERS, G.A. (1986) Bacteria associated with granular activated carbon particles in drinking water. *Applied and Environmental Microbiology*, **52**, 434–438.

CARSON, L.A., FAVERO, M.S., BOND, W.W. & PETERSON, N.J. (1972) Factors affecting comparative resistance of naturally occurring and subcultured *Pseudomonas aeruginosa* to disinfectants. *Applied Microbiology*, **23**, 863–869.

DAVIS, B.J. (1985) Whirlpool operation and the prevention of infection. *Infection Control*, **6**, 394–397.

DE JONCKHEERE, J.F. (1982) Hospital hydrotherapy pools treated with ultra violet light: bad bacteriological quality and presence of thermophilic *Naegleria*. *Journal of Hygiene (Camb.)*, **88**, 205–214.

EDLICH, R.F., BECKER, D.G., PHUNG, D., McCLELLAND, W.A. & DAY, S.G. (1988) Water treatment of hydrotherapy exercise pools. *Journal of Burn Care and Rehabilitation*, **9**, 510–515.

FAVERO, M.S. (1984) Whirlpool spa-associated infections: are we really in hot water? *American Journal of Public Health*, **74**, 653–655.

FAVERO, M.S., CARSON, L.A., BOND, W.W. & PETERSON, N.J. (1971) *Pseudomonas aeruginosa*: growth in distilled water from hospitals. *Science*, **173**, 836–838.

GORMAN, S.P. (1991) Microbial adherence and biofilm production. In Denyer, S.P. & Hugo, W.B. (eds.) *Mechanisms of Action of Chemical Biocides: Their Study and Exploitation*. Society for Applied Bacteriology Technical Series No. 27, pp. 271–295. Blackwell Scientific Publications, Oxford.

GROOTHIUS, D.G., HAVELAAR, A.H. & VEENENDAAL, H.R. (1985) A note on legionellas in whirlpools. *Journal of Applied Bacteriology*, **58**, 479–481.

HAVELAAR, A.H., DURING, M. & DELFGOU-VAN ASCH, E.H.M. (1985) Comparative study of membrane filtration and enrichment media for the isolation and enumeration of *Pseudomonas aeruginosa* from sewage, surface water and swimming pools. *Canadian Journal of Microbiology*, **31**, 686–692.

HIGHSMITH, A.K., LE, P.N., KHABBAZ, R.F. & MUNN, V.P. (1985) Characteristics of *Pseudomonas aeruginosa* isolated from whirlpools and bathers. *Infection Control*, **6**, 407–412.

JONES, F. & BARTLETT, C.L.R. (1985) Infections associated with whirlpools and spas. *Journal of Applied Bacteriology Symposium Supplement 14*, **59**, 61S–66S.

KUCHTA, J.M., STATES, S.J., McNAMARA, A.M., WADOWSKY, R.M. & YEE, R.B. (1983) Susceptibility of *Legionella pneumophila* to chlorine in tap water. *Applied and Environmental Microbiology*, 46, 1134–1139.

KUCHTA, J.M., STATES, S.J., McGLAUGHLIN, J.E., OVERMEYER, J.H., WADOWSKY, R.M., McNAMARA, A.M., WOLFORD, R.S. & YEE, R.B. (1985) Enhanced chlorine resistance of tap water-adapted *Legionella pneumophila* as compared with agar medium-passaged strains. *Applied and Environmental Microbiology*, 50, 21–26.

LeCHEVALLIER, M.W., HASSENAUER, T.S., CAMPER, A.K. & McFETERS, G.A. (1984) Disinfection of bacteria attached to granular activated carbon. *Applied and Environmental Microbiology*, 48, 918–923.

LeCHEVALLIER, M.W., BABCOCK, T.M. & LEE, R.G. (1987) Examination and characterisation of distribution system biofilms. *Applied and Environmental Microbiology*, 53, 2714–2724.

LeCHEVALLIER, M.W., CAWTHON, C.D. & LEE, R.G. (1988) Factors promoting survival of bacteria in chlorinated water supplies. *Applied and Environmental Microbiology*, 54, 649–654.

McCAUSLAND, W.J. & COX, P.J. (1972) *Pseudomonas* infections traced to motel whirlpool. *Journal of Environmental Health*, 37, 455–459.

MURACA, P., STOUT, J.E. & LU, V.L. (1987) Comparative assessment of chlorine, heat, ozone and UV light for killing *Legionella pneumophila* within a model plumbing system. *Applied and Environmental Microbiology*, 53, 447–453.

PENNY, P.T. (1991) Hydrotherapy pools of the future — the avoidance of health problems. *Journal of Hospital Infection*, 17 (Supplement A), 535–542.

POOL WATER TREATMENT ADVISORY GROUP (1990) *Pool Monitor. A Survey of Swimming Pool Water Management Standards*. Field House, Thrandeston, Diss, Norfolk IP21 4BU, UK.

PRICE, D. & AHEARN, D.G. (1988) Incidence and persistence of *Pseudomonas aeruginosa* in whirlpools. *Journal of Clinical Microbiology*, 26, 1650–1654.

RATNAM, S., HOGAN, K., MARCH, S.B. & BUTLER, R.W. (1986) Whirlpool associated folliculitis caused by *Pseudomonas aeruginosa*: report of an outbreak and review. *Journal of Clinical Microbiology*, 23, 655–659.

SEYFRIED, P.L. & FRASER, D.J. (1980) Persistence of *Pseudomonas aeruginosa* in chlorinated swimming pools. *Canadian Journal of Microbiology*, 26, 350–355.

SWIMMING POOL AND ALLIED TRADE ASSOCIATION LTD (1986) *Standards for Spa Pools: Installations, Chemicals & Water Treatment*. SPATA House, Junction Road, Andover, Hampshire SP10 3QT, UK.

TOSTI, E. & VOLTERRA, L. (1988) Water hygiene of two swimming pools: microbial indicators. *Journal of Applied Bacteriology*, 65, 87–91.

US DEPARTMENT OF HEALTH AND HUMAN SERVICES. PUBLIC HEALTH SERVICE (1981) *Suggested Health and Safety Guidelines for Public Spas and Hot Tubs*. Centers for Disease Control, Atlanta.

VAN KLINGEREN, B., PULLEN, W. & REIJNDERS, H.F.R. (1980) Quantitative suspension test for the evaluation of disinfectants for swimming pool water: experiences with sodium hypochlorite and sodium dichlorisocyanurate. *Zentralblatt für Bakteriologie. Mikrobiologie und Hygiene. I Abteilung Originale B*, 170, 457–468.

WARREN, I.C., HUTCHINSON, M. & RIDGWAY, J.W. (1981) Comparative assessment of swimming pool disinfectants. In Collins, C.H. Allwood, M.C. Bloomfield, S.F. & Fox, A. (eds.) *Disinfectants: Their Use and Evaluation of Effectiveness*. Society of Applied Bacteriology Technical Series No. 16, pp. 123–139. Academic Press, London.

WINN, W.C. (1988) Legionnaires, disease: historical perspective. *Clinical Microbiology Reviews*, 1, 60–81.

ZURA, R.D., GROSCHEL, D.H.M., BECKER, D.G., HWANG J.C.S. & EDLICH, R.F. (1990) Is there a need for state health department sanitary codes for public hydrotherapy and swimming pools? *Journal of Burn Care and Rehabilitation*, 11, 146–150.

14: Adhesion of Biofilms in Flowing Systems

M.E. CALLOW[1], R. SANTOS[2] AND T.R. BOTT[2]

[1]School of Biological Sciences and [2]School of Chemical Engineering, University of Birmingham, Edgbaston, Birmingham B15 2TT, UK

Biological fouling is a problem in water pipes and industrial cooling water systems, as well as on all submerged surfaces, unless some type of antifouling protection is employed. In industrial cooling systems, fouling is chiefly due to microbial biofilms, along with particulate matter, which reduces heat transfer efficiency and causes increases in pressure drop in the system, leading to increased operating costs. Microbial fouling of submerged metal structures, such as oil platforms, leads to corrosion and metal fatigue. Fouling of moving structures such as ships by biofilms composed chiefly of diatoms, leads to losses of operating efficiency due to increased frictional resistance and drag. Although visually insignificant compared to animal and macroalgal fouling, diatom slimes contribute significantly to drag (Lewthwaite et al., 1985). In view of the wide-ranging implications of microbial fouling for industry, coupled with increased environmental awareness resulting in the decreased use of biocides to control fouling, there has been interest in developing non-biocidal, 'non-stick' surfaces to prevent biofilm establishment. Consequently, a number of laboratory systems designed to study microbial adhesion have been developed. These can be used to monitor the attachment of microorganisms to surfaces under static or flowing conditions. This chapter will concentrate on methods relating to flowing systems. Once reliable protocols are established with defined surfaces, the effects of surface modifications, such as the use of polymer coatings, can be studied.

Three systems will be described: a flow apparatus which simulates the conditions found in an industrial cooling water system; a radial flow chamber to study attachment/detachment of cells under conditions of variable shear stress; and a flow cell for testing the adhesion of biofilms to raft panels. All three methods have been used in the authors' laboratories to study the effect of a variety of modifications of surface chemistry and polymer coatings on

Microbial Biofilms:
Formation and Control

the adhesion of cells or biofilms. All methods are applicable to freshwater and marine systems but, if marine systems are to be studied, all parts of the apparatus must be able to withstand saline conditions.

Flow Apparatus

The detailed operation of the apparatus has been described by Santos *et al.* (1991) and the layout is shown in Fig. 14.1. We used *Pseudomonas fluorescens* because it is known to be an early colonizer of surfaces immersed in freshwater or soil, and because it is one of the most common microorganisms in biofilms in cooling water systems. Any single or mixed culture of microorganisms can be used, but adjustments in culture conditions and operating parameters may be necessary. Controlled flows of nutrient, bacterial culture (*Ps. fluorescens*, NCIB 9046), and circulating water entered the mixing vessel, where conditions of pH, temperature and dissolved oxygen were carefully monitored. The mixing vessel was 'sparged' with filtered air and potassium hydroxide was added continuously to maintain the pH at 7. The flow of *Ps. fluorescens* from the fermenter to the mixing vessel was arranged to give 2×10^7 cfu/ml in the circulating water, and the flow of nutrient to provide 8 mg/l glucose at the beginning of each experiment. From the mixing vessel the solution was pumped through the test sections where biofilms developed. A feed and bleed system operated so that the excess solution was passed to drain. The residence time was either 30 or 60 min.

Water supply

Various types of make-up water can be used to supply the mixing vessel, depending on the quality of the local supply. Ideally, distilled water is used which, for optimal growth of *Ps. fluorescens*, requires the addition of ferric citrate (2.7 g/l) to give a final concentration of iron (as Fe^{3+}) of 1 mg/l in the circulating water. A continuous supply of distilled water which provides 20.5 l/h for a residence time of 1 h is essential. It may be feasible to use mains tapwater as the make-up water, but this has to be sequentially passed through activated charcoal to remove chlorine, a 5 µm pore-size cartridge filter to remove large debris, and a 1 µm pore-size cartridge filter to remove small debris and all living organisms apart from small bacteria and spores. If it is essential that all contaminating organisms are removed, a 0.22 µm pore-size cartridge filter is also necessary. Filters must be changed as soon as the water flow rate through the system declines — typically every 1–2 days.

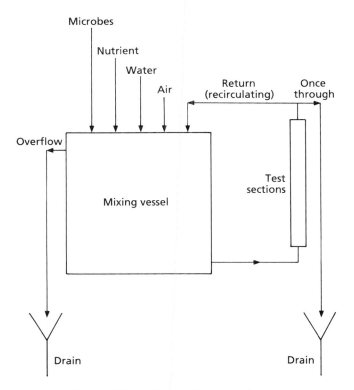

FIG. 14.1. Generalized layout of the simulated cooling water flow apparatus.

Test section and biofilm measurement

Glass test sections were chosen in order that the growth of the biofilm could be monitored continuously by absorbance, and because glass is less prone to biofilm formation than are metals under flowing conditions (Mott & Bott, 1990). However, glass test sections limit the choice of surface polymers to those that are optically clear, can be chemically anchored, and produce an even and complete surface cover.

The apparatus accomodated seven glass test sections, 765 mm in length and either 16.5 or 11.0 mm internal diameter, giving flow velocities of 0.5 or 2.5 m/s respectively, although the system can be designed to accomodate fewer or more test sections as required. Biofilm growth is continuously monitored *in situ* in each test section by infrared absorbance at 950 nm, as described by Santos *et al.* (1991). Although absorbance cannot be directly related to thickness or density of the biofilm, it provides a rapid non-destructive method for measuring biofilm accumulation. Total weight of the biofilm can be obtained by removing and draining the test sections at regular intervals (Bott & Miller, 1983). However, draining and reconnecting the flow may result in hydraulic shock, which disturbes the biofilm, particularly on surfaces where adhesion is 'weak', as well as increasing the possibility of introducing contaminating organisms. Biofilm thickness can be monitored at the completion of an experiment, provided metal test sections are used (Bott & Miller, 1983). Alternatively, the test sections can be modified to allow for the insertion of detachable metal plates, which can be removed at intervals (Harty & Bott, 1981).

Acid-washed test sections are coated with materials aimed at modifying biofilm adhesion according to the manufacturer's instructions. It is recommended that parallel studies are conducted to investigate whether the glass surface has been coated satisfactorily and whether the coating remains intact during the course of the experiment. Some of the physicochemical methods applicable to this type of study are discussed by Goodwin *et al.* (1990).

Electron microscopy

At the end of an experiment, pieces of glass test sections with adhering biofilm can be prepared for subsequent examination by scanning (SEM) or transmission electron microscopy (TEM). For SEM, pieces of glass test section with undisturbed biofilm are fixed for 1 h at 0°C in 5% glutaraldehyde in 0.1 M cacodylate buffer at pH 7.0 (Santos *et al.*, 1991). For subsequent interpretation of results, it is important to mark the direction of flow in manner which can be recognized when viewed in the SEM. After three buffer washes, specimens are post-fixed for 0.5 h in 1% osmium tetroxide in the same buffer. Specimens are dehydrated in a graded acetone series, critical point dried in liquid carbon dioxide and sputter coated with a $3-4$ nm gold film.

For TEM, if the biofilm is not sufficiently thick to be removed from the test section and survive the subsequent processing intact, it can be 'embedded' in a drop of agar cooled to 40°C. The agar gels almost instantaneously and can easily be removed from the glass. Processing is as for SEM, but dehydration in an ethanol series is preferred. Pieces are transferred from absolute ethanol into propylene oxide (two changes) and embedded in Epon 812. The extracellular polymer surrounding cells can be enhanced by including ruthenium red (Luft, 1971) in all solutions up to and including the osmium tetroxide.

Results

Infrared absorbance of biofilms developed on the sides of glass test sections during the same experiment and run at two velocities, namely 0.5 and 2.5 m/s, demonstrates that thicker biofilms are produced at the lower velocity and that these also develop more rapidly (Santos *et al.*, 1991). Examination of biofilms by SEM reveals that the biofilm is more compact at the higher velocity, and that the cells are aligned in the direction of flow (Santos *et al.*, 1991). The effect on biofilm growth of glass coated with two different chemical surface modifications, which did not alter the absorbance properties of the glass, is shown in Fig. 14.2. The surface treatments used were Glassclad HP and Glassclad IM (Petrarch Systems Silanes & Silicones, from Fluorochem Ltd). Glassclad HP is a heparin-modified siloxane resin providing hydrophilic surfaces with low thrombogenicity. Glassclad IM is a polyethyleneimine-modified resin used to provide glass surfaces with greater affinity for cell adhesion. Glassclad HP was made up as a 20% aqueous solution in water adjusted to pH 4.5 with acetic acid. Glassclad IM was made up as a 2% solution in 95% ethanol. The test sections were filled with the appropriate solutions, drained and air-dried for 24 h before use. Biofilm develops faster and becomes thicker on the Glassclad IM-coated suface than on either the glass or Glassclad HP-coated surface (Fig. 14.2). There are no differences in the growth of biofilm on the glass and Glassclad HP test sections. This result was not unexpected, because the surface properties of the two surfaces were similar.

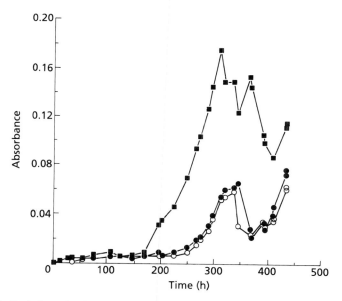

FIG. 14.2. Typical results obtained from the flow apparatus shown in Fig. 14.1. The graph shows absorbance of biofilm on the control test section (acid-washed glass) (●), and in two test sections with surface modifications: (○) Glassclad HP; (■) Glassclad IM. All test sections were run at an operating velocity of 0.5 m/s. More biofilm accumulated on the Glassclad IM. All surfaces exhibit the typical sloughing and regrowth pattern described by Bott & Miller (1983).

Radial Flow Chamber

The radial flow chamber (RFC) was developed for the study of microbial film formation under conditions of variable shear stress (Fowler & McKay, 1980). Fluid is pumped from a reservoir through the inlet pipe and flows radially between two parallel discs of known spacing via a manifold. Fluid is returned to the reservoir via a flow meter (Fig. 14.3). At constant flow, the linear fluid velocity, and hence surface shear stress, decreases radially from the centre. The apparatus was designed to investigate the shear forces that have to be overcome to allow the attachment of microorganisms. Thus, for attachment studies the reservoir is a continuous fermenter, containing bacteria in steady-

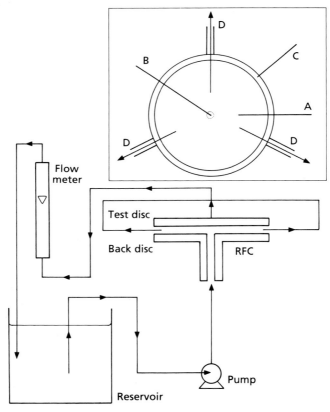

FIG. 14.3. Generalized diagram of the radial flow chamber (RFC) as used for cell detachment studies. Inset shows the RFC from above. A, Test disc; B, inlet pipe; C, outlet manifold; D, outlet pipe.

state growth (Duddridge *et al.*, 1982) and each experiment takes at least 1 week. We have used the apparatus (LH Engineering Co. Ltd) to study the shear forces required for the detachment of cells. In this mode the RFC provides a rapid method for assessing the relative strength of adhesion of cells to various substrates, each experiment taking approximately 30 min to perform.

The shear stress is calculated according to the formula given by Duddridge *et al.* (1982), where flow in the chamber is laminar ($Re_r < 2000$)

$$\tau_0 \text{ laminar } = 3Q\mu/\pi r h^2$$

or turbulent ($Re_r > 2000$)

$$\tau_0 \text{ turbulent } = 0.0288 \rho u^{-2} Re_r^{-0.2}$$

Theoretical aspects of the measurement of shear stress in the RFC have been discussed at length by Fryer *et al.* (1984, 1985). Clearly, the absolute values for critical shear stress obtained from the RFC and comparisons of values derived by other methods should only be made with caution. However, the RFC provides a relatively quick method for comparing the strength with which microorganisms adhere to various surfaces under reproducible conditions. The method has been used to study the adhesion of a number of unicellular organisms, including marine diatoms (Milne & Callow, 1985; Woods & Fletcher, 1991) and fungal spores (Hyde *et al.*, 1989).

Cell detachment studies

Acid washed 100-mm diameter plate-glass discs or discs coated with polymers can be used. If polymer coatings are used it is essential that the polymer is applied evenly over the whole disc surface, because any differences in thickness will result in aberations of flow pattern across the disc.

Cell suspensions are prepared by standard methods. It is important that single cells and not clumped cells are used for settlement. We have used *Ps. fluorescens* and the marine diatom *Amphora coffeaeformis* var. *perpusilla*. For cell attachment, discs are placed in clean perspex trays and covered with cell culture for 2 h. Discs are carefully washed in the same medium to remove unattached cells. All operations involving the movement of discs from liquid to air must be performed extremely carefully if hydrophobic coatings are used, because cells can become detached by instantaneous droplet formation as the surface passes into the air phase. Each disc is placed in the RFC apparatus which, with a 1 mm gap, typically operates at a volumetric flow of 6−10 l/min for 15 min. On removal from the apparatus, cells remaining on the discs are fixed and stained. Transparent discs are stained with 0.1% methylene blue in 7% acetic acid, and opaque discs are stained with 0.01% acridine orange for subsequent cell counting by transmitted or epifluorescent light microscopy, respectively. Each treatment is normally repeated three times. Bacteria and diatom cells are counted within a known area with ×100 or ×40 objectives, respectively, every millimetre from the centre to a radius of 40 mm along four radii at 90° to each other. The critical radius is taken as the point along the radius of the disc where 50% of the cells are detached, and the shear stress operating at this point is correspondingly termed the critical shear stress.

Results

The mean cell counts for three discs and two surfaces — acid-washed glass and a commercial silane-coupled hydrophobic coating — are presented in Fig. 14.4a and b. In Fig. 14.4a cell counts are plotted against radius from the midpoint of the disc. Disturbances in flow often cause cells to remain attached at the midpoint, and at a radius of approximately 5 mm, and these should always be ignored. Thus, the flow rate should be adjusted so that the zone where cell numbers are clearly increasing in a more or less linear manner is between 10 and 30 mm from the centre of the disc (Fig. 14.4a). Inspection of Fig. 14.4a reveals that the zone of cell detachment is less on the coated disc than on the untreated glass disc, indicating that cells adhere more tenaciously to the coated surface. Figure 14.4b shows that same data plotted as log shear stress, and critical shear stress values of 5.5 and 12.2 N/m^2 for glass and the coated surface respectively are obtained. The critical shear values for cell detachment are higher than those for attachment of cells, as discussed by Duddridge *et al.* (1982).

Flow Cell

Panels immersed from rafts are used for testing the potential performance of putative antifouling coatings (Lovegrove, 1978). Testing the potential performance of non-biocidal coatings necessitates the use of methods that measure the adhesion of organisms. Cell adhesion tests and the RFC can be used for preliminary screening of surfaces, but the field performance of surfaces requires immersion of panels from a raft prior to ship trials. Since the only forces to which biofilms are exposed on raft panels are tidal flow and water movement mediated by the wind, correlation between raft performance and ship performance is often poor when 'non-stick' surfaces are being evaluated.

The flow cell (Fig. 14.5) consists of two parallel perspex rectangular plates (0.1 × 1.5 m). A slot in the rear perspex plate holds an aluminium plate which accommodates five 100 × 100 mm panels. The front and rear perspex plates clamp together, leaving a gap of 5 mm between the plates through which water is pumped. The volumetric flow can be adjusted up to a maximum of

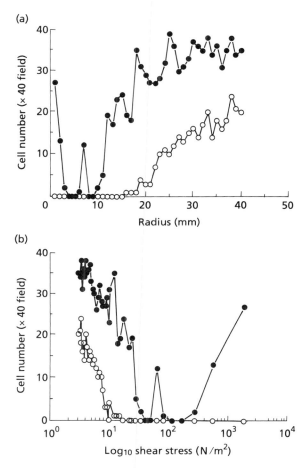

Fɪɢ. 14.4. Typical results obtained for detachment of the marine diatom *Amphora coffeaeformis* var. *perpusilla* from glass (○) and a silane-coupled hydrophobic coating (●). Each point is the mean of four cell counts taken along four 90° radii. Cell counts are plotted against radius in (a) and $\log_{10}$ shear stress in (b).

79.5 l/min, corresponding to a velocity of 2.65 m/s (Callow *et al.*, 1987). The flow cell is typically run for 10 min at maximum velocity. The dry weight of remaining adherent biofilm is measured after removal from the panel with cotton wool buds, as described by French & Evans (1986). The dry weight of panels which have not been subjected to the flow cell has also to be determined. Variation in the quantity of adherent biofilm on putative non-stick surfaces has been found to be so large that ten replicates for each treatment are required.

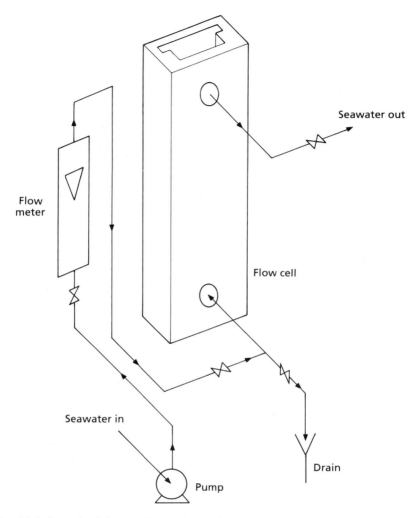

Fig. 14.5. Generalized diagram showing layout of the flow cell for testing the adhesion of biofilms to raft panels.

Results

After 8 weeks' immersion in the sea, the dry weight of the diatom slimes is less on all of the four types of low-energy silicone elastomer tested, than on the high-energy formica control. Up to 80% of the dry weight of diatom slimes are removed from low-energy silicone elastomers by the flow cell,

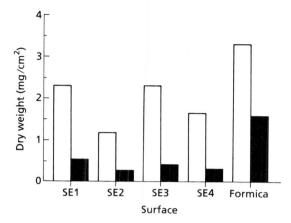

FIG. 14.6. Dry weight of fouling on (■) panels after treatment in the flow cell and (□) untreated panels. SE1−4 are four types of silicone elastomer. Untreated = formica. Each value is the mean of ten determinations.

compared with less than 50% from high-energy formica (Fig. 14.6) (Callow *et al.*, 1987).

Conclusions

The development of non-biocidal 'non-stick' surface coatings to prevent biofilm adhesion in a wide range of industrial applications, from process plant to ships' hulls, requires the development and application of laboratory methods to assess performance under dynamic conditions. In spite of the increasing number of restrictions on the use of biocides to control fouling, only a few non-biocidal 'non-stick' products have been commercialized. Methods to assess the potential of such products are available, but the development of new products seems to be slow.

Acknowledgements

Thanks are due to SERC and International Paint Ltd for financial support.

References

BOTT, T.R. & MILLER, P.C. (1983) Mechanisms of biofilm formation on aluminium tubes. *Journal of Chemical Technology and Biotechnology*, **33B**, 177−184.
CALLOW, M.E., PITCHERS, R.A. & SANTOS, R. (1987) Non-biocidal antifouling coatings. In Houghton, D.R., Smith, R.N. & Eggins, H.O. (eds.) *Biodeterioration 7*, pp. 43−48. Elsevier, London.

DUDDRIDGE, J.E., KENT, C.A. & LAWS, J.F. (1982) The effect of surface sheer stress on the attachment of *Pseudomonas fluorescens* to stainless steel under defined flow conditions. *Biotechnology and Bioengineering*, **24**, 153−164.

FRENCH, M.S. & EVANS, L.V. (1986) Fouling on paints containing copper and zinc. In Evans, L.V. & Hoagland, K.D. (eds.) *Algal Biofouling*, pp. 79−100. Elsevier, Amsterdam.

FOWLER, H.W. & McKAY, A.J. (1980) The measurement of microbial adhesion. In Berkeley, R.C.W., Lynch, J.M., Melling, J. & Rutter, P.R. (eds.) *Microbial Adhesion to Surfaces*, pp. 143−161. Academic Press, London.

FRYER, P.J., SLATER, N.K.H. & DUDDRIDGE, J.E. (1984) *On the Analysis of Biofouling Data with Radial Flow Shear Cells*. AERE Report R11295. United Kingdom Atomic Energy Authority.

FRYER, P.J., SLATER, N.K.H. & DUDDRIDGE, J.E. (1985) Suggestion for the operation of radial flow cells in cell adhesion and biofouling studies. *Biotechnology and Bioengineering*, **27**, 434−438.

GOODWIN, J.W., HARBRON, R.S. & REYNOLDS, P.A. (1990) Functionalization of colloidal silica and silica surfaces via silylation. *Colloid and Polymer Science*, **268**, 766−777.

HARTY, D.W.S. & BOTT, T.R. (1981) Deposition and growth of microorganisms on simulated heat exchanger surfaces. In Somerscales, E.F.C. & Knudson, J.G. (eds.) *Fouling of Heat Exchanger Surfaces*, pp. 335−344. Hemisphere Publishing Co., New York.

HYDE, K.D., MOSS, S.T. & JONES, E.B.G. (1989) Attachment studies in marine fungi. *Biofouling*, **1**, 287−298.

LEWTHWAITE, J.C., MOLLAND, A.F. & THOMAS, K.W. (1985) An investigation into the variation of ship skin frictional resistance with fouling. *Transactions of the Royal Institute of Naval Architects*, **127**, 269−284.

LOVEGROVE, T. (1978) Techniques for the study of mixed fouling populations. In Lovelock, D.W. & Davies, R. (eds.) *Techniques for the Study of Mixed Populations*, Society of Applied Bacteriology, Technical Series, No. 11, pp. 63−69. Academic Press, London.

LUFT, J.H. (1971) Ruthenium red and ruthenium violet. 1. Chemistry, purification, methods for use in electron microscopy and mechanism of action. *Anatomical Record*, **171**, 347−368.

MILNE, A. & CALLOW, M.E. (1985) Non-biocidal antifouling processes. In: *Polymers in a Marine Environment*, pp. 229−233. The Institute of Marine Engineers, London.

MOTT, I.E.C. & BOTT, T.R. (1990) *The Adhesion of Biofilms to Selected Materials of Construction for Heat Exchangers*. Paper 14-HX-4, 9th Heat Transfer Conference, Jerusalem.

SANTOS, R., CALLOW, M.E. & BOTT, T.R. (1991) The structure of *Pseudomonas fluorescens* biofilms in contact with flowing systems. *Biofouling*, **4**, 319−336.

WOODS, D.C. & FLETCHER, R.L. (1991) Studies on the strength of adhesion of some common marine fouling diatoms. *Biofouling*, **3**, 287−303.

15: Plasmid Exchange between Soil Bacteria in Continuous-Flow Laboratory Microcosms

L. Sun[1], M.J. Bazin[1] and J.M. Lynch[2]

[1]*Life Sciences Division, King's College London, Campden Hill Road, London W8 7AH; and* [2]*Horticulture Research International, Worthing Road, Littlehampton BN17 6LP, UK*

Genetically modified microorganisms (GEMMOs) appear to be of potential value in fields such as agriculture and pollution control (Halvorson *et al.*, 1985; Davies, 1988; Davison 1988). However, it is necessary to obtain information about their survival, competition and potential for gene exchange under natural conditions, before their release into natural ecosystems. Some studies on closed or isolated laboratory systems have been undertaken (Stotzky & Babich, 1986; Knudsen *et al.*, 1988; Rafii & Crawford, 1988; Van Elsas *et al.*, 1988, 1989; Trevors & Berg, 1989; Orvos *et al.*, 1990) but most ecosystems are thermodynamically open. In soil ecosystems, most bacteria adhere to a solid substratum, either in the form of soil particles, plant roots or terrestrial microfauna. The other major physical environmental factor that affects these bacteria is the thermodynamically open nature of soil: there is an input and output of both energy and matter to and from soil ecosystems. This contrasts with the condition in closed or isolated systems (Fig. 15.1). Essentially, bacteria in the soil grow as biofilms and are subjected to a flux of nutrient material. We have used an experimental system which has such properties, to study the physiology and genetics of soil bacteria. Here we report our observations on intraspecific plasmid exchange between *Pseudomonas cepacia* and *Enterobacter cloacae*, two frequently isolated soil bacteria.

Gene exchange between microorganisms in the natural environment may involve conjugation, transformation and transduction (Stotzky & Babich, 1986; Levy & Marshall, 1988) but plasmid transfer through conjugation is generally considered to be the most likely gene exchange process (Reanney, 1977; Trevors *et al.*, 1987). The dynamics of plasmid exchange in *Ps. cepacia* and *E. cloacae* were studied in column reactors. Bacterial growth took place on the

Microbial Biofilms:
Formation and Control

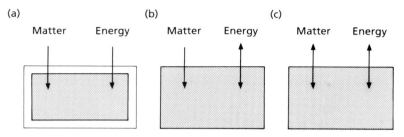

FIG. 15.1. Isolated system (a), closed system (b) and open system (c) in thermodynamics.

surface of vermiculite particles, which were irrigated with nutrient solution. Growth and any gene exchange were, therefore, in a biofilm. Analysis of the effluent from the columns indicated significant differences between *Ps. cepacia* and *E. cloacae* with respect to intraspecific plasmid exchange. Observation by scanning electron microscopy (SEM) also showed a significant difference in bacterial cell distribution with respect to distance down the column.

Column Reactor Design and Operation

In a soil column the system is heterogeneous: microbial populations adhere to soil particles and both concentrations of bacteria and growth-limiting nutrients change with distance down the column, as well as with time. For this reason, columns of four different lengths — 14, 24, 34 and 44 cm — were employed. For each length, three replicate columns were used and each column consisted of two glass tubes. The inner tube, 1.5 cm in diameter, was packed with vermiculite, which is a type of hydrated silicate. Vermiculite lamina are, in general, soft, pliable and inelastic and, when heated to 100–110°C, up to 10% water is lost (Ford, 1932). The reason for using vermiculite instead of soil as the packing material in the column was to simplify the system by providing a relatively inert solid substrate. The outer tube, 4 cm in diameter, served as a water jacket to maintain the temperature of the system at 30 ± 1°C by connecting it to a water bath (SU5 Grant, Grant Instruments Ltd, Cambridge, UK). Medium from a 5-l reservoir was pumped to the top of the columns by means of a flow inducer (Watson Marlow, Falmouth, UK) fitted with a multibranch adapter. In order to prevent bacteria growing back to the medium reservoir, anti-growback tubes were used. The system was aerated by passing air though a sterile gas filter, then through sterile water in a Dreschel bottle and finally into a manifold, to which each of the columns was connected. The air flow rate was 500 ml/min. Flasks were used to collect the effluent from each column. Details of the column construction are shown in Fig. 15.2.

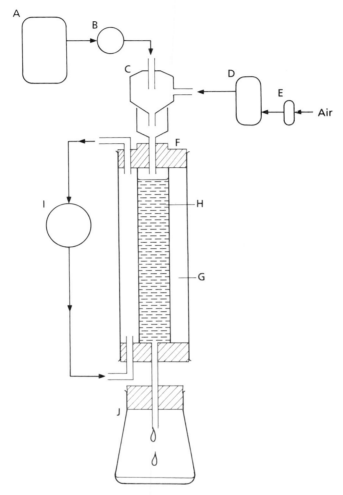

FIG. 15.2. Diagrammatic representation of the column reactor system. A, Medium reservoir; B, medium pump; C, anti-growback tube; D, Dreschel bottle; E, air filter; F, solid silicone rubber stopper; G, water jacket; H, vermiculite; I, water circulation; J, flask.

Organisms and media

Recombinant *Ps. cepacia* 120 and *E. cloacae* 67A, both containing the trans-missible plasmid R388::Tn*1721*, which carries genes encoding resistance to tetracycline and trimethoprim, were used as donor strains (Armstrong *et al.*, 1987, 1990). *Pseudomonas cepacia* 633 and *E. cloacae* 107 carrying nalidixic acid (Nal) chromosomal resistance genes were used as recipients. *Pseudomonas* *cepacia* 633 (ATCC 25416) was derived from strain PC012 (Gonzalez & Vidaver, 1979). *Enterobacter cloacae* 107 was isolated from insect faeces (Armstrong *et al.*, 1990). All bacterial strains were kindly provided by Dr Ray Seidler, United States Environmental Protection Agency, 200 SW 35th Street, Corvallis, Oregon 97333, USA.

King's B agar (King *et al.*, 1954), supplemented with 0.1 mg/ml tetracycline (Tc) and 0.05 mg/ml trimethoprim (Tp), was used to maintain and enumerate donor strains. Recipient strains were maintained and enumerated on King's B agar plates containing 0.5 mg/ml Nal. King's B agar amended with Tc, Tp and Nal was used to check for transconjugants. Defined medium was used as nutrient in the experiments, the components of which were as follows (g/l): K_2HPO_4, 1.5; KH_2PO_4, 0.5; $(NH_4)_2SO_4$, 1; $MgSO_4 . 7H_2O$, 1.2; EDTA, 12; NaOH, 2; $ZnSO_4 . 7H_2O$, 0.4; $MnSO_4 . 4H_2O$, 0.4; $CuSO_4 . 5H_2O$, 0.1; $FeSO_4 . 7H_2O$, 2; Na_2SO_4, 10; $Na_2MoO_4 . 2H_2O$, 0.1; glucose, 4. The advantage of using defined medium was that the carbon source (glucose) could be controlled to a growth-limiting concentration.

Plasmid Exchange in the Column Reactors

Donor and recipient strains were inoculated simultaneously at the top of the columns by means of a sterile hypodermic syringe. Effluent from the columns was serially diluted in 0.8% (w/v) saline spread on to plates in duplicate on appropriate selective media, and incubated for 48 h at 30°C to estimate donor, recipient and transconjugant population densities in terms of colony-forming units (cfu/ml). For each column length, results are reported as an average of the three columns employed. Defined medium was supplied to the columns at a flow rate of 0.55 ml/h.

An initial rapid increase in all strains was observed in the effluent from all of the columns. For both donor and recipient strains, the time taken to reach maximum densities of $3.5 \times 10^9 - 2.5 \times 10^{10}$ cfu/ml in the effluent appeared to be proportional to the length of the columns. Donor, recipient and transconjugant populations of *E. cloacae* tended to stabilize and come to steady state. However, in the experiment with *Ps. cepacia* only the recipient populations remained relatively stable. After reaching their maximum densities, the donor and transconjugant population densities decreased. Changes in the populations of *E. cloacae* and *Ps. cepacia* observed in the effluents from the column reactors are shown in Table 15.1.

Scanning electron microscopy

In order to obtain information about the mode of growth of the microbes on vermiculite, columns dismantled after experiments were examined by SEM.

TABLE 15.1. *Bacterial populations in the effluent from different-length columns*

	Bacterial populations (cfu/ml)			
	14 cm column		34 cm column	
	Day 5	Day 20	Day 5	Day 20
Ps. cepacia				
Donor	7.9×10^9	1.1×10^5	8.8×10^9	3.3×10^5
Recipient	9.8×10^9	9.9×10^9	9.4×10^9	8.7×10^9
Transconjugant	5.0×10^4	0.9×10^1	8.7×10^4	4.2×10^1
E. cloacae				
Donor	7.9×10^9	8.5×10^9	5.4×10^9	8.1×10^9
Recipient	7.2×10^9	7.9×10^9	5.7×10^9	8.3×10^9
Transconjugant	6.2×10^4	6.9×10^4	1.5×10^5	5.2×10^4

Donor and transconjugant populations of *Ps. cepacia* decreased after reaching their maximum densities, whereas the recipient population stabilized. All the populations of *E. cloacae* tended to be stable after reaching their maximum densities.

Vermiculite was taken from each column and placed in 2.5% glutaraldehyde in 0.1 mol/l sodium phosphate buffer at pH 7.2 for 1 h. The preparation was then gently washed with distilled water and transferred to 30% ethanol for 1 h, followed by 50% ethanol for 0.5 h, 70% ethanol for 0.5 h, 90% ethanol for 20 min, 100% ethanol for 15 min and finally, 100% acetone. Critical point drying (Emscope 750, Polaron Equipment Ltd, UK) was carried out for 1.5 h, then the dried samples were mounted on stubs and coated to a depth of 20 nm with platinum in an Polaron sputter coater (SEM Coating Unit E5100, Polaron Equipment Ltd, UK). The samples were observed under SEM (Jeol JSM-25S scanning microscope, JEOL Ltd, Tokyo, Japan) and photographs (Ilford FP4 ISO 125/22 film) were taken.

The change in cell distribution on the vermiculite surface in a 34 cm length column is illustrated in Fig. 15.3. Vermiculite from the top of the column was mostly coated with multilayered cell biofilms. Potentially, these biofilms are an ideal environment for plasmid exchange that results from cell−cell contact. Further down the column, multilayers of cell biofilms on vermiculite were rarely seen. A few cells were seen on the surface of vermiculite from the bottom layer of the column.

Conclusions

The difference in behaviour between the two bacterial species with respect to plasmid exchange is marked, and may be of considerable practical significance.

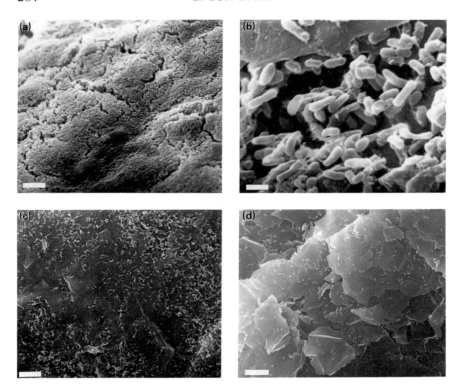

FIG. 15.3. Distribution of bacteria (*Pseudomonas cepacia*) on vermiculite taken from different layers of a 34 cm column. (a) Vermiculite from the top of the column, showing multilayered cell biofilms. (b) Slightly lower in the column than (a) and at higher magnification. (c) Middle layer from the column, showing monolayer cells. (d) Vermiculite from the bottom of the column, showing a few individual cells. Bars (a), (c) and (d), 10 μm; (b), 1 μm.

Ps. cepacia R388::Tn*1721* donors and transconjugants disappeared from the system in an exponential manner, indicating first-order decay kinetics. The identical plasmid in both donor and transconjugant cells of *E. cloacae* remained in the system, and the population densities of donor, transconjugant and recipient in the columns, as estimated from the effluent solutions, appeared to reach a steady state. One option that has been suggested for the safe release of GEMMOs is to encode the desired genes on a plasmid that behaves like R388::Tn*1721* in *Ps. cepacia*. It has been argued that, in such a case, the genetically modified message remains in the ecosystem only transiently. Thus, if a risk is associated with such a message, it may quickly disappear from the system.

Although transgeneric plasmid transfer between *Ps. cepacia* and *E. cloacae* is a rare event, we have occasionally isolated recombinant strains which indicate that it does occur. Therefore, before release of plasmid-bearing GEMMOs, with the expectation of plasmid loss, it will be advisable to check the behaviour of the plamsid in the indigenous microflora.

Acknowledgements

We thank the K.C. Wang Education of Hong Kong, the Committee of Vice-Chancellors and Principals of the Universities of the United Kingdom and Horticulture Research International for financial support to one of us (L.S.). We also wish to thank Dr D. Crawford and Dr D. Fraser (HRI, Littlehampton) for their helpful discussion of the manuscript.

References

ARMSTRONG, J.L., KNUDSEN, G.R. & SEIDLER, R.J. (1987) Microcosm method to assess survival of recombinant bacteria associated with plants and herbivorous insects. *Current Microbiology*, **15**, 229–232.

ARMSTRONG, J.L., WOOD, N.D. & PORTEOUS, L.A. (1990) Transconjugation between bacteria in the digestive tract of the cutworm, *Peridroma saucia. Applied and Environmental Microbiology*, **56**, 1492–1493.

DAVIES, J. (1988) Engineering organisms for use. In Sussman, M., Collins, C.H., Skinner, F.A. & Stewart-Tull, D.E. (eds.) *The Release of Genetically Engineered Micro-organisms*, pp. 21–28. Academic Press, London.

DAVISON, J. (1988) Plant beneficial bacteria. *Biotechnology*, **6**, 282–286.

FORD, W.E. (1932) *A Textbook of Mineralogy*, pp. 674–851. John Wiley & Sons, New York & London.

GONZALEZ, C.F. & VIDAVER, A.K. (1979) Bacteriocin, plasmid and pectolytic diversity in *Pseudomonas cepacia* of clinical and plant origin. *Journal of General Microbiology*, **110**, 161–170.

HALVORSON, H.O., PRAMER, D. & ROGUL, M. (1985) *Engineered Organisms in the Environment: Scientific Issues*. American Society for Microbiology, Washington, DC.

KING, E.O., WARD, M.K. & RANEY, D.E. (1954) Two simple media for the demonstration of pyocyanin and fluorescein. *Journal of Laboratory and Clinical Medicine*, **44**, 301.

KNUDSEN, G.R., WALTER, M.V., PORTEOUS, L.A., PRINCE, V.J., ARMSTRONG, J.L. & SEIDLER, R.J. (1988) Predictive model of conjugative plasmid transfer in the rhizosphere and phyllosphere. *Applied and Environmental Microbiology*, **54**, 343–347.

LEVY, S.B. & MARSHALL, B.M. (1988) Genetic transfer in the natural environment. In Sussman, M., Collins, C.H., Skinner, F.A. & Stewart-Tull, D.E. (eds.) *The Release of Genetically Engineered Micro-organisms*, pp. 61–76. Academic Press, London.

ORVOS, D.R., LACY, G.H. & CAIRNS, J. (1990) Genetically engineered *Erwinia carotovora*: survival, intraspecific competition, and effects upon selected bacterial genera. *Applied and Environmental Microbiology*, **56**, 1689–1694.

RAFII, F. & CRAWFORD, D.L. (1988) Transfer of conjugative plasmids and mobilization of a nonconjugative plasmid between *Streptomyces* strains on agar and in soil. *Applied and Environmental Microbiology*, **54**, 1334–1340.

REANNEY, D. (1977) Gene transfer as a mechanism of microbial evolution. *Bioscience*, **27**, 340–344.

STOTZKY, G. & BABICH, H. (1986) Survival of, and genetic transfer by, genetically engineered bacteria in natural environments. *Advanced and Applied Microbiology*, **31**, 93–138.

TREVORS, J.T., BARKAY, T. & BOURQUIN, A.W. (1987) Gene transfer among bacteria in soil and aquatic environments: a review. *Canadian Journal of Microbiology*, **33**, 191–198.

TREVORS, J.T. & BERG, G. (1989) Conjugal RP4 transfer between Pseudomonads in soil and recovery of RP4 plasmid DNA from a soil system. *Applied Microbiology*, **11**, 223–227.

VAN ELSAS, J.D., TREVORS, J.T. & STARODUB, M.E. (1988) Bacterial conjugation between pseudomonads in the rhizosphere of wheat. *FEMS Microbiology and Ecology*, **53**, 299–306.

VAN ELSAS, J.D., TREVORS, J.T., VAN OVERBEEK, L.S. & STARODUB, M.E. (1989) Survival of *Pseudomonas fluorescens* containing plasmid RP4 or pRK2501 and plasmid stability after introduction into two soils of different texture. *Canadian Journal of Microbiology*, **35**, 951–959.

16: Microbial Films
in the Light Engineering Industry

P.E. Cook[1] AND C.C. Gaylarde[2]

[1]*Department of Food Science and Technology, University of Reading, Reading RG6 2AP; and* [2]*Department of Soils, University of Rio Grande do Sul, Porto Alegre, RS, Brazil*

The light engineering industry has for many years encountered significant biodeterioration problems, which arise directly or indirectly from the growth and metabolism of microorganisms. Many of these problems relate to microbial contamination of the wide range of fluids and coolants used throughout these industries (Bennett, 1974; Rossmoore, 1986). Unstable emulsions, slimes, blocked pipework, fouled surfaces and unpleasant odours are familiar to many engineering workers. Control of these problems has usually focused on the reduction of contamination in fluids; less attention has been paid to surface films. In some cases, coolant problems may be linked to the presence of biofilms, because these can act as reservoirs for microorganisms which then recontaminate fresh coolant. Biofilms may be important causes of corrosion in metalworking fluid systems (Ortiz *et al.*, 1990) and microbial extracellular polysaccharides may help to facilitate corrosion by maintaining biofilm structure (Beech & Gaylarde, 1991).

Metalworking fluids reduce friction and power consumption as well as improving surface finish and tool life. Although primarily intended for engineering, many formulations have components that can act as carbon or nitrogen sources for the growth of microorganisms. Foxall-VanAken *et al.* (1986) demonstrated that bacteria isolated from metalworking fluids can grow on fatty acids and naphthenic petroleum oil components of non-synthetic fluids. In synthetic metalworking fluids, some fungi are capable of utilizing triethanolamine and amine borate components as their sole carbon or nitrogen sources (Prince & Morton, 1988). Metalworking fluids may also become supplemented with extrinsic sources of organic matter, which provide additional carbon and nitrogen for the growth of microorganisms (Bennett, 1972; Rossmoore, 1986).

Microbial Biofilms:
Formation and Control

A diverse range of bacteria have been isolated from metalworking fluids, including species of *Acinetobacter, Alcaligenes, Bacillus, Citrobacter, Desulfovibrio, Enterobacter, Klebsiella, Proteus* and *Pseudomonas* (Bennett, 1972, 1974; Foxall-VanAken *et al.*, 1986). Fungi also occur, particularly in synthetic coolants, and include species of *Acremonium, Aspergillus, Candida, Fusarium, Geotrichum* and *Scopulariopsis* (Bennett, 1974; Cook & Gaylarde, 1988a; Prince & Morton, 1988). On rare occasions protozoa can be found in highly contaminated fluids.

When metalworking fluids become contaminated by high populations of microorganisms the pH usually falls and the emulsion may become unstable. In addition, corrosion may occur and foul odours, such as hydrogen sulphide, may be produced. These conditions often lead to reduced efficiency and costly 'down time' for cleaning and replacement of contaminated coolants. Figure 16.1 shows the sump of a metal-cutting lathe with separation of oil on the surface of a heavily contaminated coolant. Almen *et al.* (1982) developed a high-performance liquid chromatography (HPLC) technique to study the effect of microbial contamination on the composition of emulsifiable oils.

Formation and Examination of Biofilms in Engineering Fluids

Biofilms are often difficult to locate in contaminated machines, because they may occur at sites not readily accessible during cleaning or routine maintenance. Figure 16.2 shows an extensive biofilm developing on the underside of a metal plate covering the machine sump shown in Fig. 16.1.

The composition of the biofilm may be extremely variable, depending on the coolant used and the age and design of the machine. Figure 16.3 shows bacterial and fungal cells embedded in polymeric material in a sample of biofilm from a metalworking fluid. Other biofilm components may include metal fragments, organic matter, corrosion products and oil that has separated from emulsions. A large accumulation of oil may lead to oxygen limitation in these films, particularly during machine shutdown. Fungi may be favoured by these conditions since some species, such as *Fusarium* spp., are able to grow at extremely low oxygen tensions or even under anaerobic conditions (Tabak & Cooke, 1968).

Microbiological analysis of biofilms

Plate counts can be used to isolate and count microorganisms in metalworking fluid biofilms, but it may not be possible to recover all microorganisms.

Samples can be taken by scraping a known area of biofilm with sterile razor blades, which are best used mounted in a suitable tool, such as a paint scraper. A large sample from a representative area is important, because

FIG. 16.1. Exposed sump of CNC machine showing separation of oil from the metalworking fluid emulsion.

biofilm formation is often heterogeneous. Areas which can be sampled at regular intervals are particularly important, because this will make it possible to study the effects of remedial treatments and recolonization.

Samples of biofilm from engineering workshops can be collected in sterile containers, such as Petri dishes, and returned to the laboratory, preferably on ice. For determination of viable microorganisms, 2-g samples are suspended in 20 ml of sterile 0.85% (w/v) saline and sonicated for 1 min at an amplitude of 4 μm, followed by serial dilution in saline. If samples are heavily contaminated with oil, this can sometimes be partially removed by centrifuging biofilm samples 2–3 times in saline at 12 000 g for 20 min and discarding the supernatant, which will contain the oil. Washing the biofilm by centrifugation may also reduce levels of biocides, which can interfere with recovery when cells are released from the biofilm.

Both selective and non-selective media can be used for isolation and enumeration. In general, we have found nutrient agar to be suitable for bacterial counts and malt extract agar, supplemented with 100 mg/l each of penicillin G and streptomycin sulphate, for moulds and yeasts. Cetrimide agar and *Pseudomonas*-selective medium are useful since *Pseudomonas* spp. are major contaminants in metalworking fluids and biofilms. The *Klebsiella*-selective

FIG. 16.2. The underside of the plate covering the sump shown in Fig. 16.1. Extensive development of biofilm can be seen on the surface in contact with the coolant.

medium of Tomas *et al.* (1986) may be useful, since preliminary trials have indicated that high numbers of *Klebsiella* spp. can be isolated from some non-synthetic coolants.

Care must be taken when using selective media, because cells damaged by physical or biocidal treatments may be less tolerant to some media components and may require a period of resuscitation before isolation on these media.

By means of viable plate counts it is possible to obtain information about the levels of contamination in coolant and biofilm. Table 16.1 shows the levels of different microbial groups in coolant and biofilm phases in a computerized numerical control (CNC) lathe with which a non-synthetic metalworking fluid is used. After 6 weeks' operation the coolant was heavily contaminated with bacteria. A substantial biofilm had developed on the walls of the sump and contained high numbers of bacteria. The microbial numbers were much lower after cleaning and replacement of coolant and a further 5 weeks' operation, although surface-associated microorganisms were generally present in higher numbers than in the coolant.

The magnitude of the surface microbial populations in these fluids can be demonstrated by determining the viable count in the fluid before and after disruption of the biofilm. Table 16.2 shows viable counts in the coolant of a

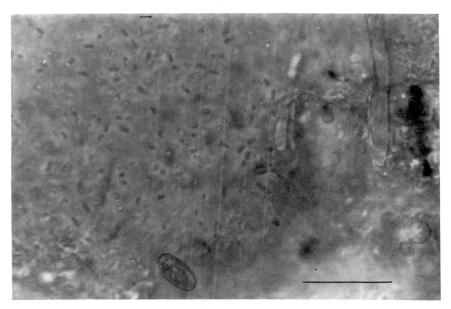

FIG. 16.3. Biofilm from metalworking fluid system, showing fungal hyphae, yeasts and bacterial cells. Bar, 20 μm.

TABLE 16.1. *Viable counts of microorganisms in the coolant and biofilm of a CNC lathe containing a non-synthetic metalworking fluid. The machine was sampled (a) after 6 weeks' operation and (b) a further 5 weeks after cleaning and introduction of fresh coolant*

	Viable counts (cfu)			
Source	Aerobic bacteria	Anaerobic bacteria	Coliform bacteria	Fungi
(a) After 6 weeks				
Coolant (per ml)	7.8×10^{10}	4.3×10^{10}	7.0×10^{10}	2.2×10^{5}
Biofilm (per cm^2)	6.0×10^{9}	5.3×10^{8}	3.2×10^{10}	1.3×10^{5}
(b) A further 5 weeks after cleaning				
Coolant (per ml)	8.3×10^{4}	5.8×10^{5}	1.0×10^{4}	2.4×10^{2}
Biofilm (per cm^2)	6.1×10^{5}	1.5×10^{5}	3.6×10^{5}	3.2×10^{3}

From Cook & Gaylarde (1988a).

TABLE 16.2. *Viable counts of microorganisms in a non-synthetic metalworking fluid in a CNC milling machine before and after disruption of the sump biofilm with a wire brush*[*]

| | Viable count (cfu/ml) | | |
Microbial group	Before brushing	After brushing	Increase
Aerobic bacteria	2.2×10^8	4.6×10^9	$\times$ 21
Anaerobic bacteria	2.0×10^8	2.2×10^9	$\times$ 11
Coliform bacteria	7.0×10^6	9.0×10^7	$\times$ 13
Fungi	2.9×10^5	4.2×10^5	$\times$ 1.5

[*] The walls of the sump were rubbed with a wire brush for 30 s and the coolant was mixed before taking a sample. From Cook & Gaylarde (1988a).

CNC milling machine before and after disruption of a biofilm in the machine sump. Bacterial counts increased by more than tenfold after disruption, whereas fungal counts increased only slightly.

Model systems for studying biofilm formation

Since biofilms are difficult to sample in engineering workshops, it is desirable to use model systems to examine biofilm formation and control. Cook & Gaylarde (1988a) developed a model circulating system to monitor biofilm formation in aqueous metalworking fluids. The experimental system is shown in Fig. 16.4. Coolant was pumped from a 12-l base tank to a header tank and allowed to flow at a rate of 1.5 l/min through a series of cylindrical glass tubes (30 × 4 cm) connected in series, each containing four rectangular mild steel coupons (4 × 1.5 cm). Twelve litres of coolant were circulated continuously for 7 h each day for 5 days, followed by 2 days when the system was not run.

The development of planktonic and biofilm populations of microorganisms in a non-synthetic metalworking fluid in the presence and absence of a triazine biocide are shown in Figs 16.5 and 16.6. Bacteria and fungal populations developed much more rapidly in the absence of biocide, but the biofilm population was less stable and viable counts declined after 6 days. When biocide was present, planktonic and biofilm numbers increased more slowly but eventually reached levels similar to those in the absence of biocide.

Model systems are particularly valuable for evaluating biocides, because differences in the coolants used, machine design and workshop practices can result in trials that are difficult to interpret in engineering workshops.

Direct microscopy of biofilms

Viable counts on agar media provide a means of isolating and quantifying microorganisms from biofilms, but the approach may lead to the underestimation

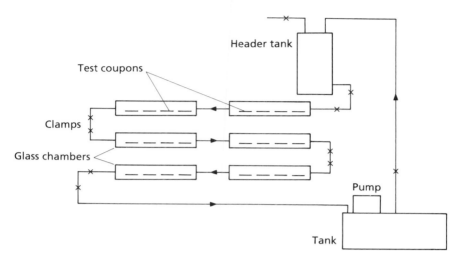

FIG. 16.4. Model circulating system for monitoring biofilm formation in metalworking fluids. From Cook & Gaylarde (1988a).

of living cells, because some microorganisms may be difficult to isolate. Direct microscopy offers an alternative approach. Appropriate staining methods can be used to differentiate between living and dead cells, or between those which are active and inactive.

Epifluorescence microscopy has been widely used to examine microorganisms in fluids, on surfaces, and in homogenates of solid substrates such as plant materials. Staining can be used to detect nucleic acids, membranes or enzyme activity. Unfortunately, fluorescence microscopy is difficult to use with engineering fluids, because some coolant components are prone to fluoresce and others may quench the fluorescence of microbial cells.

The direct epifluorescence filter technique (DEFT) (Pettipher & Rodrigues, 1981) has been used successfully to quantify bacteria in food emulsions, such as milk and milk products, and might find application for quantifying microorganisms in aqueous metalworking fluids and biofilms. Appropriate pretreatments with surfactants and enzymes would be required to remove interference from oil, emulsifiers and biofilm components.

Epifluorescence microscopy can be used to examine microbial cells and their distribution on machine surfaces where oil is absent. Cook & Gaylarde (1988b) examined the spatial dynamics of *Desulfovibrio desulfuricans* subsp. *desulfuricans* (New Jersey, NCIMB 8313) on 7 mm mild steel discs that had received different degrees of surface polishing. After exposure to *D. desulfuricans* in Postgate's Medium C (Postgate, 1984), the steel discs were rinsed gently in sterile distilled water and the attached cells stained in 0.001% (w/v) aqueous

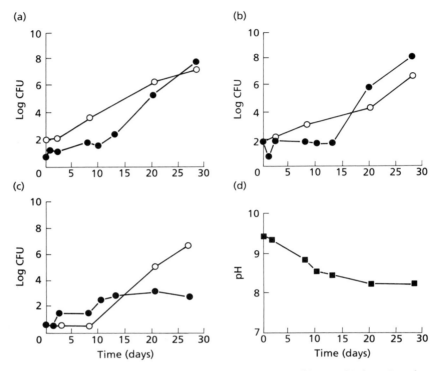

FIG. 16.5. Changes in the viable counts of (a) aerobic bacteria, (b) anaerobic bacteria and (c) fungi associated with the coolant (●) (viable counts; cfu/ml) and biofilm (○) (viable counts; cfu/cm^2) in an aqueous metalworking fluid plus triazine biocide in a model system. The pH of the coolant is shown in (d). Inoculum consisted of a mixture of contaminated metalworking fluids. From Cook & Gaylarde (1988a).

acridine orange solution for 5 min. After rinsing in sterile distilled water, the discs were examined by epifluorescence microscopy and the numbers of fluorescing cells could be counted using a 10×10 gridded graticule.

Staining with tetrazolium salts

Tetrazolium salts have been extensively used in the study of metabolically active microorganisms in soil (Smith & Pugh, 1979; Macdonald, 1980; Trevors et al., 1982), water (Tabor & Neihof, 1982) and food (Betts et al., 1989). They are also used in dip-slides for the detection of microbial contamination in fluids such as machine coolants (Hill et al., 1976). Benbouzid-Rollet et al. (1991) used 2-(p-iodophenyl)-3-(p-nitrophenyl)-5-phenyltetrazolium chloride

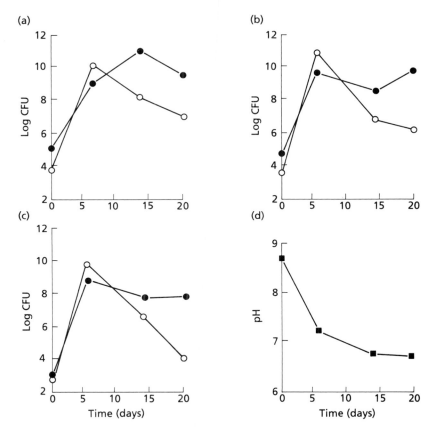

FIG. 16.6. Changes in the viable counts of (a) aerobic bacteria, (b) anaerobic bacteria and (c) fungi associated with the coolant (●) (viable counts; cfu/ml) and biofilm (○) (viable counts; cfu/cm^2) in an aqueous metalworking fluid in a model system. The pH of the coolant is shown in (d). Inoculum consisted of a mixture of contaminated metalworking fluids. From Cook & Gaylarde (1988a).

(INT) to detect active cells of *Vibrio natriegens* in a marine aerobic biofilm on stainless-steel surfaces in a laboratory tubular flow system.

MacDonald (1980) demonstrated that a number of tetrazolium salts can be used in conjunction with the agar-film technique (Jones & Mollison, 1948) to quantify active and inactive bacteria in the soil. Unfortunately, low metabolic activity results in tetrazolium formazan crystals, which are very small and diffuse. Substrates such as glucose, succinic acid, glutamic acid, NADH and NADPH are often added along with tetrazolium salts, since this can result in higher metabolic activity and increased staining. Of the tetrazolium salts,

INT and thiazolyl blue (MTT) give clearer and more intense staining of a wide range of microorganisms than triphenyl tetrazolium chloride (TTC) or tetrazolium blue (TB).

If the sample contains fungal hyphae and quantification is required, the sample should be homogenized, either before or after the incubation with tetrazolium salts. Incubation before homogenization has the advantage that the hyphal contents are not disrupted and exposed directly to osmotic and pH changes, biocides or enzyme inhibitors. In addition, tetrazolium formazan crystals are insoluble in aqueous systems and are less likely to be lost from hyphae during homogenization. A disadvantage is that, without homogenization, tetrazolium salts may not reach hyphae or bacterial cells deeply embedded in biofilms. In all cases, homogenization treatments should be standardized and optimized for a particular type of biofilm. High-speed homogenization for a short time is preferable to low speeds for extended periods. An optimum release of fungal hyphae can usually be obtained after about 1 min with a high-speed open-blade homogenizer at 25 000 rpm.

Incubation of samples should be for as long as possible, because some microbial cells may be viable but metabolically inactive and only show staining after prolonged incubation. The percentage of active fungal hyphae after incubating fungal homogenates for different periods at 20°C is shown in Fig. 16.7. Most of the hyphae were stained in the first hour. No additional staining could be detected between 3 and 7 h.

Procedure for staining with tetrazolium

1   An intact or homogenized sample of biofilm is suspended in 0.1 M Tris-HCl buffer, pH 7.6, containing 1 mg/ml NADH and 1 mg/ml NADPH.
2   The sample is mixed with an equal volume of filter-sterilized (0.45 μm) 0.1 M Tris-HCl buffer, pH 7.6, containing INT or MTT to give a final concentration of 0.2 mg/ml.
3   The sample is incubated at 20°C for 3 h and further microbial growth halted by adding 40% (v/v) aqueous formalin to give a final concentration of 4% (v/v).
4   Fixed samples can then be homogenized if necessary, and used to prepare agar films according to Jones & Mollison (1948).
5   Agar films can be examined with phase-contrast or bright-field microscopy. If required, microbial cells can be stained and mounted in phenolic aniline blue. This can be prepared by mixing 3.75 g phenol, 0.05 g aniline blue and 100 ml of 20% (v/v) acetic acid. After mixing, the stain is left to stand for 2−3 h before filtration.

Tetrazolium-stained bacteria can be counted with a gridded graticule and lengths of tetrazolium-stained fungal hyphae measured with a gridded graticule

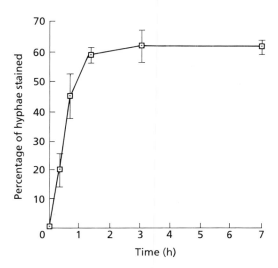

FIG. 16.7. Influence of incubation time at 20°C on the percentage of hyphae in agar films stained with 2-(*p*-iodophenyl)-3-(*p*-nitrophenyl)-5-phenyl tetrazolium chloride (INT). Figures are mean values (±SEM) based on five determinations at each incubation time.

and the intersection method of Olson (1950).

An alternative method to agar films has been used by Betts *et al.* (1989), who used INT staining to count viable bacteria and yeasts retained on filter membranes. There was a good correlation between the numbers of tetra-zolium-stained cells and viable counts of pure cultures of bacteria and yeasts. Benbouzid-Rollet *et al.* (1991) also found good agreement between counts of tetrazolium-stained cells, viable plate counts and epifluorescence counts of *Vibrio natriegens* in a model biofilm system.

Acrylic resin replicas

Birkby & Preece (1981) described a technique for the differentiation of Gram-positive and Gram-negative bacteria with transparent acrylic resin emulsion replicas from the surfaces of plants. We have used an acrylic copolymer emulsion (Spectrum Oil Colours, Horsham) to produce replicas of metal surfaces exposed to microorganisms. Microorganisms removed by the resin can be Gram stained or preincubated with tetrazolium salts before application of the resin. This can be examined microscopically and the spatial distribution of microorganisms examined. The method is unsuitable for systems containing oil, and is only suitable for studying the early stages of surface colonization.

Indirect methods for detecting microbial activity in biofilms

Many indirect methods have been proposed for the detection of microbial activity and biomass in biofilms, and Geesey & White (1990) have provided information on the range of techniques available for bacteria.

Fluorescein diacetate hydrolysis

Fluorescein diacetate (FDA) hydrolysis is a simple but sensitive method of measuring non-specific microbial activity, and assays have been used with a wide range of substrates (Schnurer & Rosswall, 1982). Marriot *et al.* (1991) used a spectrophotometric determination of FDA hydrolysis to measure microbial activity in biofilm samples from a rotating biological contactor. The assay was found to be sensitive and rapid, and FDA activity could be terminated with acetone. FDA hydrolysis increased linearly with time and with the dry-weight biomass of microorganisms.

FDA can also be used to measure microbial activity in some aqueous metalworking fluids. After incubation and termination of enzyme activity, high-speed centrifugation at $20\,000\,g$ can be used to separate metalworking fluid emulsions into aqueous and oil phases. The aqueous phase can then be used to detect released fluorescein at 492 nm. The assay for FDA hydrolysis may also be applicable to biofilms in engineering fluids. The disadvantages of using FDA are that not all microorganisms hydrolyse FDA to the same extent, and microbial cells deeply embedded in biofilms may not be exposed to the compound.

Extraction of tetrazolium salts

Although tetrazolium-stained cells can be observed and counted micro-scopically, an alternative approach is to extract the tetrazolium formazan and measure it spectrophotometrically. Smith & Pugh (1979) extracted tetrazolium formazan from soil with acetone and used absorbance at 492 nm as an indirect measure of microbial activity.

Biofilm samples that have been incubated with tetrazolium salts can also be extracted with a suitable solvent and the absorbance determined to provide an indication of microbial activity. In experiments with biofilms of sulphate-reducing bacteria on mild steel coupons (4×1.5 cm), INT-stained cells were extracted with 20 ml of acetone for 2 min after first rinsing in sterile distilled water. After filtration, the absorbance was measured at 480 nm against a blank. Appropriate controls are important, because acetone may extract a number of components from biofilms.

Quantification of fungal contamination in coolant and biofilm

In a study of microbial problems in small engineering workshops, biofilms were encountered in which fungal hyphae formed a significant component. Fungal biofilms often become detached, and the mats of hyphae can cause blockages in pipework and tanks. Fungal problems in metalworking fluids may be becoming more significant, particularly with the increasing use of synthetic fluids which are more prone to fungal attack (Prince & Morton, 1988).

Fungi are difficult to quantify by means of viable plate counts, because the number of colonies of moulds and filamentous yeasts on agar cannot readily be related to biomass, hyphal length or numbers of propagules. Alternative methods include assays for chitin and membrane components, such as ergosterol, which has the advantage of being found almost exclusively in fungi.

Seitz *et al.* (1977, 1979) have developed UV spectroscopy and HPLC methods to quantify ergosterol in fungal cultures and stored grain. Ergosterol assays have also been used to quantify fungal biomass in plant materials (Matcham *et al.*, 1985; Newell *et al.*, 1988). The following procedure was used by Cook & Gaylarde (1988a) to quantify ergosterol in biofilm and coolant samples in a non-synthetic metalworking fluid. The saponification, extraction and thin-layer chromatography (TLC) steps were carried out under subdued lighting conditions to reduce degradation of ergosterol.

Procedure for ergosterol extraction and quantification

1 The sample of biofilm is dispersed in 250 ml of sterile distilled water and centrifuged at 20 000 g for 20 min. The pellet is resuspended in sterile distilled water and the procedure repeated 3–4 times to remove as much metalworking fluid as possible. The pellet can either be used directly or freeze-dried for later analysis. When sampling metalworking fluids, these can be centrifuged directly to pellet the microorganisms and separate the fluid into oil and aqueous

TABLE 16.3. *Viable counts of fungi, ergosterol content and estimated fungal biomass in the coolant and biofilm phases of an aqueous metalworking fluid containing a triazine biocide*

Source	Viable count (cfu)	Ergosterol (µg)	Fungal biomass* (µg)
Coolant (per ml)	5.5×10^2	0.01	5.3
Biofilm (per cm²)	3.2×10^6	17.10	8555.5

* Ergosterol content assumed to be 2 µg/g mycelial dry weight. From Cook & Gaylarde (1988a).

phases. Several litres of fluid may have to be centrifuged to obtain sufficient fungal material for ergosterol assay.

2   The pellet is suspended in 20 ml of methanol and homogenized at 25 000 rpm for 2 min. After standing for 5 min the homogenate is filtered, together with an additional 10 ml of methanol used to rinse the homogenization flask.

3   The filtered homogenate is mixed with 12.5 ml of ethanol and 5 g of potassium hydroxide and saponified by refluxing for 30 min. After cooling, 12.5 ml of distilled water is added.

4   The non-saponifiable fraction is extracted three times with 25 ml of petroleum ether (bp 40−60°C), and the combined extracts evaporated to dryness under vacuum at 50−60°C in a rotary evaporator.

5   Extracts are dissolved in an appropriate volume of dichloromethane and loaded as bands on to precoated and activated TLC plates (Silica G, Camlab Ltd, Cambridge, UK). The plates are run in petroleum ether (bp 40−60°C), diethyl ether and glacial acetic acid (70:30:1). The sterol band between R_f 0.314−0.438 is scraped from the TLC plate, eluted with ethanol and filtered.

6   Ergosterol is determined by reverse-phase HPLC with a C_{18} column. The solvent system usually consists of methanol or methanol and water (99 : 1). Ergosterol can be detected at 282 nm in 6−9 min at a flow rate of 1.5 ml/min, depending on the column and HPLC system used.

A comparison between the viable fungal plate counts and the ergosterol content of coolant and biofilm in a model aqueous metalworking fluid system is shown in Table 16.3. The data show a considerable difference between the coolant and the biofilm, both in fungal plate counts and biomass estimated using ergosterol.

Control of Biofilms
in the Light Engineering Industry

Biocides are regularly used to prolong the life of machine coolants, and a wide range of agents are available (Hill *et al.*, 1976; Shennan, 1983). Although many of these are effective against planktonic microorganisms, less is known about their ability to control biofilms.

Bacteria are the most frequently encountered contaminants in water-based metalworking fluids, largely because isolation techniques are more likely to recover these than non-sporulating filamentous fungi. Moulds are less often detected because they tend to accumulate on walls, in crevices and in piping, often at the coolant/air interface. Because a wide range of contaminant bacteria and fungi can occur in machine coolants, it is important that biocides are broad-spectrum in action. The most commonly used biocides in metalworking fluids are triazines, glutaraldehyde and pyrithione derivatives. Isothiazolones are used for controlling biofilms in the offshore oil industry, although they are rapidly inactivated by nucleophilic components such as amines in metalworking fluids (Sandin *et al.*, 1991). Glutaraldehyde has been used successfully to control biofilms in some metalworking fluids.

Since biofilms can act as a microbial reservoir for reinfection of replaced coolant, it is important that biocides are targeted towards controlling surface populations of microorganisms. The regular use of biocides at levels sufficient to control planktonic microorganisms is likely to expose biofilm microorganisms to sublethal concentrations. The constant reinfection of coolants from biofilms is likely to lead to the emergence of strains resistant to biocides.

In synthetic metalworking fluids filamentous fungi present more of a problem than bacteria (Prince & Morton, 1988), and attention has recently turned to methods for their control. Antifungal agents, such as those used in agriculture, are being evaluated as possible biocides. Pohlman (1991) has investigated the fungicidal efficacy of a diiodomethyl-*p*-tolylsulphone emulsion in metalworking fluids.

Model systems should make it possible to develop strategies for the control of microbial films in the light engineering industry. With a better understanding of biofilm formation and detachment, it may be possible to develop and apply chemical and physical treatments that extend the life of engineering fluids and reduce 'down time' and corrosion.

References

ALMEN, R., MANTELLI, G., McTEER, P. & NAKAYAMA, S. (1982) Application of high-performance liquid chromatography to the study of the effect of microorganisms in emulsifiable oils. *Lubrication Engineering*, **38**, 99–103.

BEECH, I.B. & GAYLARDE, C.C. (1991) Microbial polysaccharides and corrosion. *International Biodeterioration*, **27**, 95–108.

BENBOUZID-ROLLET, N.D., CONTE, M., GUEZENNEC, J. & PRIEUR, D. (1991) Monitoring of a *Vibrio natriegens* and *Desulfovibrio vulgaris* marine aerobic biofilm on a stainless steel surface in a laboratory tubular flow system. *Journal of Applied Bacteriology*, **71**, 244–251.

BENNETT, E.O. (1972) The biology of metal-working fluids. *Lubrication Engineering*, **28**, 237–249.

BENNETT, E.O. (1974) The deterioration of metal cutting fluids. *Progress in Industrial Microbiology*, **13**, 121–149.

BETTS, R.P., BANKES, P. & BANKS, J.G. (1989) Rapid enumeration of viable micro-organsims by staining and direct microscopy. *Letters in Applied Microbiology*, **9**, 199–202.

BIRKBY, K.M. & PREECE, T.F. (1981) Differentiation of Gram-positive and Gram-negative bacteria in transparent acrylic resin emulsion replicas of surfaces of plants. *Journal of Applied Bacteriology*, **50**, 59–63.

COOK, P.E. & GAYLARDE, C.C. (1988a) Biofilm formation in aqueous metal-working fluids. *International Biodeterioration*, **24**, 265–270.

COOK, P.E. & GAYLARDE, C.C. (1988b) Spatial dynamics of surface colonising microorganisms. In Morton, L.H.G. & Chamberlain, A.H.L. (eds.) *Biofilms*, pp. 35–47. The Biodeterioration Society, Kew.

FOXALL-VAN AKEN, S., BROWN, JR, J.A., YOUNG, W., SALMEEN, I., McCLURE, T., NAPIER, JR, S.

& OLSEN, R.H. (1986) Common components of industrial metal-working fluids as sources of carbon for bacterial growth. *Applied and Environmental Microbiology*, **51**, 1165–1169.

GEESEY, G. & WHITE, D.C. (1990) Determination of bacterial growth and activity at solid–liquid interfaces. *Annual Review of Microbiology*, **44**, 579–602.

HILL, E.C., GIBBON, O. & DAVIES, P. (1976) Biocides for use in oil emulsions. *Tribology International*, **9**, 121–130.

JONES, P.C.T. & MOLLISON, J.E. (1948) A technique for the quantitative estimation of soil microorganisms. *Journal of General Microbiology*, **2**, 54–69.

MACDONALD, R.M. (1980) Cytological demonstration of catabolism in soil micro-organisms. *Soil Biology and Biochemistry*, **12**, 419–423.

MARRIOTT, N.J., SMITH, R.N. & HALL, J.J. (1991) FDA hydrolysis as a measure of microbial activity in the biofilm of a rotating biological contactor. In Rossmoore, H.W. (ed.) *Biodeterioration and Biodegradation 8*, pp. 456–458. Elsevier, London.

MATCHAM, S.E., JORDON, B.R. & WOOD, D.A. (1985) Estimation of fungal biomass in a solid-substrate by three independent methods. *Applied Microbiology and Biotechnology*, **21**, 108–112.

NEWELL, S.Y., ARSUFFI, T.L. & FALLON, R.D. (1988) Fundamental procedures for determining ergosterol content of decaying plant material by liquid chromatography. *Applied and Environmental Microbiology*, **54**, 1876–1879.

OLSON, F.C.W. (1950) Quantitative estimates of filamentous algae. *Transactions of the American Microscopical Society*, **69**, 272–279.

ORTIZ, C., GUIAMET, P.S. & VIDELA, H.A. (1990) Relationship between biofilms and corrosion of steel by microbial contaminants of cutting-oil emulsions. *International Biodeterioration*, **26**, 315–326.

PETTIPHER, G.L. & RODRIGUES, U.M. (1981) Rapid enumeration of bacteria in heat-treated milk and milk products using a membrane filtration–epifluorescent microscopy technique. *Journal of Applied Bacteriology*, **50**, 157–166.

POHLMAN, J. (1991) Fungicidal efficacy of a diiodomethyl-*p*-tolylsulfone emulsion in metal working fluid. In Rossmoore, H.W. (ed.) *Biodeterioration and Biodegradation 8*, pp. 502–503. Elsevier, London.

POSTGATE, J.R. (1984) *The Sulphate-Reducing Bacteria*. Cambridge University Press, Cambridge.

PRINCE, E.L. & MORTON, L.H.G. (1988) Fungal biodeterioration of synthetic metal working fluids. In Morton, L.H.G. & Chamberlain, A.H.L. (eds.) *Biofilms*, pp. 107–122. The Biodeterioration Society, Kew.

ROSSMOORE, H.W. (1986) Microbial degradation of water-based metal working fluids. In Moo-Young, M. (ed.) *Comprehensive Biotechnology*, pp. 249–269. Pergamon Press, Oxford.

SANDIN, M., ALLENMARK, S. & EDEBO, L. (1991) Selectivity of toxicity in alkaline liquids by alkylalkanolamines. In Rossmoore, H.W. (ed.) *Biodeterioration and Biodegradation 8*, pp. 428–430. Elsevier, London.

SCHNURER, J. & ROSSWALL, T. (1982) Fluorescein diacetate hydrolysis as a measure of total microbial activity in soil and litter. *Applied and Environmental Microbiology*, **43**, 1256–1261.

SEITZ, L.M., MOHR, H.E., BURROUGHS, R.T. & SAUER, D.B. (1977) Ergosterol as an indicator of fungal invasion in grains. *Cereal Chemistry*, **54**, 1207–1217.

SEITZ, L.M., SAUER, D.B., BURROUGHS, R.T., MOHR, H.E. & HUBBARD, J.D. (1979) Ergosterol as a measure of fungal growth. *Phytopathology*, **69**, 1202–1205.

SHENNAN, J.L. (1983) Selection and evaluation of biocides for aqueous metal-working fluids. *Tribology International*, **16**, 317–330.

SMITH, S.N. & PUGH, G.J.F. (1979) Evaluation of dehydrogenase as a suitable indicator of soil microbial activity. *Enzyme and Microbial Technology*, **1**, 279–281.

TABAK, H.H. & COOKE, W.B. (1968) Growth and metabolism of fungi in an atmosphere of nitrogen. *Mycologia*, **60**, 115–140.

TABOR, P.S. & NEIHOF, R.A. (1982) Improved method for determination of respiring individual microorganisms in natural waters. *Applied and Environmental Microbiology*, **43**, 1249–1255.

TOMAS, J.M., CIURANA, B. & JOFRE, J.T. (1986) New, simple medium for selective, differential recovery of *Klebsiella* spp. *Applied and Environmental Microbiology*, **51**, 1361–1363.

TREVORS, J.J., MAYFIELD, C.I. & INNISS, W.E. (1982) Measurement of electron transport (ETS) activity in soil. *Microbial Ecology*, **8**, 163–168.

17: Adherent Growth of *Staphylococcus aureus* from Poultry Processing Plants

C.E.R. DODD, B.J. CHAFFEY, K. DAEMS AND W.M. WAITES

Department of Applied Biochemistry and Food Science, University of Nottingham, Sutton Bonington Campus, Loughborough LE12 5RD, UK

Staphylococcus aureus can become endemic in the defeathering machinery of poultry processing plants and can increase by up to 1000-fold on carcasses during processing (Lahellec *et al.*, 1977; Gibbs *et al.*, 1978; Jacobsen, 1979; Notermans *et al.*, 1982; Adams & Mead, 1983; Thompson & Patterson, 1983; Dodd *et al.*, 1988a, b; Mead & Dodd, 1990). Such contamination is not easily removed, either by subsequent water immersion chilling or even spray-washing (Notermans *et al.*, 1982). Plasmid profiling has shown that only certain strains colonize the defeathering machinery (Dodd *et al.*, 1988a) and that these strains are resistant to cleaning and disinfection because they are almost eight times more resistant to hypochlorite than other strains (Bolton *et al.*, 1988). Such strains grow in macroscopic clumps up to 1 cm in diameter even in liquid-shake culture, and attach to the glass surfaces of culture vessels, particularly above the splash zone, as well as growing as 'sticky' colonies on agar media (Bolton *et al.*, 1988; Dodd *et al.*, 1988a; Chaffey *et al.*, 1989). Such stickiness is due to the production of extracellular polymer (Chaffey *et al.*, 1989).

This unusual growth habit in culture vessels makes the usual methods of measurement of growth, such as viable counting and turbidity readings, totally inadequate. In this chapter we examine alternative and more accurate methods of measuring growth parameters. All the strains described are isolates of *S. aureus* from a poultry processing plant (Adams & Mead, 1983) except strain BM1-1111#4, which is a plasmidless clumping derivative of the non-clumping strain BM1-1111 produced by growth in linoleic acid, according to the method of Butcher *et al.* (1976).

Microbial Biofilms:
Formation and Control

Inoculum Preparation

Since clumping strains do not produce a turbid suspension during growth, preparation of a uniform suspension is necessary to allow the use of a reproducible starting inoculum. The simplest means of producing this is by sonication of the clumps of growth. Cells are grown on brain heart infusion (BHI) agar overnight at 37°C. Tryptone soya broth (TSB; 50 ml) in 250 ml Erlenmeyer flasks is inoculated from this and incubated at 37°C with shaking at 200 rpm. The culture is centrifuged at 20 000 g for 8 min at 4°C, and the pellet resuspended in 25 ml of ice-cold phosphate-buffered saline (PBS; pH 7.3, Oxoid) by vortex mixing. Centrifugation and mixing are repeated twice and the final pellet is resuspended in 25 ml ice-cold PBS. The suspension is sonicated for 15 s in a sonicator (100 W, MSE, Loughborough, Leicestershire, UK), cooled in ice-water and the sonication repeated. This produces a single-cell suspension. This method was used by Chaffey *et al.* (1991) and Bolton *et al.* (1988) for clumping strains of *S. aureus*. The method should be applicable to strains of other genera, although the exact sonication conditions may have to be varied. These can be established by sonicating a culture of a non-clumping strain of the same species and confirming that the viable count is not reduced by more than 5%. After preparation, the production of a single-cell suspension should be confirmed by microscopy.

Impedance

Impedance is the resistance to flow of an alternating electrical current through a conducting material, and depends on the production of ions in the growth medium by the metabolism of growing microorganisms (Eden & Eden, 1984). In general, impedance changes are detectable when the number of microorganisms reaches the threshold level of 10^7 cells/ml. The time for the initial inoculum to reach the threshold level is designated as the *detection time*, and is a function of both the initial concentration and the specific growth kinetics of the organism in the given medium. The generation time (t_g) of the organism is given by the equation $t_g = \log 2/B$, where B is the slope of the regression line produced when detection times are plotted against corresponding viable counts as $\log_{10}$ cfu/ml. For *S. aureus* the following protocol (Daems *et al.*, unpublished) with a Bactometer Microbial Monitoring System M-120B (bioMérieux UK Ltd, Basingstoke, UK) was used. Serial dilutions of a single-cell suspension are made in peptone (0.1% w/v). Duplicate samples (0.1 ml) of appropriate dilutions are inoculated into the wells of modules (bioMérieux, UK) containing TSB (1 ml) and incubated at 35°C for 20 h in the Bactometer incubation chamber. Similar samples are spread on to BHI agar and incubated at 37°C overnight to obtain the corresponding viable counts. Capacitance measurements on the modules are made automatically by the Bactometer every 6 min and the detection times are obtained for each dilution. Detection times are plotted against viable counts as $\log_{10}$ cfu/ml, and the mean generation time is determined with the Bactometer programme. After 20 h, tests are carried out to confirm growth of the original inoculant strain. A sample (0.1 ml) is grown on BHI agar and the appearance of the colonies, microscopic appearance under phase contrast, and API STAPH (bioMérieux UK) profile are determined. Typical results are shown in Table 17.1. These show that generation times for the clumping strain BM1-1111#4 were very similar to those for the closely related non-clumping strain BM1-1111. This suggests that diffusion of nutrients into, and that of metabolic products out of, the centre of a clump did not reduce the growth rate. The method may be adapted for use with other genera by the use of appropriate growth media.

Radiolabel Incorporation into DNA

Measurement of the uptake of a radiolabel into the cells and its incorporation into the DNA allows estimation of their growth rate. A method with ^{3}H-deoxythymidylic acid (^{3}H-dTMP) as the source of label has been used by the authors (Chaffey *et al.*, unpublished). Cells from frozen culture are inoculated into semisynthetic nutrient medium (SNM 50 ml, De Repentigny *et al.*, 1964) in a 250 ml flask and incubated at 37°C for 18 h with shaking at 200 rpm. The cells are centrifuged, washed and sonicated to produce a single-cell suspension. A sample of suspension (0.2 ml) is used to inoculate sterile acid-washed glass scintillation vials containing SNM (5 ml) plus ^{3}H-dTMP (final concentration 7.4 kBq/ml; 7.5 μmol/l), and is incubated at 37°C with shaking at 200 rpm. Labelled dTMP is added aseptically to sterile SNM as the ammonium salt of [methyl-^{3}H]-thymidine-5'-monophosphate. At intervals, three replicate vials are harvested by placing in ice-water. The cultures are filtered (0.45 μm HVLP filter, Millipore UK), the vials rinsed twice with PBS

TABLE 17.1. *Generation times (t_g) of clumping and non-clumping strains of Staphylococcus aureus determined by impedance (capacitance) measurements*

	t_g (min)		
S. aureus strain	Sample 1	Sample 2	Mean
BM1-1111 (sonicated)	20	20	20
BM1-1111 (non-sonicated)	19	19	19
BM1-1111#4 (sonicated)	18	19	18.5
BM1-1111#4 (non-sonicated)	17	16	16.5

(10 ml) at 4°C and the washings filtered through the same filter. The filters are rinsed twice with ice-cold trichloroacetic acid (TCA; 5 ml, 5% w/v), placed in a clean scintillation vial and dried overnight. Ice-cold TCA (10 ml, 5% w/v) is added to each culture vial, left to stand for 10 min on ice and the liquid filtered (0.45 μm HVLP filter, Millipore). Optiphase-X (Fisons, Loughborough, UK) scintillation fluid (10 ml) is added to each sample and the samples are counted twice for 10 min on a Packard Tri-Carb 4640 scintillation counter (Canberra Packard Ltd, Pangbourne, UK) or until standard deviation is <2%. ^{3}H-dTMP must be dephosphorylated before it is incorporated into DNA. Dephosphorylation is the rate-limiting step so that the addition of exogenous dTMP results in a slow uptake of thymidine, which precludes the induction of thymidine phosphorylase and prevents the label being shunted into other metabolic pathways. Brietman *et al.* (1967) have shown that cells grown in the presence of ^{3}H-dTMP incorporate the label into DNA at a constant rate for extended periods of time. The authors (Chaffey *et al.*, unpublished) have used this method to examine the growth in *S. aureus.* Measurement of growth by turbidity at 540 nm is shown compared with that by incorporation of ^{3}H-dTMP for two non-clumping strains, BM1-1111 (Fig. 17.1(a)) and SV1-25 (Fig. 17.1(b)). The growth in SNM, examined by the uptake of a radiolabel, is close to exponential and similar to that measured spectrometrically, with doubling times of 1.25 and 1.09 h respectively for strain BM1-1111, and 0.69 h for strain SV1-25 by either means of measurement. Such comparisons are, of course, not possible for clumping isolates, because reproducible spectrometric readings cannot be produced. However, when the doubling time for clumping strain BM1-1111#4 is determined by ^{3}H-dTMP incorporation (Fig. 17.2(b)), the maximum observed growth rate is 1.35 h (calculated from total counts). This strain also shows a longer lag period than does the non-clumping strain (Fig. 17.2(a)), although the final cell yield is greater for the clumping phenotype.

A further use of this method is for the examination of the adhesive properties of strains, because it allows differentiation between cells adherent to the walls of the culture vessel and those free in the culture medium (Chaffey *et al.*, unpublished). When adherence of the non-clumping strain of *S. aureus* BM1-1111 to the culture vessel wall is examined in this way (Fig. 17.2(a)), it is apparent that over 90% of the counts are present in the liquid phase. In contrast, in the clumping strain, BM1-1111#4, whereas most of the counts are present in cells in the liquid phase during the first 5 h of growth (Fig. 17.2(b)), more than 90% of the label is subsequently recovered from cells attached to the glass. Furthermore, the apparent growth rate of adherent cells of the clumping strain is a doubling time of 0.67 h, in contrast to 1.35 h for overall counts during the same period. This suggests that the cells grew faster when attached. However, there is evidence that 'capture' of non-

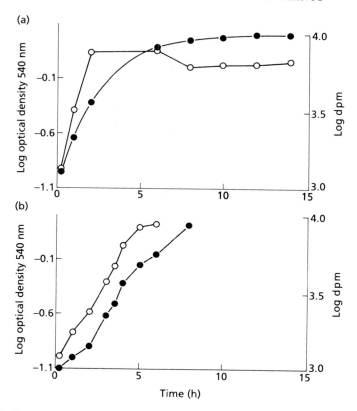

FIG. 17.1. Growth of *Staphylococcus aureus* strain BM1-1111 (a) and strain SV1-25 (b) in batch culture measured by following changes in optical density at 540 nm (●) or ³H-dTMP incorporation (○), as described in text.

adherent cells by attached cells also occurred, making it premature to suggest that attached cells have a growth advantage over unattached cells.

Differences in metabolism between closely related strains of *Staphylococcus* have not previously been reported, although Christensen *et al.* (1982) showed that slime production of *Staphylococcus epidermidis* did not affect growth rate. The production of large amounts of extracellular material by the clumping strain used by Chaffey *et al.* (1989, 1991) may be sufficient to reduce the rate of growth in a semisynthetic medium such as SNM, but not in a more nutrient-rich medium, such as TSB.

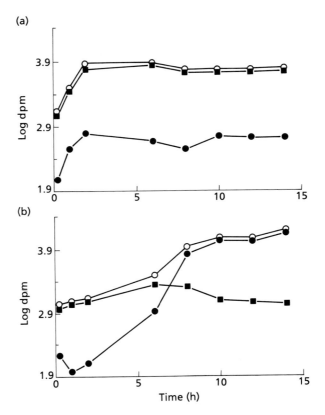

FIG. 17.2. Partitioning of *Staphylococcus aureus* in batch culture between adherent and non-adherent cells, determined by ^{3}H-dTMP incorporation as described in text. (a) Non-clumping strain BM1-1111 grown in SNM supplemented with 0.025 mol/l D-glucose. (b) Clumping strain BM1-1111#4 grown in SNM supplemented with 0.025 mol/l D-glucose. Total radioactivity counts (○) calculated from the sum of adherent counts (●) and non-adherent counts (■).

Conclusions

For organisms such as strains of *S. aureus*, which display a clumping habit and attach to the walls of culture vessels, the usual methods of measuring growth by viable counts and turbidity fail to provide accurate estimates of the extent or rate of growth. The use of impedance or the incorporation of ^{3}H-dTMP allows growth rates to be determined accurately. In addition, measurement of ^{3}H-dTMP incorporation allows the proportion of attached cells present in the population to be estimated.

Acknowledgements

We thank Mrs Sheila Godber and Mr D. Fowler for skilled secretarial and technical assistance; the AFRC and MAFF for financial support and bioMérieux UK Ltd. for the provision of equipment. Katja Daems was an Erasmus exchange student from the University of Ghent.

References

ADAMS, B.W. & MEAD, G.C. (1983) Incidence and properties of *Staphylococcus aureus* associated with turkeys during processing and further processing operations. *Journal of Hygiene, Cambridge,* **91**, 479–490.

BOLTON, K.J., DODD, C.E.R., MEAD, G.C. & WAITES, W.M. (1988) Chlorine resistance of strains of *Staphylococcus aureus* isolated from poultry processing plants. *Letters in Applied Microbiology,* **6**, 31–34.

BRIETMAN, T.R., BRADFORD, R.M. & CANNON, W.D. (1967) Use of exogenous deoxythymidylic acid to label the deoxyribonucleic acid of growing wild-type *Escherichia coli. Journal of Bacteriology,* **93**, 1471–1472.

BUTCHER, G.W., KING, G. & DYKE, K.G.H. (1976) Sensitivity of *Staphylococcus aureus* to unsaturated fatty acids. *Journal of General Microbiology,* **94**, 290–296.

CHAFFEY, B.J., DODD, C.E.R. & WAITES, W.M. (1989) Chemical and physical characteristics of the cell surface of adhesive *Staphylococcus aureus. Journal of Applied Bacteriology,* **67**, xiv.

CHAFFEY, B.J., DODD, C.E.R., BOLTON, K.J. & WAITES, W.M. (1991) Adhesion of *Staphylococcus aureus*: its importance in poultry processing. *Biofouling,* **5**, 103–114.

CHRISTENSEN, D.G., SIMPSON, W.A., BISNO, A.L. & BEACHEY, E.H. (1982) Adherence of slime-producing strains of *Staphylococcus epidermidis* to smooth surfaces. *Infection and Immunity,* **37**, 318–326.

DE REPENTIGNY, J., SONEA, S. & FRAPPIER, A. (1964) Differentiation by immunodiffusion and by quantitative immunofluorescence between 5-fluorouracil-treated and normal cells from a toxinogenic *Staphylococcus aureus* strain. *Journal of Bacteriology,* **88**, 444–448.

DODD, C.E.R., CHAFFFEY, B.J. & WAITES, W.M. (1988a) Plasmid profiles as indicators of the source of contamination of *Staphylococcus aureus* endemic within a poultry processing plant. *Applied and Environmental Microbiology,* **54**, 1541–1549.

DODD, C.E.R., MEAD, G.C. & WAITES, W.M. (1988b) Detection of the site of contamination by *Staphylococcus aureus* within the defeathering machinery of a poultry processing plant. *Letters in Applied Microbiology,* **7**, 63–66.

EDEN, R. & EDEN, G. (1984) *Impedance Microbiology.* John Wiley and Sons, New York.

GIBBS, P.A., PATTERSON, J.T. & THOMPSON, J.K. (1978) The distribution of *Staphylococcus aureus* in a poultry processing plant. *Journal of Applied Bacteriology,* **44**, 401–410.

JACOBSEN, C. (1979) Stafylokokker på et fjerkraeslagteri. *Dansk Veterinaertidsskrift,* **62**, 817–824.

LAHELLEC, C., COLIN, P. & MEURIER, C. (1977) Origine et caractérisation des souches de *Staphylococcus aureus* présentes sur les carcasses de volailles et dans les produits transformés. *Bulletin d'Information, Station Expérimentale d'Aviculture de Ploufragan,* **17**, 84–98.

MEAD, G.C. & DODD, C.E.R. (1990) Incidence, origin and significance of staphylococci on processed poultry. In Jones, D., Board, R.G. & Sussman, M. (eds.) *Staphylococci,* pp. 81–91. Society for Applied Bacteriology Symposium Series No. 19. Blackwell Scientific Publications, Oxford.

NOTERMANS, S., DUFRENNE, J. & VAN LEEUWEN, W.J. (1982) Contamination of broiler chickens by *Staphylococcus aureus* during processing: incidence and origin. *Journal of Applied Bacteriology*, **52**, 275–280.

THOMPSON, J.K. & PATTERSON, J.T. (1983) *Staphylococcus aureus* from a site of contamination in a broiler processing plant. *Record of Agriculture Research (Belfast)*, **31**, 45–53.

18: Microbial Adherence to Food Contact Surfaces

A. GILMOUR[1,2], A.B. WILSON[2] AND T.W. FRASER[1]

[1]*Department of Agriculture for Northern Ireland; and* [2]*The Queen's University of Belfast, Agriculture and Food Science Centre, Newforge Lane, Belfast BT9 5PX, UK*

Food contact surfaces are subject to bacterial adherence and colonization. These are of importance in the subsequent contamination of product, and are therefore of great significance in relation to food spoilage and safety (Stanley, 1983). To investigate the nature of the adhering microflora or the degree of proliferation of bacteria on surfaces under various environmental conditions, generally requires either removal of the organisms or the application of *in situ* methods to simplified experimental systems. This chapter describes some of the methods currently available for the study of bacterial adherence and proliferation on dairy or food contact surfaces, such as glass, rubber and stainless steel.

Removal by Rinsing

In tests involving the recovery of bacteria from disinfected surfaces, greater efficiency in recovery was achieved when rubber-bladed squeegee devices were used in conjunction with a rinse solution (Neave & Hoy, 1947; Cousins, 1963). Lisboa (1959) devised small circular squeegees for sampling the internal surfaces of stainless steel tubes. In our laboratory, a squeegee rinse procedure consisting of an initial rinse followed by six successive double squeegee rinses, has been used to assess the nature of the microbial population that adheres to the internal surfaces of milk-soiled glass, stainless steel and rubber pipe sections after removal from an experimental milking plant.

Pipe sections and end closures

Sections can be prepared from glass, rubber and stainless steel pipe (Fig. 18.1) obtainable from any dairy supply company. The glass and stainless steel

Microbial Biofilms:
Formation and Control

FIG. 18.1. Pipe sections, end caps/bungs and squeegee.

sections we have used are 33 cm long, whereas rubber sections are 37 cm long to compensate for the different end closures used. All have an internal diameter of 3.1 cm. The glass and stainless steel pipe sections are sealed with rubber end-caps (Fullwood and Bland Ltd, Ellesmere, Shropshire, UK), and the rubber pipe sections are sealed with 3 cm diameter rubber bungs introduced 2 cm into each end.

Conditioning

New Pipe sections should be conditioned by scrubbing with warm detergent solution, 0.75% (w/v) sodium carbonate and 0.05% (w/v) sodium sulphite (Cousins *et al.*, 1960), then rinsed with cold water and autoclaved at 115°C for 10 min. The pipe sections are then soiled with aged raw milk, by incubation at 22°C for 6 h, closed, incubated overnight at 22°C, washed again and sterilized. This soiling and cleaning process is carried out three times before the conditioned pipe sections are individually sterilized ready for use.

Soiling

Pipe sections are inserted one at any one time into an experimental milking installation, such as the Fullwood three-point system (Fig. 18.2), which consists of three teat cup clusters, three 27.3-l recorder jars, a receiver vessel and associated milk and vacuum stainless steel pipelines.

Before inserting each pipe section, the milking installation is cleaned as follows. Caustic detergent (e.g. Farm Cold Cleaner, Diversey (Ireland) Ltd, Glasnevin, Dublin, Republic of Ireland) is dissolved in 45 l water at 85°C to give a 0.5% (w/v) solution, and hypochlorite is added to a concentration of 300 parts per million available chlorine. This solution is then circulated round

FIG. 18.2. Experimental milking installation.

the plant for 10 min before being discharged to waste. The plant is finally flushed with 45 l cold water.

A sterile pipe section can now be inserted into the milk transfer pipeline, immediately before the receiver jar, as shown in Fig. 18.3. Two stainless steel connector pieces are used to connect rubber pipe sections. As shown (Fig. 18.3(a)), these are inserted 2 cm into each end of the pipe section and to the same depth into an adjacent rubber elbow joint on one side and a rubber T-piece on the other side. On removal of these rubber pipe sections, and until sampling can take place, rubber closures are inserted 2 cm into each end. Glass and stainless steel sections, however, are sealed with rubber end-caps on removal, until sampling can take place.

In our laboratory, milk of various descriptions has been used to soil the equipment, including the inserted pipe section. For each soiling, the milk (22 l) is introduced into the equipment via the clusters to the recorder and receiver jars; the latter is emptied by the milk pump, which returns the milk to the churn. This operation is carried out twice for each cluster, that is, six times in all. After each soiling, 22 l of sterile deionized water is drawn into the machine under vacuum and flushed to waste. Normally this soiling procedure is carried out twice per day and continued for 5 days, after which the inserted pipe section is removed.

(a)

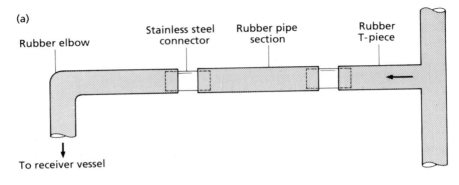

(b)

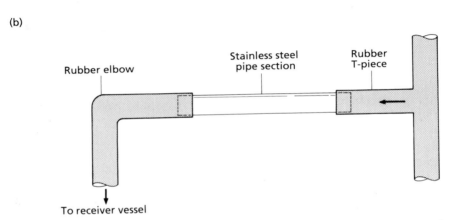

FIG. 18.3.(a) Method of insertion of rubber pipe sections. (b) Method of insertion of glass and stainless steel pipe sections. The arrows show direction of milk flow.

Sampling

The squeegee sampler (Fig. 18.1) consists of a metal rod to which a 3.25 cm diameter 3 mm thick neoprene rubber disc is attached, with a 3 mm diameter drilled hole in each quadrant, to allow the passage of rinse solution. Squeegees are individually sterilized ready for use.

A suitable rinse solution is quarter-strength Ringer's solution containing 0.1% (w/v) of a 1:4 dilution of Lissapol N (BDH Chemicals Ltd, Poole, UK).

After removing the top closure of the soiled pipe sections, 25 ml of warmed rinse solution (44 ± 1°C) is added and the closure replaced. The pipe section is then rolled horizontally six times, first in one direction, then the other, three times each way across a distance of 45.7 cm. This rolling procedure should take about 6 s. The entire rolling procedure is repeated and the pipe section inverted twice before the top closure is removed and the initial rinse allowed to drain for 1 min into a sterile screw-capped bottle for subsequent bacteriological examination.

This initial rinse is followed by six successive double squeegee rinses. For each of these, after the addition of the rinse solution and subsequent initial rolling, the top closure is removed and, while the pipe section is held vertically, a sterile squeegee is inserted and pushed to the bottom of the pipe section. The squeegee is then drawn up the pipe section until the neoprene washer just reaches the top rim, or 2 cm from the top rim in the case of rubber pipe sections. This operation is repeated before the squeegee is removed and the final rolling, inversion and collection of the rinse solution carried out.

The initial rinse and the six squeegee rinses can be analysed separately or combined before bacteriological examination by an appropriate plating procedure.

Comments

It is well known that it is impossible to remove all adhering microorganisms from soiled food contact surfaces. The technique described was, however, found to be more effective in achieving this than other previously evaluated scraping and swabbing procedures. With aged raw milk as the soiling medium, the procedure described was found to remove 99.99% of the total removable bacterial numbers, as determined by applying 15 double squeegee rinses (Speers *et al.*, 1984a). An initial rinse is included to count bacteria associated more loosely with the surfaces and therefore more easily removed.

With this technique we were able to determine both numbers and types of bacteria associated with glass, rubber and stainless steel (Speers & Gilmour, 1982; Lewis & Gilmour, 1987) when these surfaces were exposed to milks with differing microflora. In addition, it was possible to show that, although bacteria and residues tend to accumulate in areas that are difficult to clean, such as crevices, joints and dead-ends, areas of relatively heavy contamination can also occur in straight pipe sections. Although we have described the squeegee technique in relation to dairy applications, it can be adapted for wider use in the study of microbial adherence in relation to other food contact surfaces.

In situ Epifluorescence Miroscopy

Epifluorescence microscopy can be used to determine numbers of bacterial cells that adhere to food contact surfaces under a variety of environmental conditions. A great advantage of this technique over conventional transmitted light microscopy is that it is easily possible to visualize microorganisms attached to the surface of opaque materials. Many fluorescent stains are available, including euchrysine, fluorescein and rhodamine, but the most commonly used is acridine orange. It was originally thought that viable cells stained with acridine orange fluoresce orange-red, whereas non-viable cells show green fluorescence due to the interaction of the stain with the RNA and DNA of the cell. However, Pettipher *et al.* (1980) found that, under certain conditions, some green-fluorescing cells are still active, thereby casting doubt on this hypothesis. It is now thought that it is only under the rigorous conditions of the direct epifluorescent filtration technique (DEFT) that viability can be determined with acridine orange.

We have used epifluorescence microscopy to assess the adherence of a number of milk-associated bacteria to inert surfaces in the presence of various attachment solutions.

Surfaces

Experimental surfaces of glass (conventional microscope slides), rubber (Avon Industrial Polymer Ltd, Melksham, UK) and stainless steel (Stainless Steel Fabrications Ltd, Belfast, UK) were obtained as microscope slide-like pieces, 75 mm long and 25 mm wide. Before each experiment the surfaces were thoroughly washed in a detergent solution (5% Decon; Decon Laboratories Ltd, Hove, UK), rinsed well with deionized water and autoclaved at 115°C for 10 min.

Attachment

Each slide is placed for 1 h at 25°C in a Petri dish containing 20 ml of suspending solution inoculated with $1-2 \times 10^6$ bacteria/ml. These suspending solutions are formulated to include the major constituents of raw milk. The slides are then removed with sterile forceps and washed gently by passing them three times through 200 ml sterile distilled water. Finally, the slides are drained for 10 s on sterile filter paper and allowed to air-dry.

Staining

Before staining, the attached cells are fixed for 3 min in Kilpatrick's fixative (ethanol, 60 ml; chloroform, 30 ml; formalin, 10 ml) and the surface allowed to dry in air. The cells are stained with 0.025% w/v acridine orange (Difco) which had previously been filtered. An optical brightener such as Uvitex AN (0.025% w/v) (Tinopal AN; Ciba-Geigy Ltd, Cambridge, UK) can also be included in the stain to enhance the bacterial fluorescence (Paton & Jones, 1973). If this is used, the commercial compound is purified by dissolving in ethanol, filtering through Whatman Grade 1 filter paper and allowing evaporation of the alcoholic extract to form pale-yellow needle-shaped crystals (Paton, personal communication). After 2 min the surface is rinsed with distilled water and allowed to dry in air.

Viewing

To prepare a slide for examination, a drop of sterile water is placed on its surface and a coverslip marked with a circle (area 1 cm^2) is located on the water drop. The coverslip is sealed with a smear of Vaseline. The specimen is best viewed with a ×100 oil immersion objective and a non-fluorescent immersion oil.

Enumeration

Cells may either be counted automatically with an image analyser, or manually. When counting manually, cells that are separated by a distance less than the widest diameter of the other constituent cells, are regarded as one clump. It has been shown that counts produced by both manual and automated methods correspond closely to conventional plate counts. Twenty microscope fields, or 200 organisms/clump — whichever is reached first — are counted on each surface. Fields are selected at equidistant intervals throughout the delineated area, according to the method of the Modified Breed Smear Count (Anon, 1984). The number of adherent cells per field of view is determined from the mean count of the fields counted. To calculate the number of cells per mm^2, it is first necessary to estimate the area of the field of view. For modern microscope eyepieces the field of view is designated in millimetres as the field number (FN). For a ×100 objective, the eyepiece FN ×10 equals the actual diameter of the field of view expressed in micrometres. In this case, an eyepiece of FN20, 1 mm^2 is represented by 31.84 fields of view. For eyepieces of unknown field number a stage micrometer must be used.

Comments

We have used this technique to study the ability of various milk-associated bacteria to adhere to inert surfaces and to study the effects of various milk components on adherence (Speers & Gilmour, 1985). Results have shown that both the bacterial type and the nature of the attachment surface influence adherence. It has also been shown that substances absorbed on to the attachment surface can affect adherence. The greatest adherence was observed with a

Moraxella-like species on stainless steel with lactose and non-casein protein as the preabsorbed component.

In situ Bioluminescence Measurements

Adenosine triphosphate (ATP) is present in all living cells and can be measured with the enzyme luciferase and its substrate luciferin which, in the presence of ATP, is converted to oxyluciferin. This reaction generates light which can be accurately measured by means of a luminometer. The reaction is as follows:

$$\text{Luciferin} + \text{ATP} \rightarrow \text{Oxyluciferin} + CO_2 + \text{AMP} + \text{Light}$$

It has been shown that the concentration of ATP present in broth cultures is proportional to bacterial numbers, between 10^3 and 10^8 bacteria per gram or millilitre (Stanley, 1989).

Before the ATP can be assayed it must be extracted from the bacterial cells. Extraction agents include acids, boiling buffers and steam, as well as commercially available products. Trichloroacetic acid is the most commonly used extractant but, because it inhibits luciferase at higher concentrations, a range of these should be tested for the maximum yield of ATP with the minimum luciferase inhibition. Assay conditions are also extremely important: the optimum temperature for enzyme activity is 25°C and the optimum pH is 7.75. Most buffer systems can be used, but 0.1 M buffers are best because the enzyme is inhibited by high ionic strength. Ethylenediamine tetra-acetic acid (EDTA) and magnesium ions must also be present, as stabilizer and cofactor, respectively. Finally, oxygen, which can be a limiting factor, is extremely important for the reaction and most luminometers are fitted with a mixer to ensure adequate oxygenation. The luciferin/luciferase cocktails available commercially, contain all the necessary components for the reaction.

A method developed to study the adherence and proliferation of *Listeria monocytogenes* on the surfaces of square glass coverslips $(18 \times 18 \, \text{mm})$ is described below.

Preparation of coverslips

Coverslips are first soaked in 20% (w/v) NaOH for 3 days to clean the attachment surface. They are then thoroughly rinsed, first in 0.02 mol/l HCl and then in distilled water, and finally autoclaved at 121°C for 15 min.

Medium

A diluted tryptone soya broth (TSB) devised by Frank & Koffi (1990) is used to grow the attached cells. This is prepared by adding 2 g TSB (Oxoid) and 8 g glucose to 1 l distilled water.

Method

Diluted TSB (250 ml) is inoculated with 5 ml of an overnight culture of *L. monocytogenes* CRA 433 (in TSB (Oxoid) with 0.6% (w/v) yeast extract) giving an initial count of 1×10^7 cells/ml; 10 ml volumes are dispensed into 30 ml sterile disposable universal containers (Sterilin) and one coverslip is added to each bottle. These are incubated at 30°C and sampled at regular intervals over a suitable time period. In our case this was every 2 h for 30 h.

At each sampling time three coverslips are removed and tested. One coverslip is used for an internal standardization procedure, which takes into account changes in enzyme activity due to instability or interference by other components of the assay. The internal standardization procedure is carried out by adding to the sample 100 µl of an ATP standard of known concentration. The concentration of this ATP standard and its light-output value are included in the formula supplied by the manufacturer to calculate the concentration of ATP present in the sample. The other two coverslips are used as duplicate samples for the following bioluminescence assay.

Each of the three coverslips is removed aseptically from the medium with a pair of tweezers, drained for 10 s by touching edgewise to sterile filter paper, and washed gently by passing three times through 10 ml sterile distilled water. The coverslips are then drained as before and broken into four pieces with a sterile glass cutter, leaving the attached film of cells intact. The pieces of coverslip are placed in a cuvette (Clinicon, Billingshurst, UK) to which 200 µl of sterile distilled water is added, followed by a further 200 µl of the ATP-releasing agent (Enzymatix, Cambridge, UK). After 2 min, which is sufficient for the releasing agent to extract the ATP, 100 µl is removed and added to 100 µl of buffer in another cuvette, which is used for the bioluminescence assay. A volume of 100 µl of the luciferin/luciferase cocktail (Enzymatix, Cambridge, UK) is added to this cuvette and the relative light unit (RLU) output over 10 s is recorded in a luminometer (e.g. Bio-Orbit 1251, Hungerford, UK).

Comments

Typical results are shown in Fig. 18.4, which reveals a close relationship between the number of bacteria as determined respectively by a plating procedure carried out on the crushed glass coverslip and the ATP concentration by the bioluminescence assay. This shows that bioluminescence can

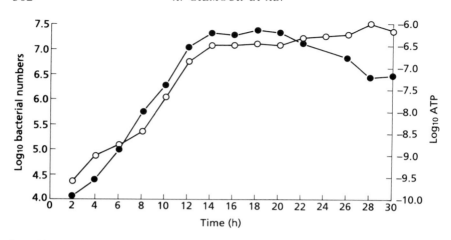

F<small>IG</small>. 18.4. Numbers of *Listeria monocytogenes* determined using both a plating procedure and ATP concentration on glass surfaces. (●) Log_{10} bacterial numbers; (○) log_{10} ATP concentration (relative light units).

be used to assess the attachment and proliferation of *L. monocytogenes* over a 30-h period.

Scanning Electron Microscopy

Although scanning electron microscopy (SEM) offers an attractive and direct method for the visualization of microbial biofilms, it is important to appreciate that the results achieved by conventional methods of preparation, although often spectacular, may not portray the specimen in a lifelike condition. With the exception of the recently introduced environmental and low-vacuum SEMs, the high-vacuum operating conditions of conventional SEM requires that the water content of hydrated biological samples is either removed or immobilized, to avoid deterioration of the vacuum and to allow the samples to withstand the hostile environment. Most of these procedures result in the creation of artefacts. A further requirement for SEM is that samples are rendered conductive, usually by the application of a film of inert heavy metal such as gold or platinum, to prevent charge build-up by interaction of the electron beam with the specimen. Heavy metal coating also has the advantage of improving secondary electron emission and hence image quality. Microbial biofilms for study in the SEM may either be obtained naturally or can be prepared artificially in the laboratory. Although the former are frequently desirable, experimentally prepared specimens provide wider scope for the study of biofilms by electron microscopy. For instance, the effects of temperature, substrate and inoculum on biofilm development can be studied in a carefully controlled manner.

Specimen preparation for SEM

Either cryo methods or conventional chemical methods can be used.

Cryopreparation methods

Deep-freezing or cryofixation, although generally accepted as providing the most lifelike biological specimens for SEM, suffers from certain practical limitations as well as limited availability. To ensure lifelike specimens inherently limits the scope for the study of some important aspects of natural or simulated biofilms, particularly where some degree of soiling is involved. Although Fig. 18.5 provides a realistic view of milk-soil contaminated with a *Micrococcus* spp., it offers little or no microscopic detail of the substrate or the micro-organisms. Such problems can be avoided under experimental conditions by culturing the biofilms in a residue-free culture medium such as brain heart infusion. A further consideration is that, because of the immediacy of the cryo technique, batches of specimens cannot be prepared in advance of viewing, unlike with conventional SEM. Nevertheless, cryo-SEM provides standards against which subsequent conventionally prepared material can be judged (Figs 18.6–18.8) and should, where possible, be included in any SEM study. The cryofixation process requires that the specimen is deep-frozen, usually by plunging it into subcooled liquid nitrogen at less than $-196°C$, followed by warming to $-80°C$ under vacuum. This is to allow sublimation of any unbound water and then recooling to approximately $-170°C$ on the cold stage of the

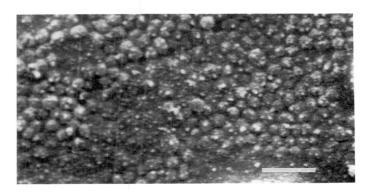

Fig. 18.5. Cryofixation. Milk-soiled stainless steel surface contaminated with a *Micrococcus* sp. (*cf.* Fig. 18.12). Bar = 5 μm.

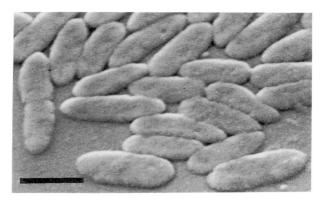

FIG. 18.6. *Pseudomonas fragi* grown on plastic tissue-culture coverslips in brain–heart infusion, a soil-free medium. Cryofixation. Note the close contact between the microorganisms and lack of fibrils. Bar = 2 µm.

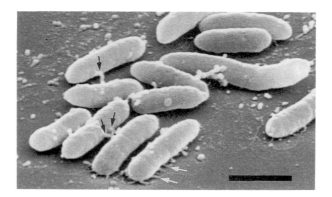

FIG. 18.7. *Pseudomonas fragi* grown on plastic tissue-culture coverslips in brain–heart infusion, a soil-free medium. Glutaraldehyde fixation, ethanol dehydration followed by critical point drying. Note the separation of the microorganisms and the appearance of attachment fibrils (arrows). Bar = 2 µm.

preparation chamber, before sputter coating and subsequent transfer under vacuum to the SEM cold stage (−160°C) for viewing. The complete process is carried out under vacuum in a dedicated cryopreparation unit, a number of which are commercially available.

A problem associated with the cryopreparation of wet biofilms — like those encountered on many food contact surfaces — can be premature air-drying of the specimen while mounting it on to the cryostub, before freezing. To avoid

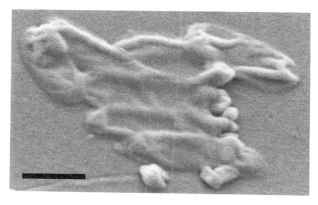

FIG. 18.8. *Pseudomonas fragi* grown on plastic tissue-culture coverslips in brain–heart infusion, a soil-free medium. Air drying. Although collapsed, the organisms bear resemblance to those of the cryoprepared specimen. Bar = 2 μm.

a similar problem when studying suspensions of microorganisms from aqueous culture, the following simple technique was devised (Fraser & Gilmour, 1986).

A 1 cm square of 0.45 μm porosity (hydrophilic) Millipore membrane (Millipore SA, Molsheim, France) is attached to the cryostub with Tissue Tec 11 Oct compound (Lab-Tek Division, Miles Laboratories Inc, Naperville, Ill, USA). An 8 mm square of 0.4 μm porosity hydrophobic Nuclepore poly-carbonate membrane (Nuclepore Corporation, Pleasanton, CA, USA), is then similarly attached directly on top of the Millipore membrane. Immediately before freezing, a droplet (*ca.* 20 μl) of the culture is placed on to the Nuclepore membrane. After a few moments the droplet will be completely drawn through into the Millipore membrane layer, which acts as an absorbent, leaving the organisms on the Nuclepore membrane layer, fully hydrated but protected by the moist Millipore membrane beneath. At this point the cryostub should be plunged into the liquid nitrogen slush. The method is also suitable for samples saturated with ethanol (Fig. 18.9; see Fig. 18.11).

Conventional chemical preparation methods

In the absence of — or as supplement to — cryopreparation, conventional SEM preparation methods can be applied to the study of biofilms. All such methods involve the three distinct processes of chemical fixation, solvent dehydration and then removal of the dehydration solvent ('drying'). Varying these processes, particularly at the fixation or dehydration stages, can greatly affect the final

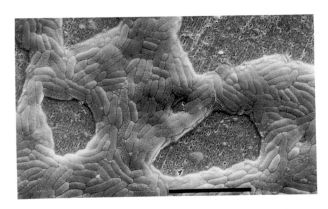

FIG. 18.9. Cryofixation. Aqueous suspension of *Pseudomonas fragi* on Nuclepore membrane examined by cryo-SEM. Bar = 10 μm.

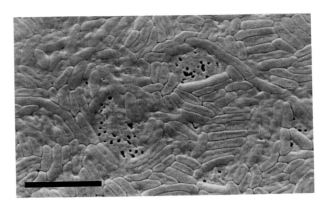

FIG. 18.10. Air drying. Aqueous suspension of *Pseudomonas fragi* on Nuclepore membrane air-dried and examined by cryo-SEM. Bar = 10 μm.

appearance of the specimen. This flexibility to influence the final appearance of the specimen can be of great value to the electron microscopist.

As a preliminary to chemical procedures it is useful to examine the biofilm in the air-dried state, followed by sputter coating. Exceptions are films rich in fats or oils, which would contaminate the SEM and be difficult to sputter coat. *Pseudomonas fragi* grown on Thermanox plastic tissue-culture coverslips (Miles Laboratories Inc, Naperville, Ill, USA) in brain—heart infusion and rinsed in distilled water before air-drying, show a considerable degree of collapse (Fig. 18.8) compared with cryospecimens (Fig. 18.6) and chemically fixed, dehydrated and critical point dried specimens (Fig. 18.7). However, the

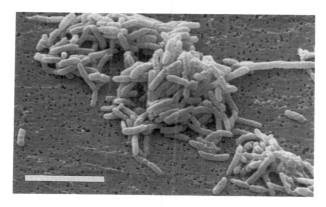

FIG. 18.11. Glutaraldehyde fixation and ethanol dehydration. Ethanolic suspension of *Pseudomonas fragi* on Nuclepore membrane examined by cryo-SEM. Bar = 10 μm.

specimen provides a reliable guide to the number and distribution of the organisms. Suspensions of *Ps. fragi*, similarly air-dried (Fig. 18.10), appear more lifelike than chemically fixed and dehydrated suspensions (Fig. 18.11), when compared with cryotreated specimens (Fig. 18.9).

Fixation is the first step in the chemical preparation process, the purpose of which is to preserve the specimens in as lifelike a condition as possible. Buffered glutaraldehyde is routinely employed as the primary fixative for SEM, and acts by cross-linking proteins to increase the mechanical rigidity of microbial structures. The value of osmium tetroxide post-fixation, invariably included in transmission electron microscopy (TEM) routines, is debatable when applied to SEM. However, osmium tetroxide cross-links and stabilizes proteins and lipids by oxidizing double bonds. Its effectiveness as a secondary fixative in any particular SEM situation can best be assessed by comparative studies.

The next stage in the preparation procedure is the removal of water (dehydration) from the specimen. This is achieved by substituting the water with a polar organic solvent, usually either ethanol or acetone. Although little morphological difference can be detected between biological specimens dehydrated by graded concentrations of either ethanol or acetone, ethanol is 'kinder' to some synthetic substrates, such as nitrile rubber (Speers *et al.*, 1984b) and to the seals of the critical point dryer, which is used at the next stage. Acetone-dehydrated samples should be transferred to amyl acetate before critical point drying. Complete substitution of dehydration solvent by liquid carbon dioxide in the critical point dryer is indicated when the exhausted carbon dioxide no longer smells of the solvent. Critical point drying is most commonly used to remove the dehydration solvent from SEM specimens because it completely prevents surface tension artefacts. Briefly, the dehydrated samples are transferred to the critical point drying apparatus and infiltrated with liquid carbon dioxide under pressure. The temperature is then raised until the critical point for carbon dioxide is reached and the density of its vapour is the same as that of its liquid phase. Venting off the vapour at this stage results in a dry sample free from any surface tension artefacts.

For preliminary studies the following schedule is suggested:

1   4% glutaraldehyde in 0.025 mol/l phosphate buffer pH 7.0 for 2−4 h at 4°C;
2   0.025 mol/l phosphate buffer pH 7.0 for 10 min;
3   10% v/v ethanol for 10−20 min;
4   30% v/v ethanol for 10−20 min;
5   50% v/v ethanol for 10−20 min;
6   70% v/v ethanol for 10−20 min;
7   90% v/v ethanol for 10−20 min;
8   100% ethanol for 10−20 min;
9   100% absolute ethanol for 10−20 min.

Ethanol dried over anhydrous sodium sulphate or a similar dehydrating agent must be avoided, to prevent particulate contamination of the sample.

The above holding times in each reagent should be regarded as the minimum necessary, but it should be appreciated that prolonged dehydration times will increase chemical extraction by the dehydration solvent, producing disproportionate loss of the least well fixed components of the specimen. The samples are transferred to the critical point dryer in 100% ethanol. Before viewing by SEM the dried samples are mounted on suitable specimen stubs and rendered conductive by the deposition of a thin layer of either gold, gold−palladium alloy or platinum, usually by sputter coating. These metals are favoured for their complete resistance to oxidation, excellent secondary electron emission and ease of sputter coating. Of the three, platinum produces the finest grain and is the most suitable for high-resolution imaging.

Brain−heart infusion-cultured cells of *Ps. fragi* prepared by this procedure are shown in Fig. 18.7. Compared with cryopreparation (Fig. 18.6), some degree of shrinkage is evident — an inevitable consequence of chemical preparation techniques; most of the shrinkage has been shown to occur at the dehydration stage (Fraser & Gilmour, 1986). The collapse associated with air drying (Fig. 18.8), however, has been avoided.

Critical point drying may adversely affect biofilm attachment surfaces such as nitrile rubber. In such cases the technique described by Lamb & Ingram (1979), of argon replacement-induced drying (ARID) from ethanol, can be used to dry biofilms without appreciably altering the specimen appearance. This technique simply involves the gentle evaporation from the specimen of the 100% ethanol of the final dehydration step in a stream of dry argon gas in a suitable container. The results obtained with this technique are illustrated in Figs 18.12 and 18.13.

The biofilm microorganisms encountered in the food industry are invariably associated with residues or matrices, such as milk residues. Whereas cryo-SEM preserves such substrates in their entirety (see Fig. 18.5), the use of glutaraldehyde as the sole or primary fixative has the tendency to modify these, sometimes to the extent of removing them completely (Fig. 18.12).

FIG. 18.12. Glutaraldehyde fixation, ethanol dehydration, argon replacement-induced drying (ARID). Milk-soiled stainless steel surface contaminated with a *Micrococcus* sp. The ethanol dehydration was prolonged to achieve extraction of the milk soil (*cf.* Fig. 18.5). Bar = 5 μm.

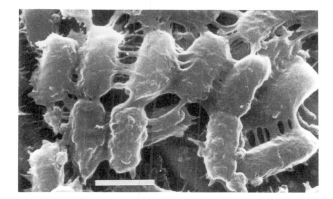

FIG. 18.13. Glutaraldehyde fixation, ethanol dehydration, argon replacement-induced drying (ARID). Milk-soiled stainless steel surface contaminated with an *Acinetobacter* sp. The strandlike appearance of the extracellular polymer can probably be attributed to shrinkage during the chemical preparation procedure. Bar = 1 μm.

FIG. 18.14. Chicken skin surface, standard critical point drying procedure. Bar = 10 μm.

FIG. 18.15. Chicken skin surface, standard critical point drying procedure but including diethyl ether rinse. Bar = 10 μm.

Indeed, if the purpose of specimen preparation is to visualize the surface to which the biofilm is adherent, or to examine by SEM the morphology of those organisms attached to the surface, the standard glutaraldehyde method should be employed, or even modified by the inclusion of a diethyl ether wash between the first and second absolute alcohol stages. In studies of the heavily contaminated skin of commercially prepared chicken carcasses the inclusion of a 30-min diethyl ether wash proved particularly effective at removing soiling to expose the underlying skin surface (Figs 18.14 and 18.15).

Contaminating surface layers and deposits often support many of the microorganisms of the biofilm, so that the removal of these layers and deposits will also reduce the apparent number of microorganisms. This was graphically illustrated by McMeekin *et al.* (1979) in a study of chicken skin. By exposing

FIG. 18.16. Chicken skin surface contaminated with *Pseudomonas* spp. Osmium tetroxide vapour prefixation, glutaraldehyde post-fixation, ethanol dehydration and critical point drying. Note the increased retention of surface soil and microorganisms when compared with glutaraldehyde fixation alone. Bar = 5 μm.

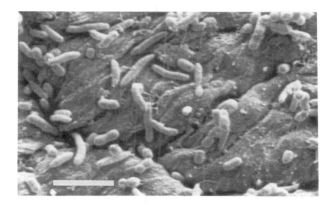

FIG. 18.17. Chicken skin surface contaminated with *Pseudomonas* spp. Glutaraldehyde fixation, ethanol dehydration and critical point drying. Bar = 5 μm.

some skin samples to osmium tetroxide vapour for 24 h before secondary fixation in 4% glutaraldehyde, these workers were able to demonstrate more complete preservation of the adherent biofilm (Fig. 18.16) than when glutaraldehyde alone was used as the fixative (Fig. 18.17). McMeekin *et al.* (1979) proposed that the osmium tetroxide vapour stabilized the lipid fraction of the surface film, which is normally removed by immersion in aqueous glutaraldehyde. More recently, the method has been shown to be effective for stabilizing milk soil (Lewis, 1985) (Fig. 18.18). Osmium tetroxide vapour fixation is easily achieved by sealing the samples into a glass or plastic Petri dish, together with approx 1 ml of 4% (w/v) osmium tetroxide solution in a

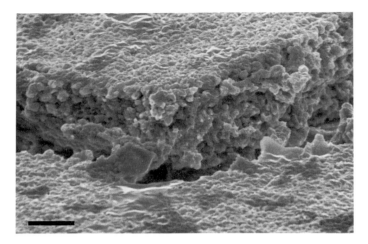

FIG. 18.18. Osmium tetroxide vapour prefixation can be included in chemical preparation schedules to maximize the preservation of milk soil (*cf.* Figs 18.5 and 18.12). Bar = 20 μm.

small watch glass or other inert vessel, with some means of preserving 100% humidity such as a small plug of cotton wool soaked in distilled water.

Comments

SEM has played an integral and important part in our studies of adherence, proliferation and exopolymer production by microorganisms inoculated on to food-related surfaces. SEM revealed that proliferation and exopolymer production occurred on both glass and stainless steel, but not on rubber. It also identified possible influencing factors, such as imperfections, including microchannels in stainless steel and milk deposits (Speers *et al.*, 1984b). SEM has also been used to monitor the development of individual cells inoculated by a single-cell inoculation technique (Lewis *et al.*, 1987). Cryo-SEM has been used to explore the changes that take place during the chemical fixation and dehydration procedures leading to critical point drying. The results showed that the ethanol dehydrating stage of the procedure is responsible for the appearance of attachment fibrils and fibrillar polymer which are now generally considered to be artefacts due to shrinkage during dehydration (Fraser & Gilmour, 1986).

References

ANON. (1984) *Microbiological Examination for Dairy Purposes.* BS4285 (Section 2.5). British Standards Institution, London.

COUSINS, C.M. (1963) Methods for the detection of survivors on milk-handling equipment with reference to the use of disinfectant inhibitors. *Journal of Applied Bacteriology*, **26**, 376–386.

COUSINS, C.M., HOY, W.A. & CLEGG, L.F.L. (1960) The evaluation of surface-active disinfectants for use in milk production. *Journal of Applied Bacteriology*, **23**, 359–371.

FRANK, J.F. & KOFFI, R.A. (1990) Surface adherent growth of *Listeria monocytogenes* is associated with increased resistance to surfactant sanitizers and heat. *Journal of Food Protection*, **53**, 550–554.

FRASER, T.W. & GILMOUR, A. (1986) Scanning electron microscopy preparation methods: their influence on the morphology and fibril formation in *Pseudomonas fragi* (ATCC 4973). *Journal of Applied Bacteriology*, **60**, 527–533.

LAMB, J.C. & INGRAM, P. (1979) Drying of biological specimens for scanning electron microscopy directly from ethanol. *Scanning Electron Microscopy*, **3**, 459–464.

LEWIS, S.J. (1985) *Factors Affecting the Adherence of Microorganisms to Dairy Equipment Surfaces*. PhD Thesis, The Queen's University of Belfast.

LEWIS, S.J. & GILMOUR, A. (1987) Microflora associated with the internal surfaces of rubber and stainless steel milk transfer pipeline. *Journal of Applied Bacteriology*, **62**, 327–333.

LEWIS, S.J., GILMOUR, A., FRASER, T.W. & MCCALL, R.D. (1987) Scanning electron microscopy of soiled stainless steel inoculated with single bacterial cells. *International Journal of Food Microbiology*, **4**, 279–284.

LISBOA, N.P. (1959) A tube test for evaluating agents possessing both detergent and sterilizing properties. *15th International Dairy Congress London*, **3**, 1816–1821.

MCMEEKIN, T.A., THOMAS, C.J. & MCCALL, D. (1979) Scanning electron microscopy of microorganisms on chicken skin. *Journal of Applied Bacteriology*, **46**, 195–200.

NEAVE, F.K. & HOY, W.A. (1947) The disinfection of contaminated metal surfaces with hypochlorite solutions. *Journal of Dairy Research*, **15**, 24–54.

PATON, A.M. & JONES, S.M. (1973) The observation of microorganisms on surfaces by incident fluorescence microscopy. *Journal of Applied Bacteriology*, **36**, 441–443.

PETTIPHER, G.L., MANSELL, R., MCKINNON, C.H. & COUSINS, C.M. (1980) Rapid membrane filtration–epifluorescent microscopy technique for direct enumeration of bacteria of raw milk. *Applied and Environmental Microbiology*, **39**, 423–429.

SPEERS, J.G.S. & GILMOUR, A. (1982) A quantitative and qualitative comparison of microbial flora associated with glass, rubber and stainless steel components of a milking plant. *Brief Communications of the XXI International Dairy Congress* 1, Book 2, 368–369.

SPEERS, J.G.S. & GILMOUR, A. (1985) The influence of milk and milk components on the attachment of bacteria to farm dairy equipment surfaces. *Journal of Applied Bacteriology*, **59**, 325–332.

SPEERS, J.G.S., LEWIS, S.J. & GILMOUR, A. (1984a) Bacteriological sampling of glass, rubber and stainless steel pipe sections. *Journal of Dairy Research*, **51**, 547–555.

SPEERS, J.G.S., GILMOUR, A., FRASER, T.W. & MCCALL, R.D. (1984b) Scanning electron microscopy of dairy equipment surfaces contaminated by two milk-borne microorganisms. *Journal of Applied Bacteriology*, **57**, 139–145.

STANLEY, P.M. (1983) Factors affecting the irreversible attachment of *Pseudomonas aeruginosa* to stainless steel. *Canadian Journal of Microbiology*, **29**, 1493–1499.

STANLEY, P.E. (1989) A concise beginner's guide to rapid microbiology using adenosine triphosphate (ATP) and luminescence. In Stanley, P.E., McCarthy, B.J. & Smither, R. (eds.) *ATP Luminescence – Rapid Methods in Microbiology*, SAB Technical Series No. 26, pp. 1–10. Blackwell Scientific Publications, Oxford.

19: The Statistical Evaluation of Adherence Assays

A.D. WOOLFSON

School of Pharmacy, Medical Biology Centre, The Queen's University of Belfast, Belfast BT9 7BL, UK

The most commonly used *in vitro* methods for quantifying the number of bacteria that adhere to a given substrate are measurement of the radiation from adhering radiolabelled bacterial cells (Wyatt *et al.*, 1990), and direct counting, with light microscopy, of the number of adherent organisms per host cell (Gorman *et al.*, 1987). In both cases, one or more candidate anti-adherence treatments is compared with the results obtained from a control experiment.

Visual counting by microscopy gives direct information about the mean number of adhering organisms per epithelial cell (Reid & Brooks, 1985). By comparison, adherence assays based on radiolabelled organisms (Schaeffer *et al.*, 1981; Kozody *et al.*, 1985) or electronic particle counting techniques (Gorman *et al.*, 1986) give only an overall view of adherence, without information about the variation in adhesion from cell to cell. However, irrespective of the experimental model used, the interpretation of data obtained in adherence assays is critically dependent on application of the appropriate statistical testing in order to evaluate the level of significance, if any, of the corresponding antiadherence effect.

The choice of statistical test for the analysis of adherence assay results depends on both the experimental design and the pattern of distribution of organisms adhering to the substrate. Often, but not always, interpretation of adherence assays involves the statistical comparison of two sets of observations, where one data set is independent of the other. The statistical treatment applicable to this situation is different from that applicable to an experimental design involving more than two treatments, or from one which compares the effect of more than two treatments on more than one substrate.

The first decision to be made in applying a statistical procedure to adherence assay results concerns the nature of the distribution pattern observed. For

Microbial Biofilms:
Formation and Control

normally distributed data a parametric test may be used. If the data are not normally distributed — the commonest situation on adherence assays — a non-parametric method is usually required. Data obtained by light-microscopic assay consist of counts of the number of organisms adhering per epithelial cell, with perhaps 100 randomly selected cells counted in each data set. The number of organisms adhering per cell varies greatly. Some cells have none or only a few adherent organisms, whereas others may have relatively large numbers attached. The resulting frequency distribution will therefore be discontinuous. Consequently, the distribution of adherence data observed in adherence assays will be Poissonian rather than normal (Rosenstein *et al.*, 1985).

Parametric and Non-Parametric Hypothesis Testing

The statistical interpretation of adherence assay data is an example of hypothesis testing. Generally, a hypothesis is formulated with regard to the effect of a given compound or treatment on microbial adherence to substrate cells. The assumption is always that the compound or treatment has no effect on microbial adherence. Thus, the hypothesis to be tested is always a null hypothesis.

There are two forms of hypothesis testing. In one case an assumption is made in respect of the distribution pattern of the data. This fixes a particular parameter for the procedure and such tests are, therefore, known as *parametric tests*. The commonly assumed distribution pattern is the normal distribution. Consequently, a parametric hypothesis test involves the actual or assumed knowledge that the data to be examined are normally distributed. When the assumption of normality in the distribution is invalid, a *non-parametric test* is indicated. Such a test makes no underlying assumption as to the distribution of the data.

Asymmetric distribution (skewness)

The distribution of data from adherence assays tends to be skewed rather than normally distributed. In a skewed or asymmetric distribution there is an excess of data in one side or tail of the distribution. If a unimodal distribution has a longer tail extending towards lower values of the variate (observation) the distribution shows negative skewness. Where a longer tail extends towards higher values of the variate the distribution is positively skewed. In adherence assays the distribution tends to be positively skewed (Rosenstein *et al.*, 1985).

A measure of the extent to which the distribution of adherence data deviates from normality can be made by calculating a skewness coefficient for the data. In its simplest form, this gives a measure of the asymmetry of the distribution as the difference between the mean and the mode values, where the mode is that value of the variate possessed by the greatest number of members of the population. A dimensionless measure of skewness can be obtained from

$$\frac{(\bar{x} - \text{mode})}{s}$$

where $\bar{x}$ is the mean value of the data, the mode is the most frequently occurring value and s is the standard deviation of the data. However, not all distributions have a modal value and some may actually be bimodal. A bimodal distribution of adherence assay data may indicate the presence of two independent adherence factors.

A more powerful measure of skewness is the third-moment coefficient of skewness, where the term moment is the mean value of a given power of the variate, in this case the third power. The third-moment coefficient of skewness is defined as

$$\{\Sigma[(x - \bar{x})/s]^3\}/n$$

where n is the number of values in the data set. A normally distributed population has a third-moment skewness coefficient of zero. A value greater than unity indicates that the population (of data) deviates markedly from the normal distribution.

Student's t-test and the Mann–Whitney U-test (Wilcoxon rank sum test)

An unpaired t-test is a parametric test based on the t-distribution — the theoretical distribution of a small sample drawn from a normally distributed population. It may be used to compare the means of two independent groups. When the population variance is unknown, as is the case with adherence assays, it is calculated by pooling the data from the two groups, with the assumption that both data sets are drawn from the same normally distributed population. A more sensitive design involves a paired t-test, in which the comparisons of two means are made on related samples. The unpaired test is, however, appropriate for adherence assays, since the data are usually obtained from a single substrate rather than from several different substrate pairs.

The non-parametric equivalent of the unpaired t-test is the Mann–Whitney U-test, also known as the Wilcoxon rank sum test. Here, observations from the two data sets are pooled and then ranked in numerical order. Tied data are given a rank equal to the average of the ranks. The basis of the test is the sum of the ranks of the smaller sample. Another ranking test, the Wilcoxon signed rank test, must be used for the distribution-free comparison of two treatments in a paired design, and is therefore the non-parametric alternative to a paired t-test.

Adherence Assays:
Comparison of Results in Two Independent Groups
by Parametric and Non-Parametric Tests

Many statistical treatments of adherence assay data have been by a two-sided unpaired t-test, irrespective of the degree of skewness in the population (King et al., 1980; Lee & King, 1983; Kozody et al., 1985; McCourtie et al., 1986). The accepted level of significance for these tests is 5%. Although the application of a parametric test to adherence assay data is statistically incorrect, in most cases the interpretation of the results by either method will be different only in respect of the level of significance, if at all. Furthermore, the Mann—Whitney U-test is not without problems when applied to adherence assay data. Here, the rank orders are analysed rather than the data themselves. In a ranking test of this type it is generally assumed that ties (two or more numerically identical data in the ranking) result only from experimental limitations, since the data are assumed to be continuous. This is clearly not the case in adherence assays, since it is perfectly reasonable for several epithelial cells to have identical numbers of adhering organisms, irrespective of the measurement technique. Thus, the large number of ties inevitably present in adherence assay data may reduce, to some extent, the power of the Mann—Whitney U-test for this application.

An extensive comparison of the parametric and non-parametric tests for the comparison of the means of two independent groups has been made by Woolfson et al. (1987). The data in Table 19.1 consist of 17 assays of the antiadherence capacity of the antimicrobial agent Taurolin (Gorman et al., 1987), in which buccal epithelial cells were the substrate and water treatment was the control. Pretreatment of either host cells or organisms with Taurolin was used (see Chapter 8).

For a level of significance of 5% ($p < 0.05$) there was little apparent difference when results were analysed by either a parametric or a distribution-free method. Thus, in Table 19.1, for tests A—C, and also test K, neither method indicated a statistically significant difference between Taurolin treatment and water controls. In test H, the non-parametric method yielded significance at the 5% level, whereas the t-test indicated that there was no significant difference. This was the only case in which the significance of the result was in doubt between the parametric and non-parametric methods. It is of interest that this case involved a concentration of the antimicrobial agent well below that recommended for clinical use. Thus, the antiadherence effect could reasonably be expected to be marginal at this level. In all other tests the conclusions reached from application of either statistical method were in agreement. Thus, in tests D, E, N, P and Q both methods gave results significant at a maximum level ($p < 0.001$). In the remaining tests, although the

TABLE 19.1. *Comparative statistical evaluation of adherence assay results*

Test*	Organism	Host	Pretreatment	N†	Statistical evaluation t‡	U§
	Candida albicans blastospores	Buccal epithelial	Organisms			
A	Water			5.99 ± 0.79		
	Taurolin 0.005%			6.58 ± 0.92	NS	NS
B	Water			5.99 ± 0.79		
	Taurolin 0.05%			6.95 ± 1.02	NS	NS
C	Water			5.99 ± 0.79		
	Taurolin 0.1%			4.29 ± 0.44	$p < 0.1$	NS
D	Water			5.87 ± 0.74		
	Taurolin 1%			2.15 ± 0.29	$p < 0.001$	$p < 0.001$
E	Water			5.87 ± 0.74		
	Taurolin 2%			2.01 ± 0.19	$p < 0.001$	$p < 0.001$
	Candida albicans blastospores	Buccal epithelial	Host			
F	Water			5.39 ± 0.56		
	Taurolin 0.5%			3.39 ± 0.63	$p < 0.02$	$p < 0.001$
G	Water			5.39 ± 0.56		
	Taurolin 2%			3.47 ± 0.36	$p < 0.005$	$p < 0.01$
	Staphylococcus saprophyticus	Buccal epithelial	Organisms			
H	Water			22.62 ± 1.84		
	Taurolin 0.05%			18.36 ± 1.75	$p < 0.1$	$p < 0.001$
I	Water			22.62 ± 1.84		
	Taurolin 0.5%			18.36 ± 1.75	$p < 0.01$	$p < 0.001$
J	Water			22.62 ± 1.84		
	Taurolin 2%			15.93 ± 1.74	$p < 0.01$	$p < 0.001$
	Staphylococcus saprophyticus	Buccal epithelial	Host			
K	Water			27.71 ± 2.29		
	Taurolin 0.05%			22.72 ± 1.76	$p < 0.1$	NS
L	Water			27.71 ± 2.29		
	Taurolin 0.5%			20.36 ± 1.37	$p < 0.01$	$p < 0.05$
M	Water			27.71 ± 2.29		
	Taurolin 2%			17.65 ± 1.10	$p < 0.001$	$p < 0.01$
	Escherichia coli	Buccal epithelial	Organisms			
N	Water			24.10 ± 1.74		
	Taurolin 0.5			15.29 ± 1.14	$p < 0.001$	$p < 0.001$
O	Water			24.10 ± 1.74		
	Taurolin 2%			17.02 ± 1.42	$p < 0.02$	$p < 0.001$

continued on p. 320

TABLE 19.1. (*continued*)

Test*	Organism	Host	Pretreatment	N†	*t*‡	U§
					Statistical evaluation	
	Escherichia coli	Buccal epithelial	Host			
P	Water			20.99 ± 1.26		
	Taurolin 0.5%			12.51 ± 0.96	$p < 0.001$	$p < 0.001$
Q	Water			20.99 ± 1.26		
	Taurolin 2%			12.37 ± 1.05	$p < 0.001$	$p < 0.001$

* Treated for 30 min at 37°C with either water or Taurolin solution.
† Mean number of organisms ± standard deviation adhering per epithelial cell.
‡ Unpaired, two tailed *t*-test.
§ Mann–Whitney U-test.
NS, not significant.
Microbial adherence to host buccal epithelial cells was determined by light microscopy following pretreatment of either host cells or organisms with a Taurolin solution, as specified, or a water control.
Reproduced from Woolfson *et al.* (1987) with permission of the copyright owner.

levels of significance differed slightly in some cases, there was agreement that the antiadherence effect of Taurolin was highly significant.

Skewness coefficients were calculated (Table 19.2) for each data set in Table 19.1. Tests A–E indicated a reduction in the skewness of the distributions at all Taurolin concentrations, when compared with the corresponding controls. The greatest reduction occurred with the greatest concentration of antimicrobial agent. A reduction in the skewness coefficient with increasing concentration of antimicrobial agent was clearly seen in grouped tests (F, G), (H, I, J), (K, L, M) and (N, O).

Calculation of the skewness coefficient gives information about the actual data obtained in the adherence assay, whereas the parametric or non-parametric analysis of these data simply gives information on the differences between means in control and treated groups. Reduction in the skewness of the distribution indicates a shift towards normality in the population as a consequence of the effect of the antimicrobial agent. This was probably due to a reduction in the number of epithelial cells with large numbers of adherent organisms, and a shift towards increasing numbers of cells with either zero or few adherent organisms. The number of cells with zero adherent organisms appeared to increase with increasing concentration of the antimicrobial agent. Thus, the skewness of the distribution was reduced.

The statistical analysis of the adherence assay results (Table 19.1) suggests that there was little practical difference in the conclusions reached, irrespective of the test used, provided a high enough level of significance was chosen.

TABLE 19.2. *Skewness coefficients of adherence assay data from Table 19.1*

Test*	Skewness coefficient	
	Water control	Taurolin treatment
A	2.11	1.32
B	2.11	1.46
C	2.11	3.28
D	2.43	2.12
E	2.43	1.05
F	3.35	2.48
G	3.35	1.56
H	2.20	2.06
I	2.20	1.34
J	2.20	0.61
K	1.36	1.89
L	1.36	1.03
M	1.36	0.47
N	1.95	1.81
O	1.95	0.89
P	1.60	1.30
Q	1.60	1.45

* As designated in Table 19.1.

Thus, conclusions obtained in previous studies based on a *t*-test are not necessarily invalid. However, a test of normality by calculation of the skewness coefficient for each data set, and the use of a non-parametric test if the data deviate significantly from normality, is advisable.

Data transformation

A possible solution to the problem of a non-normal distribution in adherence is to transform the variate (observation) such that the distribution approximates normality. A parametric test may then be applied. The most common method of transformation is to take the logarithm of the number of adhering bacteria. However, this will not in every case achieve transformation of the data to a normal distribution (Rosenstein *et al.*, 1985). Alternative methods of transferring a variate Y have been reported. These include the transformations $Y \to \sqrt{Y}$ (Gilsdorf *et al.*, 1989) and $Y \to (Y + 0.5)^{-1/2}$ (Persi *et al.*, 1985).

Statistical Analysis of Multiple Comparisons

When the means of more than two groups of data are to be compared, it is not statistically valid to use a *t*-test or the Mann−Whitney U-test by selecting from the data set any two groups and then performing several two-group comparisons. This can lead to contradictory conclusions, not least because different variances are being used for different comparisons. Therefore, in more complex experimental designs of this type it is necessary to use a powerful statistical tool, the analysis of variance (ANOVA). When ANOVA is to be used for statistical evaluation of the data, the experiment must be designed with this in mind, since the experimental design determines which of several ANOVA procedures is the correct choice for a particular application. In this context, the *t*-test is simply a special case of ANOVA in which only two means are compared. A further complication is that, as with the two-group comparisons, both parametric and non-parametric ANOVA procedures are available and the same considerations will apply here as to the two-group tests.

One-way and two-way ANOVA (parametric)

If the data are normally distributed, or can be transformed so as to approximate a normal distribution, the simplest experimental design is a one-way ANOVA, In this procedure the equality of treatment means is tested in experiments in which two or more treatments are randomly assigned to different experimental units. An example of this type of experimental design, applied to adherence assays, has been given by Roberts *et al.* (1990), in which the adherence of a given bacterial strain to three different urethral catheters was tested. A more complex design is a two-way ANOVA, sometimes known as a randomized block design. This is an extension of the paired *t*-test to more than two treatments.

Multiple-range tests

An ANOVA result that indicates a significant difference between treatments may not, in itself, be a satisfactory conclusion to draw from an adherence assay. It is reasonable to seek an ordering of treatments, from best to worst, with respect to antiadherence capacities. This type of statistical procedure is known as a multiple-range test, and can only be performed after a significant difference between treatments has been established by ANOVA.

Several multiple-range tests are available, including Tukey's test, Scheffe's method and the Newman−Keuls multiple-range test (Snedecor & Cochran, 1980). The Newman−Keuls test has been applied by Gilsdorf *et al.* (1989) to adherence assay data which had initially been analysed by ANOVA after square-root transformation.

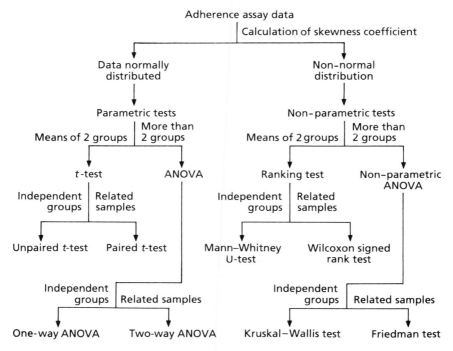

FIG. 19.1. Statistical procedures for the analysis of adherence assay data.

One-way and two-way ANOVA (non-parametric)

When a non-normal distribution of adherence assay data has been established by calculation of the skewness coefficient, a non-parametric ANOVA should be used for experimental designs involving multiple comparisons. The Kruskal—Wallis test is a non-parametric one-way ANOVA and is, essentially, an extension of the Wilcoxon rank sum test to more than two treatments. A significant difference obtained by this procedure indicates that at least two of the treatments compared are different. When the experimental design follows that for a two-way ANOVA, the non-parametric Friedman test should be used. It may be regarded as an extension to the signed rank test, though the computation procedures are different. These methods are treated in detail by Hollander & Wolfe (1973).

An application of the Kruskal—Wallis test to adherence assay data can be found in Wyatt *et al.* (1990). A one-way ANOVA design was used to compare the adherence of a *Staphylococcus aureus* strain to six different batches of a particular cell monolayer. Rosenstein *et al.* (1985) used the two-way non-parametric Friedman test statistically to analyse adherence assay data from a more complex design. In this example, the adhesion of three *Escherichia coli* strains to uroepithelial cells from five healthy female volunteers was investigated.

Selection and Computation Procedures

Although numerous additional statistical methods can be applied to adherence assays, those outlined in this chapter form the basis of the statistical evaluation of adherence assay data. The flowchart in Fig. 19.1 provides a guide to selection of the correct statistical procedure in order to ensure that the conclusions drawn from adherence assays are reliable, and that published work in this area is comparable.

The detailed computational procedures for performing the statistical tests discussed are outside the scope of the present chapter. However, many worked examples are provided in several standard statistical texts, including those by Bolton (1984) and Zar (1974). The procedures may conveniently be performed, with suitable software, on a microcomputer.

References

BOLTON, S. (1984) *Pharmaceutical Statistics*. Marcel Dekker, New York.

GILSDORF, J.R., WILSON, K. & BEALS, T.F. (1989) Bacterial colonization of intravenous catheter materials *in vitro* and *in vivo*. *Surgery*, **106**, 37–44.

GORMAN, S.P., McCAFFERTY, D.F. & ANDERSON, L. (1986) Application of an electronic particle counter to the quantification of bacterial and *Candidal* adherence to mucosal epithelial cells. *Letters in Applied Microbiology*, **2**, 97–100.

GORMAN, S.P., McCAFFERTY, D.F., WOOLFSON, A.D. & JONES, D.S. (1987) Reduced adherence of microorganisms to human mucosal epithelial cells following treatment with Taurolin, a novel antimicrobial agent. *Journal of Applied Bacteriology*, **62**, 315–320.

HOLLANDER, M. & WOLFE, D.A. (1973) *Non-Parametric Statistical Methods*. Wiley, New York.

KING, R.D., LEE, J.C. & MORRIS, A.L. (1980) Adherence of *Candida albicans* and other *Candida* species to mucosal epithelial cells. *Infection and Immunity*, **27**, 667–674.

KOZODY, N.L., HARDING, G.K.M., NICOLLE, L.E., KELLY, K. & RONALD, A.R. (1985) Adherence of *Escherichia coli* to epithelial cells in the pathogenesis of urinary tract infection. *Clinical and Investigative Medicine*, **8**, 121–125.

LEE, J.C. & KING, R.D. (1983) Characterization of *Candida albicans* adherence to human vaginal epithelial cells *in vitro*. *Infection and Immunity*, **41**, 1024–1030.

McCOURTIE, J., MacFARLANE, T.W. & SAMARANAYAKE, L.P. (1986) A comparison of the effects of chlorhexidine gluconate, amphotericin B and nystatin on the adherence of *Candida* species to denture acrylic. *Journal of Antimicrobial Chemotherapy*, **17**, 575–583.

PERSI, M.A., BURNHAM, J.C. & DUHRING, J.L. (1985) Effect of carbon dioxide and pH on adhesion of *Candida albicans* to vaginal epithelial cells. *Infection and Immunity*, **50**, 82–90.

REID, G. & BROOKS, H.J.L. (1985) A fluorescent antibody staining technique to detect bacterial adherence to urinary tract epithelial cell. *Stain Technology*, **60**, 211–217.

ROBERTS, J.A., FUSSELL, E.N. & KAACK, M. (1990) Bacterial adherence to urethral catheters. *Journal of Urology*, **144**, 264–269.

ROSENSTEIN, I.J., GRADY, D., HAMILTON-MILLER, J.M.T. & BRUMFITT, W. (1985) Relationship between adhesion of *Escherichia coli* to uroepithelial cells and the pathogenesis of urinary infection: problems in methodology and analysis. *Journal of Medical Microbiology*, **20**, 335–344.

SCHAEFFER, A.J., JONES, J. & DUNN, J.K. (1981) Association of *in vitro Escherichia coli* adherence to vaginal and buccal epithelial cells with susceptibility of women to recurrent urinary tract infections. *New England Journal of Medicine*, **304**, 1062–1066.

SNEDECOR, G.W. & COCHRAN, W.G. (1980) *Statistical Methods*, 7th edn. State Unviersity Press, Ames, I.

WOOLFSON, A.D., GORMAN, S.P., McCAFFERTY, D.F. & JONES, D.S. (1987) On the statistical evaluation of adherence assays. *Journal of Applied Bacteriology*, **63**, 147–151.

WYATT, J.E., POSTON, S.M. & NOBLE, W.C. (1990) Adherence of *Staphylococcus aureus* to cell monolayers. *Journal of Applied Bacteriology*, **69**, 834–844.

ZAR, J.H. (1974) *Biostatistical Analysis*. Prentice-Hall, New Jersey.

Index

Acetone, as a dehydration solvent, 307
Acridine orange, 298
Acrylic resin replicas, 277
Adenylate distribution, 62
Adenylate energy charge, 62
Adenylate status, of biofilm, 79
Adherence,
    by CNS, 138–144
    to food contact surfaces, 293–312
Adherence assay, 125–126, 148–152, 159,
    315–324
Adhesion,
    control of, 147–163
    medical device-associated, 133–144
Adhesion number tests, 13–15
Adhesive properties, of *Staphylococcus aureus*
    strains, 288
Aeromonads, in drinking water, 219
Aeromonas hydrophila, 219–225
Agar diffusion, 176
Agar dilution method, 176, 177
Amidinocillin, 36
Ammonium compounds, 235
Amoebae, 201, 206, 234
Amphora coffeaeformis, 253
Anionic ion exchange resin, retention on, 22
ANOVA procedures, 322–324
Antiadhesive agents, 175
Antibacterial agents, sensitivity to, 187–198
Antibiotic effect, measurement of, 190
Antibiotic penetration, of polysaccharide, 54
Antibiotic susceptibility, 39
Antibiotics,
    effects on chemostat-grown biofilms, 193
    penetration through biofilms, 9
Antimicrobial agents, effect of antimicrobial
    chemotherapy on adherence of
    microorganisms, 147–163
Antimicrobial methods, to evaluate plaque
    control agents, 178
Antimicrobials, response to, 82
'Antiplaque', definition of, 175
Argon replacement-induced drying, 308
'Artificial mouths', 169
Asymmetric distribution, 316–318
ATP, concentration of, 300–301

Attachment, bacterial, 13–16, 109–129
    inhibition of, 175
    on teeth, 167–170
Attenuated total reflectance, 6–9

Bacillus subtilis, 61
Bactericidal activity, tests of, 177
Bacteriological sampling, 243
Bacteriological water quality, 236–238
Bacterionema
    fragilis, 97, 110–129
    loeschii, 84
    matruchotii, 84
    melaninogenicus, 169
Bactometer Microbial Monitoring System,
    286–287
Batch culture growth, use of MRD with,
    3–4
Batch-culture studies, of CNS biofilms on
    biomedical materials, 190–192
BATH test, 17, 24
Beads, immunomagnetic, 119–120
Beer–Norton function, 8
Biocides,
    evaluation of, 209–211
    for metalworking fluids, 280
    for recreational waters, 230–243
Bioelectric effect, 198
Bioluminescence, *in situ* measurements of,
    300–302
Bioluminescence assay, 160
Biomaterials, biofilms on, 190–192
Bitumen-painted mild steel, 222–224
Bleeding on probing, 174
Bromine, 209, 228, 236
2-2′-bromo-chloro-dimethylhydantoin, 211
Broth dilution methods, 176
Buccal retention test, 178
Buffered bovine albumin solution, 240

^{13}C nuclear magnetic resonance, 9
Calorifier, microflora from, 202
Candida albicans, 148
Carbol fuchsin, 111

Carbon dioxide, effect on CNS adherence, 141–142
Carbon limitation, 55
Caries, 167, 202
Catheters,
  discs of material from, 192
  Foley latex urinary, 155–157
  indwelling, 69
  pathogens on, 133, 144
  peritoneal, 98–106
Caulobacter, 55, 57
Caustic detergent, 294
CDFF, 63–84
Cefamandole, 36
Cell distraction, 13
Cellulose nitrate membrane filters, 18–19
Chemostat growth,
  under low-iron conditions, 192–193
  model of, 202–206
  use of MRD with, 4–6
Chemostats, 38–46, 65, 169
'Chick–Watson law', 210
Chloramines, 235
Chlorhexidine, 36, 160–162, 177, 178–179, 181, 187
Chlorine, 209–210, 217–225, 228–243
Chlorine demand free water, 241
Chlorine resistance, 241
Closed growth systems, 31–32
Clumping strains, of *Staphylococcus aureus*, 286–290
CNS, *see* Coagulase-negative staphylococci
Coaggregation, 85
Coagulase-negative staphylococci, 133–144
Code of Practice, for institutional buildings, 208–209
Coliform biofilms, 217–225
Colony on filter method, 187–189
Column reactor, 260–263
Computer-enhanced microscopy, 10
Concanavalin A, 137
Concentration gradients, 58
Conditioning, of pipe sections, 294
Conditioning films, 56
Confocal laser microscopy, 87
Confocal laser scanning microscopy, 95–106
Constant-depth biofilm, 63
Contact angle measurements, 18–19, 23
Contact time, 210
Continuous ambulatory peritoneal dialysis, 97–101
Continuous culture, of oral bacteria, 169
Continuous-culture model, 219–220
Continuous perfused biofilms, 44, 193–194
Conventional SEM preparation methods, 305–312

Cooling towers, hospital, 209–210
Cooling water systems, 247–257
Copiotrophic bacteria, 55, 61
Copper, colonization of, 205
Cornified epithelial cells, 150
Counting, of milk-associated bacteria, 299
Counting techniques, 315
Counts, for spa water quality, 236
Coverslips, glass, bacterial, adherence to, 300–302
Critical force tests, 13–15
Critical point drying, 307–308
Cryofixation, for SEM, 303–305
Cryosectioning, of biofilms, 74
CSF shunts, 155–157
Cyanobacteria, 201
Cystic fibrosis, 32
Cytokine release, 110

Dark-field microscopy, 171
Debris index, 172
Defeathering machines, *Staphylococcus aureus* from, 285–290
DEFT, 273–274
Dehydration solvent, removal of, 305–308
Density, of biofilm, 59
Density gradient centrifugation, 117–119
Dental plaque, 39, 69, 202
  area modification of Shaw & Murray stain index, 172
  control trials, 181
  identification of, 171–172
  indices, 172
  study of, 167–181
[3]H-deoxythymidylic acid, 287–288
Dermatitis, 231
Desulfovibrio desulfuricans, 273
Detachment, 57, 253–254
Device-associated infection, diagnosis of, 134–136
Devices,
  implantable, 133–144
  indwelling, 147–163
Dextran, as a cryoprotectant, 79
Diarrhoea, 219
Diatom slimes, 247, 253, 256
Differential interference contrast microscopy, 209
Differential labelling, 139–140
Direct microscopy, 272–274
Disclosing solutions, 171–172
Disinfectants, residual, 217–225
Disinfection concentration, 210
Distribution, Poissonian, 316
DLVO theory, 15

DNA, incorporation of radiolabel in, 287–289
Doppler shift frequency, 21–22
Drinking water, biofilms in, 209–213, 217–225, 240
Drop shape analysis, 19
Dry eosin/carbol fuchsin method, 111–112
Dutch Standard Suspension Test, 240
Dynamic adhesion, 13

Ecology, of legionellas, 207–209
Effluent, from column reactor, 262–263
Elastomeric materials, 206
Electrophoretic mobility, 21–23
Elemental composition, of biofilms, 51–52
End closures, 293–297
Engineering fluids, *see* Metalworking fluids
Enterobacter cloacae, 259–265
Enterobacteriaceae, 16
Environmental SEMs, 302
Epifluorescence microscopy, 154, 156, 213, 253–254, 273–274, 297–300
Epithelial cells,
   adherence to, 125–127
   adhesion to by antimicrobial chemotherapy, 147–163
   collection of, 125
Eppendorf centrifuge, 126
Ergosterol, 279–280
Erosion, 57
Erosion rate, 58–59
Erythrosine, 171, 172
ESCA, 22–23
Escherichia coli, 9, 30, 32, 33, 36, 60, 110, 118, 128, 133, 140, 159–160, 193–195, 217–225, 234, 237, 239, 324
Ethanol, as a dehydration solvent, 307–309
Evanescent wave, 6
Exfoliated epithelial cells, 148–151
Experimental plaque methods, 179
Extracellular matrix, 53–55
Extracellular polymer substances, 52–55, 218
Extraction agents, for ATP, 300
Eyepieces, microscope, 299

Faecal contamination, of water, 217–225
Fermenters,
   constant-depth, 198
   continuous-culture, 37–39
Fibronectin-binding proteins, 16
Filters, colonies on, 187–189, 239
Fimbriae, 109–110, 112, 115, 128, 233

mannose-binding, 16
FITC conjugate, 117
Fixation, chemical, 305–307
Flagella, 109, 115
Flavobacteria, 201, 221
Flotation pools, 231
Flow apparatus, 247–251
Flow cell, 254–257
Flow rate, 254
'Flow-through' system, 5
Flowing systems, adhesion of biofilms in, 247–257
Fluid shear forces, 39, 40
Fluid velocity, 204
Fluids, metalworking, 267–281
Fluorescein diacetate hydrolysis, 278
Fluorescein isothiocyanate, 213
Fluorescent antibody techniques, 238
Fluorescent stains, 298
Folliculitis, 231–232
Food contact surfaces, adherence to, 293–312
Forces,
   critical, 13–15, 254
   interfacial, 15–16
Formica, slimes from, 257
Fouling, in industrial cooling systems, 247–257
Fourier transform infrared spectroscopy, 6–9, 21
Freezing, of biofilms, 78–79
Fungi, in metalworking fluids, 268, 279, 281
Fusarium, 268
Fusiforms, 169
Fusobacterium nucleatum, 85

Gas production, 58
Gas tension, 141–142
Gene exchange, 259–265
Genetically modified organisms, 259–265
Giemsa stain, 126, 150
Gingival crevicular fluid flow measurements, 173–174
Gingival index, 173–174
Gingivitis, 167, 173–175, 180–181
Gingivitis methods, 180
Gingivitis scoring, 173–175
Glass, adherence to, 293–312
Glassclad surface treatments, 251
Glass slides method, 195–197
Glass test sections, for flow apparatus, 249
Glass tiles, 222–223
Glow discharge treatment, 152–153, 155
Glutaraldehyde, 281, 307, 309, 311
Gradient centrifugation, 117–119

Gradostat, 65
Growth dynamics, of biofilm, 70
Growth rate-controlled biofilms, 43, 193–194
Guidelines, for testing of oral hygiene products, 175

Haemagglutination, 123–125
Haemophilus influenzae, 33
Hair follicles, 230
Heat treatment, 209
Helmstetter & Cummings technique, 43
HEp2 cells, 151
Herbert device, 65
Heterogenicity, 56–58
Hexadecane, 17
Hickman and Broviac catheters, 141
High-performance liquid chromatography, 268
Homogeneous populations, preparation of, 117–120
Homogenization method, 241
Hospital cooling towers, 209–210
Hospital water supply system, 240
Hot water systems, legionellas in, 208
Hydraulic shock, 250
Hydrocarbons, adherence to, 122
Hydrocarbons (BATH), bacterial adherence to, 17, 24
Hydrogel coatings, 155
Hydromer coating, 155, 156–157
Hydrophobic interaction assay, 18, 22, 24
Hydrophobic interaction chromatography (HIC), 17, 122
Hydrophobicity, 16–18, 23–24, 121–122, 155
Hydrotherapy pools, 239
Hydroxyapatite adherence assay, 168–169
Hyphomicrobium, 57, 86
Hypobromous acid, 236
Hypochlorite,
    cleaning of milk installations with, 294
    resistance to, 285
Hypochlorous acid, 210, 235
Hypothesis testing, 316–324

Image analyser, 299
Image analysis, 152, 155, 157, 197–198
Immunofluorescence, 95, 116
Immunogold electron microscopy, 116–118
Immunomagnetic beads, 120–121
Impedance changes, 286–287

Implants, medical, 133–144
Indicator organisms, 236
Indole production, 217
Inert solid/liquid interfaces, 30
Infrared absorbance, 250–251
Inhibition,
    of adherence, 128–129
    tests of, 176
Inoculum,
    for clumping strains of *Staphylococcus aureus*, 286
    for two-stage model, 220
Internal reflectance element, 6–9
Intrauterine contraceptive devices, 157–162
Invasive potential, of bacteria, 97
Irreversible adhesion, 15
Isopycnic density centrifugation, 118
Isothiazolones, 210, 280–281

Kill curves, 177–178
Klebsiella, 269–270
Klebsiella pneumoniae, 59
Kruskal–Wallis test, 323

Laboratory methods, for biofilm production, 29–46
β-lactams, penetration of through biofilms, 8–9
Lactose, 217
Laser confocal scanning microscopy, 197–198
Lathe, contaminated coolant from, 270–271
Lawns, bacterial, 18–19
Legionella, 39, 201–213, 230–243
    bozemanii, 208–209
    pneumophila, 201–213, 232, 237
Legionnaire's disease, 201–213, 232
Light engineering industry, 267–281
Light microscopy, 111–116
Listeria monocytogenes, 300–301
Location, of biofilm development, 30–37
Low-vacuum SEMs, 302
Luciferin, 300

Macromolecules, 218
Malvern Zetasizer II, 21
Mann–Whitney U-test, 127, 317–321
Mannitol, 217
Mannose-specific lectins, 137
Media,
    used in study of CNS slime and

adherence, 142
used with metalworking fluid organisms, 269–270
Membrane filtration, 237, 241
Metabolic activity, 59–62
Metal corrosion, 52
Metal test sections, 250
Metalworking fluids, 69, 267–281
Methicillin-resistant bacteria, 197
Micrococcus, spp., 303
Microcolonies, exposure of to antibiotics, 195–196
Microscopy, characterization of surface structures by, 110–116
'Microslicer', 59
Microtitre tray adherence assay, 138–139
Milk, as a soiling medium, 295–297
Milking plant, microbial adherence to, 293–312
Millipore membrane, 305
Mini-cells, 55
Minimum bactericidal concentrations, 177
Minimum inhibitory concentration, 177
Mixing vessel, 248–251
Model systems, 272–273
Modified Breed Smear Count, 299
Modified microtitre tray adherence assay, 141
Modified Robbins device, 2–6, 11, 40, 189–190
Monochloramine, 209–210, 221–225
Monoclonal antibodies, 116–117, 120, 128, 212–213
Monofilaments, PHEMA-coated, 158–162
Monolayers, epithelial, 150–151
Moraxella, 300
Morphology, within a biofilm, 55–56
Mortality figures, for CNS infection, 133
Most probable number method, 237
Mouthwashes, testing of, 177–180
Mucosal epithelial cells, 148–151

Naegleria
fowleri, 239
lovaniensis, 239
Neisseria, 58
Newman–Keuls multiple-range test, 322
Nitrile rubber, attachment to, 308
Nitrobacter, 61
Nitrogen limitation, 55
Nitrosomonas europaea, 61
Non-parametric test, 316–324
Non-stick surfaces, 255–257
Nuclepore polycarbonate membrane, 305

Nutrient availability, 29–30

Oligosaccharide moieties, in mammalian systems, 128
Oligotrophic bacteria, 55
Open continuous culture, 37–43
Open growth systems, 31
Optical sectioning, 96
Oral bacteria, biofilms of, 79–82
Osmium tetroxide fixation, 112, 307, 311
Oxygen depletion, 58
Ozone, 209, 237

Paints, antifouling, 211
Paraformaldehyde fixation, 114
Parametric tests, 316–324
Partitioning materials, 20
'Patchiness', 59
Pathogenicity, link with adherence, 147
Pelvic inflammatory disease, 157
Percoll, preparation of, 117–120
Perfused-biofilm method, 194–195
Periotron, 174
Peritoneal catheters, 97
Peritoneal dialysis fluid, as a culture medium, 142–143
Peritonitis, 97, 135, 138
Perspex adherence model, 169–170
pH
gradients, 30
of metalworking fluids, 268
for swimming pools, 242–243
PHEMA-coated monofilaments, bacterial adhesion to, 158
Philadelphia outbreak, the, 232
Pili, 109
Pipe sections, 293–297
Placebos, for potential plaque inhibitors, 181
Planktonic bacteria, 158
Planktonic populations, 1, 204–205, 233, 238–241
Plaque, *see* Dental plaque
Plasmin formation, 109
Plastic surfaces, 206, 238
Plasticizers, 140
Platinum–gold shadowing, 115–116
Plumbing materials, biofilms on, 205–206, 240–241
Pluronic surface treatment, 153–155
pmS 110-broth, 136–137
Polymer surfaces, adherence to, 140–141
Polymers, model, 152–155
Polysaccharides, bacterial, 53–55

Polystyrene, 20, 23, 152−155, 233
Polyurethane catheters, 141
Polyvinylchloride, 140
Polyvinylpyrrolidone, 155
Pontiac fever, 201−202, 232
Pool Water Treatment Advisory Group, 227, 229, 235, 237, 242−243
Porphyromonas gingivalis, 128
Poultry plants, *Staphylococcus aureus* from, 285−290
Premature air-drying, 304
Proportional integral derivative controllers, 204
Prostheses, 133−144, 189
Protozoa, in biofilms, 207−212
Pseudomonas, 57, 269
  aeruginosa, 9, 32−33, 46, 54, 61, 62−63, 69, 72, 75−76, 83−85, 140, 142, 187, 189, 193, 231−243
  cepacia, 54, 259−265
  fluorescens, 5, 10, 222
  fragi, 306−308
  paucimobilis, 222
  putida, 86
PVC, biofilms on, 206

Quantitative suspension test, 240
Quasielastic laser light scattering, 21
Quigley & Hine plaque index, 172

Radial flow chamber, 14−15, 252−254
Radial flow reactor, 40
Radiolabel, incorporation into DNA, 287−289
Recirculating systems, 6, 229−231
Recombinant strains, of soil bacteria, 265
Recreational waters, control of biofilm in, 227−243
Retention studies, 178
Reversible adsorption, 15
Reynolds number, 204
Rheogoniometer, 54
Rinsing, removal of bacteria by, 293−294, 296−297
River water, 69
Robbins device, 2−6, 11, 40−43, 63−64, 241
Rocked-tile method, 123
Role-plate technique, 135−136
Rotating disc method, 35
Roto Torque system, 39, 63
Rubber, adherence to, 293−312
Ruthenium red staining, 112−114, 115

Saliva/plaque levels with time, 178
Salivary bacterial count, 178−179
Salmonella, 109
Salmonella typhimurium, 233
Salt aggregation tests, 19−20, 24
Sample discs submerged, 34−35
Saprospira, 57
Scanning confocal laser microscopy, 10
Scanning electron microscopy, 72, 96, 99−102, 139−140, 250, 260, 262−263, 302−310
Scanning optical microscopy, 96
Scheffe's method, 322
Scoring,
  of gingivitis, 173−175
  of plaque, 172
Selective enrichment, 120
SEM, *see* Scanning electron microscopy
Septicaemia, 133, 144
Serratia marcescens, 140
Sessile drop technique, 18
Sessile population, 1
Shadow imaging, 100
Shear stress, 14−15, 35, 58, 252−254
Shearing forces, 30, 105
Ships, fouling of, 247−254
Short-term plaque regrowth studies, 179
Silicone elastomers, fouling on, 256−257
Silicone rubber, CNS adherence to, 141
Slide agglutination assay, 137
Slime production,
  by CNS, 136−140
  of *Staphylococcus epidermidis*, 289
Sloughing, 57, 58, 212
Soft tissues, infections of, 40, 45
Soil bacteria,
  plasmid exchange between, 259−265
  staining of, 274
Soiling, of an experimental installation, 294−295
Solid/air interfaces, 30
Solid nutrient/liquid interfaces, 30, 33, 34
Solute gradients, 74
Solvent dehydration, 305−312
Spas, 227−243
Spatial heterogeneity, 58
Spectrometer, double-beam, 8
Spitz−Holter shunts, colonization of, 136
Sputter−cryo techniques, 102, 306, 308
Squeegee rinse procedure, 293, 296−297
Stabilizers, 140
Staining methods, 111, 112
Stainless steel, adherence to, 293−312
Stainless steel pipe sections, 293−297
Staphylococcus

aureus, 16, 33, 101, 133, 140, 197, 237, 285–290
epidermidis, 23, 33, 46, 101, 103, 137, 152–155, 191, 197
hominis, 191
hyicus, 191
Starvation, 53
Starvation survival processes, 55
Static settling method, 123–125
Statistical evaluation, of adherence assays, 315–324
'Sticky' colonies, 285
Streptococci, 169, 236
Streptococcus
oralis, 169
sanguis, 84–85, 168, 187
Structures, surface, of bacteria, 109–128
Student's *t*-test, 127, 317–321
Study designs, for plaque control trials, 181
Submerged test pieces, 37–39
Substantivity, 178–179
Substrate utilization, 60
Sugars, addition of to haemagglutination assay, 128
Sulcus bleeding index, 174
Sulphate-reducing bacteria, 52, 278
Sulphides, 52
Surface charge, 20–21
Surface chemical analysis, 22–23
Surface structures, of bacteria, 109–128
Surfaces, stimulation of metabolic activity by, 61
Surfactants, pluronic, 153–155
Swimming pools, 209, 227–243

Tap water, biofilms in, 219
Taurolin, 318–320
TEM, *see* Transmission electron microscopy
Temperature, of recreational waters, 229–231, 235
Tenckhoff catheters, 98–106, 138
Tetrazolium salts, 274–276, 278
Thermal conductivity, of mixed-culture biofilm, 54
Tissue-chamber model, 196–197
Tobramycin, 189
Toothbrushing, home-use methods, 180–181
Toothpastes, testing of, 177–180
Topology, biofilm, 212–213

Total dissolved solids, 242–243
Transmission electron microscopy, 72, 99–102, 139–140, 250, 307
Treatment plant failure, 218
Trichloroacetic acid, 300
TRITC conjugate, 117
Trophic state, 55
Tube assay, for slime production, 137
Tukey's test, 322
Two-step immunolabelling procedure, 116–117

Uterus, devices in, 157–162
UV spectroscopy, 279

Variance, analysis of, 322–324
Veillonella, 58
Vermiculite, 260–263
Viability, across a biofilm profile, 79
Viable counts,
of batch-cultured biofilms, 191–192
of colonies on filters, 188
of coolant organisms, 270–272
Vibrio, 55, 56
natriegens, 275, 277
parahaemolyticus, 33
Volatile fraction, of a biofilm, 52

Wall growth, in glass vessels, 34
Wastewater treatment plant, 63
Water chemistry, effect on legionella ecology, 208
Water Quality and Public Health laws, 217
Water systems, growth of bacteria in, 201–213, 217–225
Wet India ink method, 111
Wetted air/solid interfaces, 35

X-ray photoelectron spectroscopy, 22–23

Yersinia enterocolitica, 233

Zeta potential, of bacterial cells, 21–24
Zinc selenide crystals, 6–9

THE SOCIETY FOR APPLIED BACTERIOLOGY TECHNICAL SERIES

Book Series Editors: F.A. Skinner (1966–1992)
M. Sussman (1992–)

1 Identification Methods for Microbiologists Part A *1966*
   Edited by B.M. Gibbs and F.A. Skinner
   Out of print
2 Identification Methods for Microbiologists Part B *1986*
   Edited by B.M. Gibbs and D.A. Shapton
   Out of print
3 Isolation Methods for Microbiologists *1969*
   Edited by D.A. Shapton and G.W. Gould
4 Automation, Mechanization and Data Handling in Microbiology *1970*
   Edited by Ann Baillie and R.J. Gilbert
5 Isolation of Anaerobes *1971*
   Edited by D.A. Shapton and R.G. Board
6 Safety in Microbiology *1972*
   Edited by D.A. Shapton and R.G. Board
7 Sampling — Microbiological Monitoring of Environments *1973*
   Edited by R.G. Board and D.W. Lovelock
8 Some Methods for Microbiological Assay *1975*
   Edited by R.G. Board and D.W. Lovelock
9 Microbial Aspects of the Deterioration of Materials *1975*
   Edited by R.J. Gilbert and D.W. Lovelock
10 Microbial Ultrastructure: The Use of the Electron Microscope *1976*
   Edited by R. Fuller and D.W. Lovelock
11 Techniques for the Study of Mixed Populations *1978*
   Edited by D.W. Lovelock and R. Davies
12 Plant Pathogens *1979*
   Edited by D.W. Lovelock
13 Cold Tolerant Microbes in Spoilage and the Environment *1979*
   Edited by A.D. Russell and R. Fuller
14 Identification Methods for Microbiologists (2nd edn) *1979*
   Edited by F.A. Skinner and D.W. Lovelock
15 Microbial Growth and Survival in Extremes of Environment *1980*
   Edited by G.W. Gould and Janet E.L. Corry
16 Disinfectants: Their Use and Evaluation of Effectiveness *1981*
   Edited by C.H. Collins, M.C. Allwood, Sally F. Bloomfield and A. Fox
17 Isolation and Identification Methods for Food Poisoning Organisms *1982*
   Edited by Janet E.L. Corry, Diane Roberts and F.A. Skinner
18 Antibiotics: Assessment of Antimicrobial Activity and Resistance *1983*
   Edited by A.D. Russell and L.B. Quesnel

19 Microbiological Methods for Environmental Biotechnology *1984*
   Edited by J.M. Grainger and J.M. Lynch
20 Chemical Methods in Bacterial Systematics *1985*
   Edited by M. Goodfellow and D.E. Minnikin
21 Isolation and Identification of Micro-organisms of Medical and Veterinary Importance *1985*
   Edited by C.H. Collins and J.M. Grange
22 Preservatives in the Food, Pharmaceutical and Environmental Industries *1987*
   Edited by R.G. Board, M.C. Allwood and J.G. Banks
23 Industrial Microbiological Testing *1987*
   Edited by J.W. Hopton and E.C. Hill
24 Immunological Techniques in Microbiology *1987*
   Edited by J.W. Grange, A. Fox and N.L. Morgan
25 Rapid Microbiological Methods for Foods, Beverages and Pharmaceuticals *1989*
   Edited by C.J. Stannard, S.B. Petitt and F.A. Skinner
26 ATP Luminescence: Rapid Methods in Microbiology *1989*
   Edited by P.E. Stanley, B.J. McCarthy and R. Smither
27 Mechanisms of Action of Chemical Biocides *1990*
   Edited by S.P. Denyer and W.B. Hugo
28 Genetic Manipulation: Techniques and Applications *1991*
   Edited by J.M. Grange, A. Fox and N.L. Morgan
29 Identification Methods in Applied and Environmental Microbiology *1992*
   Edited by R.G. Board, Dorothy Jones and F.A. Skinner
30 Microbial Biofilms: Formation and Control *1993*
   Edited by S.P. Denyer, S.P. Gorman and M. Sussman
31 New Techniques in Food and Beverage Microbiology
   Edited by R.G. Kroll, A. Gilmour and M. Sussman